METHODS IN MOLECULAR BIOLOGY

Series Editor
John M. Walker
School of Life and Medical Sciences
University of Hertfordshire
Hatfield, Hertfordshire, AL10 9AB, UK

For further volumes:
http://www.springer.com/series/7651

Viral Vectors for Gene Therapy

Methods and Protocols

Edited by

Fredric P. Manfredsson

Translational Science and Molecular Medicine, Michigan State University, Grand Rapids, MI, USA

Matthew J. Benskey

Translational Science and Molecular Medicine, Michigan State University, Grand Rapids, MI, USA

Humana Press

Editors

Fredric P. Manfredsson
Translational Science
and Molecular Medicine
Michigan State University
Grand Rapids, MI, USA

Matthew J. Benskey
Translational Science
and Molecular Medicine
Michigan State University
Grand Rapids, MI, USA

ISSN 1064-3745 ISSN 1940-6029 (electronic)
Methods in Molecular Biology
ISBN 978-1-4939-9064-1 ISBN 978-1-4939-9065-8 (eBook)
https://doi.org/10.1007/978-1-4939-9065-8

Library of Congress Control Number: 2018968117

Cover Illustration: AAV5 CAG-GFP (green) transduced fibers of passage traversing tyrosine hydroxylase (red) dopamine neurons of the substantia nigra in the rat brain. Courtesy of Fredric P. Manfredsson and Ivette M. Sandoval.

This Humana Press imprint is published by the registered company Springer Science+Business Media, LLC, part of Springer Nature.
The registered company address is: 233 Spring Street, New York, NY 10013, U.S.A.

Preface

Gene therapy, on a basic level, is defined as the delivery of nucleic acids to a cell in order to produce a desired effect. There are many methods that can be utilized in order to deliver nucleic acids to cells; however, the most efficient and adaptable method is the use of viral vectors. Viruses have naturally evolved over millennia to develop elegant mechanisms used to evade host immunity, gain entry to a cell, deliver their genetic material, and hijack host cell machinery in order to produce progeny virions. Viral vector-based gene therapy harnesses this awesome power of nature in order to efficiently deliver a desired genetic payload to cells of interest. The fundamental concept of a viral vector is relatively simple. First, the genes from the viral genome that are responsible for viral replication or untoward host response (i.e., disease) are removed, leaving only the genetic information that is absolutely essential for viral assembly. Next, the desired genetic payload is inserted into the modified viral genome. Finally, the resulting recombinant viral vector, containing the desired genetic material, is assembled in cultured cells and purified.

In its current state, viral vector gene therapy has become commonplace in both the laboratory and the clinic. Within the laboratory, viral vectors are commonly used as efficient genetic shuttles. The ability of viral vectors to infect a myriad of cells (both dividing and nondividing) and deliver various nucleic acids, which can then be integrated into the host genome or remain episomal, makes them extremely adaptable and powerful tools in biomedical research. Additionally, viral vectors are used in the lab to both model disease and research potential therapeutics. Further down the translational pipeline, viral vectors are being explored in the clinic to treat a wide range of diseases utilizing vastly different therapeutic approaches. This includes viral vectors in the treatment of cancers, neurodegenerative disease, and pulmonary disorders. Moreover, gene therapy is not limited to simple genetic overexpression, but can also accomplish a wide range of modalities such as CRISPR/CAS genome editing or manipulation of the expression of endogenous proteins. The wide range of problems to which viral vectors have been successfully applied underscores not only the tremendous potential of this tool but also how far the technology of viral vectors has grown and continues to grow [1–3]. Indeed, 2017 saw the first viral gene therapy product gain FDA approval (Luxturna©), paving the way for future efforts.

The idea of utilizing a virus to deliver a desired set of nucleic acids to a cell dates back to the early 1970s when researchers discovered that retroviruses were capable of acquiring cellular genes, giving proof in principle to the idea that viruses could be used in order to deliver nonviral genetic information to cells [4]. Not long after this observation, researchers were able to successfully generate recombinant viruses that were also capable of delivering nonviral genetic material to target cells [5–10]. Since these early landmarks in viral vector history, the field of viral vector gene therapy has grown exponentially. With advances in the fields of both molecular biology and virology, researchers have "vectorized" an ever-increasing number of different viruses with different capabilities. Further, researchers are constantly modulating every step of the viral life cycle in order to improve viral genetic delivery, while also expanding the potential repertoire of functions that a virus can perform. In this way science is constantly pushing the envelope of what a virus is able to achieve, with the end result being an unprecedented level of control over an extremely powerful tool.

However, as the maxim states, with great power comes great responsibility. Indeed, in order to achieve optimal results using gene therapy, it is the responsibility of the researcher to control for every aspect of the experiment. For years, the requisite knowledge necessary to properly control and conduct a successful gene therapy experiment has been the sole privilege of a handful of highly specialized laboratories or clinicians around the world. However, as the overall success of gene therapy has grown, so has the availability of these powerful tools. Today, the wide availability of viral vectors has made them accessible to virtually any scientist with the desire.

In spite of this increase in the availability of vectors themselves, the requisite knowledge that is absolutely essential to conducting a successful gene therapy experiment has not been made equally available. To successfully utilize viral vectors to their full potential, a large number of decisions must be made; in some instances prior to even obtaining the vector itself. It is the goal of this book to provide a comprehensive list of theoretical knowledge and detailed protocols necessary for researchers, clinicians, and students to successfully utilize viral vectors for a wide range of gene therapy applications. To begin, an introductory chapter will provide an overview of basic gene therapy modalities. Subsequent chapters will delve more deeply into specific protocols, ranging from vector production to delivery methods, which can be used as step-by-step instructions to successfully execute your desired gene therapy application.

Grand Rapids, MI, USA

Matthew J. Benskey
Fredric P. Manfredsson

References

1. Tuszynski MH, Thal L, Pay M, et al. (2005) A phase 1 clinical trial of nerve growth factor gene therapy for Alzheimer disease. Nature Medicine 11:551–555. doi: 10.1038/nm1239
2. Rainov NG, Grp GIS (2000) A phase III clinical evaluation of herpes simplex virus type 1 thymidine kinase and ganciclovir gene therapy as an adjuvant to surgical resection and radiation in adults with previously untreated glioblastoma multiforme. Hum Gene Ther 11:2389–2401. doi: 10.1089/10430340075003849 9
3. Flotte TR, Zeitlin PL, Reynolds TC, et al. (2003) Phase I trial of intranasal and endobronchial administration of a recombinant adeno-associated virus serotype 2 (rAAV2)-CFTR vector in adult cystic fibrosis patients: A two-part clinical study. Hum Gene Ther 14:1079–1088. doi: 10.1089/104303403322124792
4. Finer M, Glorioso J (2017) A brief account of viral vectors and their promise for gene therapy. Gene Ther 24:1–2. doi: 10.1038/gt.2016.71
5. Shimotohno K, Temin HM (1981) Formation of infectious progeny virus after insertion of herpes-simplex thymidine kinase gene into DNA of an avian retrovirus. Cell 26:67–77.
6. Wei CM, Gibson M, Spear PG, Scolnick EM (1981) Construction and Isolation of a Transmissible Retrovirus Containing the Src Gene of Harvey Murine Sarcoma-Virus and the Thymidine Kinase Gene of Herpes-Simplex Virus Type-1. Journal of Virology 39:935–944.
7. Tabin CJ, Hoffmann JW, Goff SP, Weinberg RA (1982) Adaptation of a retrovirus as a eucaryotic vector transmitting the herpes simplex virus thymidine kinase gene. Mol Cell Biol 2:426–436
8. Samulski RJ, Berns KI, TAN M, Muzyczka N (1982) Cloning of adeno-associated virus into Pbr322—rescue of intact virus from the recombinant plasmid in human-Cells. PNAS 79:2077–2081

9. Hermonat PL, Muzyczka N (1984) Use of Adeno-associated virus as a mammalian DNA cloning vector—transduction of neomycin resistance into mammalian tissue-culture cells. PNAS 81:6466–6470
10. Yu SF, Rüden von T, Kantoff PW, et al. (1986) Self-inactivating retroviral vectors designed for transfer of whole genes into mammalian cells. PNAS 83:3194–3198

Contents

Contributors

Eric Adams • *Northern Biomedical Research, Norton Shores, MI, USA*
Ramesh Akkina • *Department of Microbiology, Immunology and Pathology, Colorado State University, Fort Collins, CO, USA*
Patrick Aldrin-Kirk • *Molecular Neuromodulation, Wallenberg Neuroscience Center, Lund University, Lund, Sweden*
Ian E. Alexander • *Gene Therapy Research Unit, Children's Medical Research Institute, The University of Sydney, Faculty of Medicine and Health and Sydney Children's Hospitals Network, Westmead, NSW, Australia; The University of Sydney, Sydney Medical School, Discipline of Child and Adolescent Health, Westmead, NSW, Australia*
Krystof Bankiewicz • *Department of Neurological Surgery, University of California San Francisco, San Francisco, CA, USA*
Karim Benabdellah • *Centre for Genomics and Oncological Research (GENYO), Pfizer/University of Granada/Andalusian Regional Government, PTS Granada, Granada, Spain*
Rachid Benchaouir • *University of Versailles Saint-Quentin en Yvelines, Montigny-le-Bretonneux, France; Centre Scientifique de Monaco, Monaco, Monaco; SQY Therapeutics SARL, Noisy-le-Roi, France*
Matthew J. Benskey • *Department of Translational Science and Molecular Medicine, Michigan State University, Grand Rapids, MI, USA*
Tomas Björklund • *Molecular Neuromodulation, Wallenberg Neuroscience Center, Lund University, Lund, Sweden*
Florie Borel • *Gene Therapy Center, University of Massachusetts Medical School, Worcester, MA, USA*
Sakereh Carter • *Stem Cell Program and Institute for Regenerative Cures, University of California, Davis, Sacramento, CA, USA; Department of Neurology, University of California Davis, Sacramento, CA, USA*
S. Cherqui • *Division of Genetics, Department of Pediatrics, University of California, San Diego, La Jolla, CA, USA*
Wade W. Chien • *National Institute on Deafness and Other Communication Disorders/National Institutes of Health, Bethesda, MD, USA; Department of Otolaryngology-Head and Neck Surgery, Johns Hopkins School of Medicine, Baltimore, MD, USA*
Marién Cobo • *Centre for Genomics and Oncological Research (GENYO), Pfizer/University of Granada/Andalusian Regional Government, PTS Granada, Granada, Spain*
Timothy J. Collier • *Department of Translational Science & Molecular Medicine, College of Human Medicine, Michigan State University, Grand Rapids, MI, USA; Mercy Health Saint Mary's, Grand Rapids, MI, USA*
Steven J. Conrad • *Department of Microbiology and Immunology, University of Arkansas for Medical Sciences (UAMS), Little Rock, AR, USA*
Sharon C. Cunningham • *Gene Therapy Research Unit, Children's Medical Research Institute, The University of Sydney, Faculty of Medicine and Health and Sydney Children's Hospitals Network, Westmead, NSW, Australia*

PETER DENG • *Stem Cell Program and Institute for Regenerative Cures, University of California, Davis, Sacramento, CA, USA; Genome Center, MIND Institute, and Biochemistry and Molecular Medicine, University of California, Davis, Davis, CA, USA; Department of Neurology, University of California Davis, Sacramento, CA, USA*

DONGSHENG DUAN • *Department of Veterinary Pathobiology, College of Veterinary Medicine, The University of Missouri, Columbia, MO, USA; Department of Molecular Microbiology and Immunology, School of Medicine, The University of Missouri, Columbia, MO, USA; Department of Neurology, School of Medicine, The University of Missouri, Columbia, MO, USA; Department of Bioengineering, The University of Missouri, Columbia, MO, USA; Department of Biomedical Sciences, College of Veterinary Medicine, The University of Missouri, Columbia, MO, USA*

CAROLYN A. FAIRBANKS • *Department of Pharmaceutics, University of Minnesota, Minneapolis, MN, USA; Department of Neuroscience, University of Minnesota, Minneapolis, MN, USA; Department of Pharmacology, University of Minnesota, Minneapolis, MN, USA*

KYLE FINK • *Stem Cell Program and Institute for Regenerative Cures, University of California, Davis, Sacramento, CA, USA; Department of Neurology, University of California Davis, Sacramento, CA, USA*

AYSEGUL GEZER • *Department of Translational Science and Molecular Medicine, Michigan State University, Grand Rapids, MI, USA*

CHADY H. HAKIM • *Department of Molecular Microbiology and Immunology, School of Medicine, The University of Missouri, Columbia, MO, USA; National Center for Advancing Translational Sciences, NIH, Rockville, MD, USA*

BRIAN L. HOH • *Department of Neurosurgery, College of Medicine, University of Florida, Gainesville, FL, USA*

KOJI HOSAKA • *Department of Neurosurgery, College of Medicine, University of Florida, Gainesville, FL, USA*

SHUANG HU • *Department of Medical Microbiology & Immunology, University of California, Davis, CA, USA*

CRISTHIAN J. ILDEFONSO • *Department of Molecular Genetics and Microbiology, University of Florida College of Medicine, Gainesville, FL, USA; Department of Ophthalmology, University of Florida College of Medicine, Gainesville, FL, USA*

KEVIN ISGRIG • *National Institute on Deafness and Other Communication Disorders/National Institutes of Health, Bethesda, MD, USA*

ANDRANIK KAHRAMANIAN • *Department of Neurosurgery, Brain Tumor Research Center, Massachusetts General Hospital, Harvard Medical School, Boston, MA, USA*

KELLEY F. KITTO • *Department of Neuroscience, University of Minnesota, Minneapolis, MN, USA*

NATHAN C. KUHN • *Department of Translational Science and Molecular Medicine, Michigan State University, Grand Rapids, MI, USA*

RASHMI KUMARI • *Department of Comparative Pathobiology, Purdue Institute for Inflammation, Immunology, and Infectious Disease, College of Veterinary Medicine, Purdue University, West Lafayette, IN, USA*

TOSHIHIKO KURODA • *Department of Neurosurgery, Brain Tumor Research Center, Massachusetts General Hospital, Harvard Medical School, Boston, MA, USA*

ALFRED S. LEWIN • *Department of Molecular Genetics and Microbiology, University of Florida College of Medicine, Gainesville, FL, USA; Department of Ophthalmology, University of Florida College of Medicine, Gainesville, FL, USA*

Mingjie Li • *Department of Neurology and Hope Center for Neurological Disorders, Washington University School of Medicine, St Louis, MO, USA*
Jia Liu • *Department of Microbiology and Immunology, University of Arkansas for Medical Sciences (UAMS), Little Rock, AR, USA; The Center for Microbial Pathogenesis and Host Inflammatory Responses, University of Arkansas for Medical Sciences, Little Rock, AR, USA*
Noelia Maldonado-Pérez • *Centre for Genomics and Oncological Research (GENYO), Pfizer/University of Granada/Andalusian Regional Government, PTS Granada, Granada, Spain*
Fredric P. Manfredsson • *Department of Translational Science and Molecular Medicine, College of Human Medicine, Michigan State University, Grand Rapids, MI, USA; Mercy Health Saint Mary's, Grand Rapids, MI, USA*
Francisco Martín • *Centre for Genomics and Oncological Research (GENYO), Pfizer/University of Granada/Andalusian Regional Government, PTS Granada, Granada, Spain*
Michael T. Massengill • *Department of Molecular Genetics and Microbiology, University of Florida College of Medicine, Gainesville, FL, USA*
Kathryn Miller • *Department of Translational Science and Molecular Medicine, Michigan State University, Grand Rapids, MI, USA*
Suresh K. Mittal • *Department of Comparative Pathobiology, Purdue Institute for Inflammation, Immunology, and Infectious Disease, College of Veterinary Medicine, Purdue University, West Lafayette, IN, USA*
Christian Mueller • *Gene Therapy Center, University of Massachusetts Medical School, Worcester, MA, USA; Department of Pediatrics, University of Massachusetts Medical School, Worcester, MA, USA*
Kousaku Ohno • *Department of Neurological Surgery, University of California San Francisco, San Francisco, CA, USA*
Cristina D. Peterson • *Department of Neuroscience, University of Minnesota, Minneapolis, MN, USA*
Kelsey R. Pflepsen • *Department of Pharmaceutics, University of Minnesota, Minneapolis, MN, USA*
Lilith Reeves • *Cincinnati Children's Hospital Medical Center, Cincinnati, OH, USA*
Karima Relizani • *University of Versailles Saint-Quentin en Yvelines, Montigny-le-Bretonneux, France; SQY Therapeutics SARL, Noisy-le-Roi, France*
C. J. Rocca • *Division of Genetics, Department of Pediatrics, University of California, San Diego, La Jolla, CA, USA*
Sabina Sánchez-Hernández • *Centre for Genomics and Oncological Research (GENYO), Pfizer/University of Granada/Andalusian Regional Government, PTS Granada, Granada, Spain*
Lluis Samaranch • *Department of Neurological Surgery, University of California San Francisco, San Francisco, CA, USA*
Waldy San Sebastian • *Department of Neurological Surgery, University of California San Francisco, San Francisco, CA, USA*
Ivette M. Sandoval • *Department of Translational Science and Molecular Medicine, College of Human Medicine, Michigan State University, Grand Rapids, MI, USA; Mercy Health Saint Mary's, Grand Rapids, MI, USA*
Quentin Sandro • *University of Versailles Saint-Quentin en Yvelines, Montigny-le-Bretonneux, France*

Ekramy E. Sayedahmed • *Department of Comparative Pathobiology, Purdue Institute for Inflammation, Immunology, and Infectious Disease, College of Veterinary Medicine, Purdue University, West Lafayette, IN, USA*

Rhyomi L. Sellnow • *Department of Translational Science and Molecular Medicine, Michigan State University, Grand Rapids, MI, USA*

William Swaney • *Cincinnati Children's Hospital Medical Center, Cincinnati, OH, USA*

Kathy Szczur • *Cincinnati Children's Hospital Medical Center, Cincinnati, OH, USA*

Stuart Tinch • *Cincinnati Children's Hospital Medical Center, Cincinnati, OH, USA*

María Tristán-Manzano • *Centre for Genomics and Oncological Research (GENYO), Pfizer/University of Granada/Andalusian Regional Government, PTS Granada, Granada, Spain*

Roslyn Vashon • *Department of Translational Science and Molecular Medicine, Michigan State University, Grand Rapids, MI, USA*

Lucy Vulchanova • *Department of Neuroscience, University of Minnesota, Minneapolis, MN, USA*

Hiroaki Wakimoto • *Department of Neurosurgery, Brain Tumor Research Center, Massachusetts General Hospital, Harvard Medical School, Boston, MA, USA*

Lakmini P. Wasala • *Department of Veterinary Pathobiology, College of Veterinary Medicine, The University of Missouri, Columbia, MO, USA*

George L. Wilcox • *Department of Neuroscience, University of Minnesota, Minneapolis, MN, USA; Department of Pharmacology, University of Minnesota, Minneapolis, MN, USA; Department of Dermatology, University of Minnesota, Minneapolis, MN, USA*

Scott R. Witting • *Cincinnati Children's Hospital Medical Center, Cincinnati, OH, USA; Department of Pediatrics, University of Cincinnati College of Medicine, Cincinnati, OH, USA*

N. Nora Yang • *National Center for Advancing Translational Sciences, NIH, Rockville, MD, USA*

Brianna M. Young • *Department of Ophthalmology, University of Florida College of Medicine, Gainesville, FL, USA*

Yongping Yue • *Department of Molecular Microbiology and Immunology, School of Medicine, The University of Missouri, Columbia, MO, USA*

Part I

Introduction

Chapter 1

Basic Concepts in Viral Vector-Mediated Gene Therapy

Matthew J. Benskey, Ivette M. Sandoval, Kathryn Miller, Rhyomi L. Sellnow, Aysegul Gezer, Nathan C. Kuhn, Roslyn Vashon, and Fredric P. Manfredsson

Abstract

Today any researcher with the desire can easily purchase a viral vector. However, despite the availability of viral vectors themselves, the requisite knowledge that is absolutely essential to conducting a gene therapy experiment remains somewhat obscure and esoteric. To utilize viral vectors to their full potential, a large number of decisions must be made, in some instances prior to even obtaining the vector itself. For example, critical decisions include selection of the proper virus, selection of the proper expression cassette, whether to produce or purchase a viral vector, proper viral handling and storage, the most appropriate delivery method, selecting the proper controls, how to ensure your virus is expressing properly, and many other complex decisions that are essential to performing a *successful* gene therapy experiment. The need to make so many important decisions can be overwhelming and potentially prohibitive, especially to the novice gene therapist. In order to aid in this challenging process, here we provide an overview of basic gene therapy modalities and a decision tree that can be used to make oneself aware of the options available to the beginning gene therapist. This information can be used as a road map to help navigate the complex and perhaps confusing process of designing a successful gene therapy experiment.

Key words Viral vector, Gene therapy, Adeno-associated virus, Lentivirus, Adenovirus, Herpes-simplex virus

1 Introduction

Viral vector-based gene therapy was originally conceived in order to accomplish a simple goal, to transfer genetic material to a target cell. Although simple, achievement of this goal produced profound results. The ability to manipulate gene expression within any desired cell revolutionized the biomedical field. However, with the continual improvement of viral vectors, expression cassettes, and delivery methods, gene therapy has evolved far beyond the ability to simply transfer a foreign gene to a cell, now enabling researchers and clinicians to accomplish an astounding number of sophisticated cellular and molecular manipulations. For example,

Fredric P. Manfredsson and Matthew J. Benskey (eds.), *Viral Vectors for Gene Therapy: Methods and Protocols*, Methods in Molecular Biology, vol. 1937, https://doi.org/10.1007/978-1-4939-9065-8_1,

among many other possibilities, gene therapy can be used to monitor intracellular physiology, manipulate the activity of excitable cells, deliver oncolytic genes to tumors, or to permanently alter the genome of a target cell. With so many options, taking the initial steps to performing gene therapy can be a daunting task. To aid the novice gene therapist, Fig. 1 depicts a decision tree that can be used to help make the initial decisions which will ultimately lead to a successful gene therapy experiment.

The three major decisions that must be made in order to conduct a successful gene therapy experiment are: (1) which viral vector to utilize, (2) the type of expression cassette to utilize, and (3) the mode of vector delivery. These issues will be discussed in turn, however, it must be mentioned that these choices must be made in concert, as each of these decisions will reciprocally affect the others. For example, the type of viral vector one chooses will ultimately affect the size of the expression cassette that can be packaged. Alternatively, if a gene therapy experiment necessitates spatially restricted gene expression, this can be accomplished through the choice of either a specific expression cassette, a specific virus, a specific delivery method, or alternatively all three strategies combined. Thus, although these issues will be discussed individually, they must be considered in aggregate.

2 Choosing the Appropriate Viral Vector

Prior to beginning any gene therapy application, one of the first major decisions to be made is the selection of the appropriate viral vector. When selecting a viral vector for gene therapy the ultimate goal is to find a nonreplicating, recombinant vector that remains highly infective in the absence of toxicity to the host cell or organism. Further, an ideal viral vector should be amenable to genetic manipulation, and easily and reliably produced at sufficiently high titers for use in the laboratory or the clinic. In order to maximize these desired characteristics, recombinant viral vectors are engineered to remove all of the wild-type viral genome, except those portions which are required *in cis* to effectively package the desired genetic material. Removing portions of the viral genome decreases the possibility of a host-immunogenic response to encoded viral proteins, and also results in a replication incompetent virus, both of which drastically increase the safety profile of the vector while maintaining infectivity. This strategy has been successfully utilized to create an ever-growing repertoire of viral vectors, all with unique characteristics. However, the most widely used recombinant viral vectors include: (1) recombinant adeno-associated virus (rAAV), (2) adenovirus (Ad), (3) Lentivirus (LV), and (4) herpes-simplex virus (HSV). When selecting the appropriate viral vector for your application, consideration should be paid to the carrying capacity of

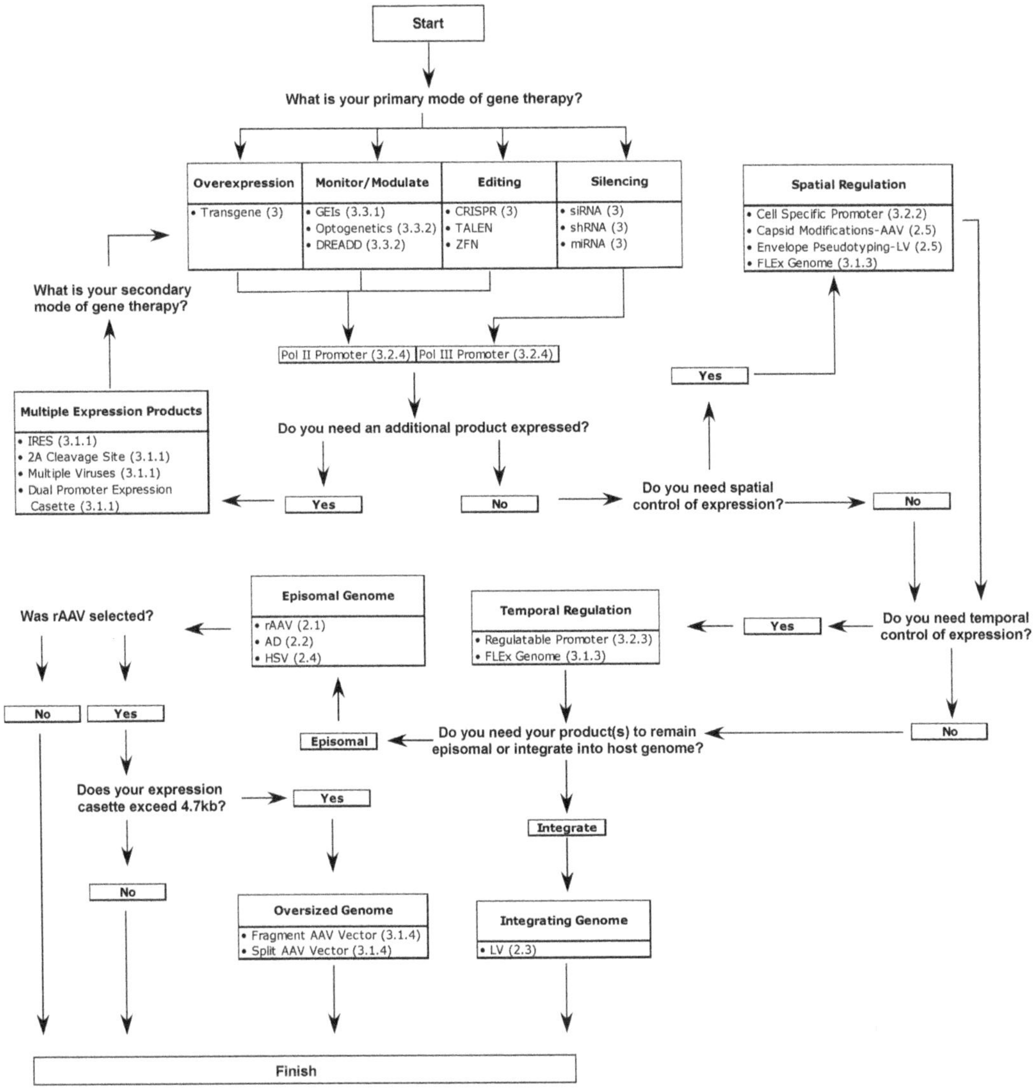

Fig. 1 Decision Tree to Facilitate the Selection of an Appropriate Expression Cassette and Viral Vector. Listed in this decision tree are some of the most important questions that must be considered when designing a gene therapy experiment using a viral vector. Begin at the top of the tree in the "Start" position and use the questions to guide you to the selection of an appropriate expression cassette (including the promoter and expression product(s)) as well as the appropriate viral vector. As you work through the decision tree, options to each answer are listed within the appropriate box subsequent to the respective question. Further information pertaining to each listed option can be found in the respective sections of this introductory chapter, which are denoted by the numbers in parentheses. Keep in mind that the selection of the appropriate expression cassette and viral vector is a decision that must be made in concert with your desired method of delivery. Upon reaching the "Finish" position, you should have an informed idea of the type of expression cassette and viral vector that you will utilize for your desired gene therapy application

the viral vector, whether the viral genome integrates into the host genome or remains episomal, the persistence of gene expression, and the toxicity/immunogenicity profile of the vector. These characteristics for each class of vector are overviewed below (also *see* Table 1), and expanded upon later in this book.

2.1 Recombinant Adeno-Associated Virus (rAAV)

Recombinant versions of AAV were originally produced in the early 1980s [1] and have since been studied and utilized more than other viral vector options. AAV is a non-enveloped virus, containing a single-stranded DNA genome of approximately 4.7 kilobases (kb) that is packaged into a small icosahedral capsid. rAAV has an ideal safety profile with little to no immunogenic or toxic response observed following high-dose administration. Production techniques have been optimized to manufacture high-titer rAAV on both small and large scales, making this vector highly amenable to virtually every application, ranging from laboratory experiments in cells or small organisms, to more elaborate experimentation or clinical trials involving systemic delivery [2] of rAAV to large organisms. The tropism of rAAV is also very versatile (*see* Subheading 2.5 and [3]) and rAAV can infect both dividing and nondividing cells. Combining the large number of naturally occurring AAV serotypes with capsid manipulations in the laboratory, such as rational capsid engineering or directed evolution, has yielded an ever-growing repertoire of AAV capsid variants, each with a unique tropism that can be harnessed to tailor viral infectivity to the researchers needs. Drawbacks to rAAV include the relatively small carrying capacity of the genome (~5 kb), limiting the amount of genetic material that can be packaged into the capsid. However, this problem has been circumvented using techniques such as split genomes (*see* Subheading 3.1.4). Further, following infection, rAAV remains episomal. This feature renders rAAV susceptible to genetic dilution and a corresponding loss of transgene expression within rapidly dividing cells. On the other hand, within nondividing cells, such as neurons, rAAV vectors provide persistent transgene expression without the need to integrate into the host cell's genome [4]. Additional drawbacks include the ubiquitous sero-prevalence of AAV in humans and animals [5]. As such, attention must be paid to the presence of neutralizing antibodies in the target host organism, as these may result in significantly reduced transduction efficacy. This is a lesser concern in rodents; nevertheless, a researcher needs to be aware that neutralizing antibodies against AAV may be present even in mice and rats [5], and may thus alter the outcome of experimentation, specifically in systemic administration paradigms. Finally, care must be taken when an experiment calls for the repeated administration of AAV, as this may lead to an immune response to the capsid [6, 7], potentially decreasing transduction efficacy or even causing a toxic response in the host.

Table 1
Basic overview of common recombinant vectors

Viral vector	Packaging capacity (kb)	Toxicity	Ease of production	Sustainability of expression	Benefits	Drawbacks
rAAV	5	Potential immunogenicity depending on capsid/low toxicity	Very labor intensive	Lifetime in nondividing cell. Subject to genomic dilution with cell division.	Extremely versatile tropism	Small carrying capacity
AD	8–35	Significant immunogenicity/ some cytotoxicity	Very labor intensive	Rapid onset, but infection often self-limiting due to immunogenicity.	High-level transgene expression	Preexisting immunity
HSV	150	Significant immunogenicity/ some cytotoxicity	Moderately labor intensive	Rapid onset, but infection often self-limiting due to immunogenicity.	Massive carrying capacity	Transgene expression can be transient
LV	8	Some immunogenicity	Easy	Persistent in dividing and nondividing cells	Prolonged transgene expression	Risk of insertional mutagenesis

2.2 Adenovirus (Ad)

Ad has had a tumultuous history in the field of gene therapy, and there are both pros and cons to Ad as a viral vector. There are many advantages to Ad: Ad is a non-enveloped double-stranded DNA virus that has a large packaging capacity and can effectively package and deliver up to 35 kb of genetic material. Ad can be grown to high-titer stocks using common laboratory practices, and Ad can very efficiently infect both dividing and nondividing cells where the delivered genetic cargo remains episomal. The major drawback to Ad is its high toxicity profile. Ad is fairly immunogenic, and a host immune response may cause adverse effects, including decreased transduction efficacy or limiting the longevity of experimentation that may be performed. Moreover, there is a broad preexisting immunity in the general population. This not only limits the serotypes of Ad that may be used in clinical trials but also prohibits the use of repeated administration in order to avoid toxicity. Despite these inherent drawbacks, current efforts aim to reengineer Ad vectors to limit their immunogenicity and toxicity, making for a safer and more targeted vector. Latest generation "gutless" Ad vectors are devoid of any viral coding sequences and exhibit a reduced immune response, although significant immunogenicity still remains [8]. Conversely, the strong immunogenicity and natural toxicity of Ad vectors makes them ideal candidates for oncolytic or vaccination gene therapy applications. Accordingly these vectors have been utilized in hundreds of clinical trials in an attempt to selectively target and destroy tumor cells or vaccinate against pandemic diseases [9].

2.3 Lentivirus (LV)

LV is a genus of retrovirus and as such contains an RNA genome that is reverse transcribed following infection. Subsequent to reverse transcription, LV DNA integrates into the host genome, making it the preferred vector for stable gene expression. For example, LV is commonly used for *ex vivo* manipulation of stem cells or the creation of stable cell lines. Recombinant LV can package and deliver ~8 kb of genetic material, has relatively low immunogenicity, and can infect both dividing and non-dividing cells. LV is very easily produced in the laboratory for both large (*see* Chapter 8) and small [10] scale applications. Additionally, LV is an enveloped virus, and as such the tropism of LV can be easily manipulated by packaging the virus using heterologous envelope proteins from different viral families (*see* Subheading 2.5, Chapter 7, and [11, 12]). A potential concern with LV is the risk of insertional mutagenesis. The LV genome will randomly integrate into areas of high transcriptional activity within the host genome following infection. This is a cause of concern due to possible disruption of the host's essential genomic material. However, newer generations of LV, known as non-integrating lentiviral vectors, lack the enzyme integrase, resulting in a virus that can maintain extra-genomic vector DNA either in a linear or circular capacity without disrupting the host genome.

2.4 Herpes-Simplex Virus (HSV)

HSV is an enveloped virus with a double-stranded DNA genome that spans an impressive 152 kb. The massive size of the HSV genome makes HSV vectors ideal for packaging and delivering large amounts of foreign DNA. HSV vectors grow well in culture and high-titer stocks are commonly obtained using standard laboratory equipment. HSV vectors are naturally neurotropic and can be further subdivided into replication-competent (attenuated) and replication-defective (amplicons) vectors. Replication-defective HSV vectors are created by removing all portions of the wild-type viral genome necessary for replication. This strategy decreases or eliminates the expression of viral proteins within target cells, greatly reducing cytotoxicity that is normally observed with HSV. Replication-defective HSV vectors are commonly used to deliver genes of interest to a wide variety of cells [13]. Notably, the naturally neurotropic nature and efficient retrograde axonal transport of HSV vectors make them ideal for transducing neurons distal to the site of application following infection of axon terminals [14]. Replication-competent HSV vectors are often used as oncolytic gene therapy tools by combining the naturally lytic nature of HSV and the delivery of immunomodulatory genetic material. HSV vectors remain episomal, which decreases the risk of insertional mutagenesis but also decreases the potential of long-term stable gene expression. Potential downsides of HSV vectors include toxicity and vector latency. The life cycle of HSV entails a latent phase, a characteristic that can be problematic if constitutive high-level transgene expression is desired. However, novel viral promoters that are not silenced during viral latency are being developed to address this issue. One final note in regard to the use of HSV; given the large genomic size of the recombinant genome, non-canonical cloning methods are required to create new genomes. This includes the use of recombination methodology and/or bacterial artificial chromosome (BAC) technology [15] (*see* Chapter 10).

2.5 Altering Viral Tropism

Of the four vector systems discussed above, arguably the two most commonly utilized vectors are rAAV and LV. The reasons for this observed preference is multifactorial. As mentioned, every viral vector has both pros and cons, and there is no perfect vector ideally suited to every gene therapy application. However, one of the most important considerations for gene therapy is the ability to deliver genetic material selectively and specifically to a target cell population. As such, the ability to easily alter the tropism of both rAAV and LV has increased the versatility of these vector systems, greatly expanding the range of applications to which these vectors are suited. Here we will briefly describe the process and benefits of vector pseudotyping for LV and rAAV, a topic which will be covered in detail in Chapter 7 for LV vectors.

Both LV and AAV gain entry to the host cell through receptor mediated endocytosis.

As such, viral tropism is dictated through interactions between host cell receptors and capsid proteins (AAV) or envelope proteins (LV). Accordingly, modifying the capsid or envelope proteins that rAAV or LV express can drastically modify the tropism of the recombinant vector. This is the basis of vector pseudotyping, the process of producing viral vectors with heterologous capsids or envelope proteins.

In the LV vector system, the vesicular stomatitis virus glycoprotein (VSV-G) is by far the most popular glycoprotein for pseudotyping HIV-1-based LV vectors. VSV-G pseudotyped LV vectors exhibit a very wide range of tropism and also result in a highly stable viral particle that is amenable to purification through ultracentrifugation. LV vectors have also been successfully pseudotyped using envelope proteins from the retrovirus, paramyxovirus, orthomyxovirus, rhabdovirus, arenavirus, and alphavirus families/genera [12, 16]. Each differently pseudotyped LV vector displays unique tropic characteristics ranging from highly selective- to broad range tropism. For example, LV vectors pseudotyped with rabies virus glycoprotein are neurotropic and display efficient retrograde transport [17], whereas LV vectors pseudotyped with Filoviral glycoproteins efficiently transduce airway epithelium of the lung [18]. There are an ever expanding number of possibilities for novel LV pseudotypes, each with advantages and disadvantages depending on the target organ or cell types [19].

Recent advances in our understating of AAV biology as well as improvements in rAAV vector engineering and production methods have drastically increased the versatility, and the corresponding potential, of rAAV. As previously mentioned, AAV tropism is dictated by the viral capsid, which differs between AAV serotypes. To date, more than one hundred wild-type AAV serotypes have been identified [20], offering a naturally existing library of capsid options. This capsid library has been further expanded through several different techniques that produce capsid variants. One of the most powerful approaches to engineer novel rAAV capsids, which alter tropism, is through the process of directed evolution. A major benefit to the directed evolution method, is that it does not require extensive knowledge regarding the biology of the target cell population, but rather relies on a random approach to produce a large number of capsid variants [21]. There are a variety of methodologies utilized for the directed evolution of rAAV capsids, including DNA shuffling, StEP PCR, and random display peptide insertion [22–24]. More targeted approaches are also successfully used to generate pseudotyped AAV variants. For example, recombining specific portions of capsids from multiple different serotypes can be used in order to produce a hybrid capsid. Alternatively, ligands can be inserted into the capsid to target a specific cell surface receptor [24, 25], or chemical modification of the capsid can be used to alter cell specificity [26, 27]. Other rational capsid bioengineering techniques have also been utilized successfully to increase AAV transduction efficiency. For example, altering specific amino acids in the capsid can modify the ability of the host cell to phosphorylate or ubiquitinate the viral vector following infection [28], increasing the likelihood that the genetic cargo will make it to the nucleus, thus enhancing transduction [29, 30].

3 Choosing the Appropriate Expression Cassette

Proper expression cassette selection is just as paramount in importance as selecting your viral vector. The expression cassette utilized must be able to sufficiently address the gene therapist's needs, and so an understanding of the current technologies and capabilities of different genomes used for gene therapy is essential before beginning a study. Overexpression of a single gene, or even multiple genes, is commonly used in research studies and clinical trials alike. Here gene expression can be controlled using either ubiquitous or tissue-specific promoters to achieve the desired expression patterns. Conversely, gene knockdown/knockout approaches are also widely used, employing either RNA interference (RNAi) techniques, such as short hairpin RNA (shRNA), small interfering RNAs (siRNA), or microRNAs (miRNA), or using gene editing technologies such as the CRISPR (clustered regularly interspaced short palindromic repeat) system. Other techniques, such as optogenetics and DREADDS (designer receptor exclusively activated by designer drugs), allow for precise control of cellular activity, while genetically encoded indicators can be used to monitor cellular processes [31]. This wide array of possibilities is mediated by the genetic material encoded and packaged into the viral vector. Thus, selecting the proper genome for a study is imperative for achieving scientifically sound results.

3.1 Genomes

3.1.1 Multicistronic Genomes

In some cases, it is necessary to express more than one gene for a particular gene therapy application. While delivering multiple separate vectors containing separate genes of interest is possible, this technique does not ensure that all cells will be transduced with both viruses, and the desired dual-expression may not be achieved. Numerous approaches for dual-expression have been developed to circumvent this problem. These "multicistronic" genomes ensure dual-expression in all transduced cells by employing bidirectional or dual promoters, internal ribosome entry sites (IRESs), or 2A cleavage sites (reviewed in [32]).

Bidirectional or dual promoters can be a useful way to control two different genes within the same vector genome. Bidirectional promoters exploit the ability of RNA Polymerase II (Pol II)-type promoters to initiate transcription both in the forward and reverse direction. The use of a strong promoter—such as the CMV promoter--joined back-to-back with other enhancer and promoter elements allows for efficient transcription and expression of two genes from a single vector [33, 34]. It is important to note that Pol II type promoters are not suited for expression of certain products such as shRNA or guide RNAs unless these are embedded in a longer transcript (these typically require a short RNA pol III promoter). This problem can be circumvented by using two independent promoters regulating each genetic product separately. The dual-promoter approach is useful for driving two genes independently in a tissue-specific manner, as well as in gene editing therapies. A common example is the use of single expression cassette containing a pol II promoter that controls the expression of a fluorescent reporter transgene and a pol III promoter controlling the expression of a shRNA [35]. Moreover, separate promoters can be used to drive multiple transgenes, however, this technique may not be ideal if packaging size is a concern, as the addition of a second promoter and gene may be too large for some delivery vehicles. However, many small versions of promoters have been developed specifically for use in gene therapy in order to obtain strong gene expression with the smallest size promoters [36, 37]. Further, the use of two identical promoters within a bicistronic genome should be avoided if possible, to decrease the risk of recombination between the homologous promoter sequences.

IRESs (internal ribosome entry sites) are RNA structures found in the 5′UTR of mRNA. In gene therapy, an IRES can be placed between two genes in order to achieve individual translation of both gene products. The IRES inhibits ribosome scanning between the two genes, and acts as an entry site for the translational machinery at the second gene [38]. Commonly used IRES sequences are ~450 bp long, which may be more suitable than a dual-promoter approach if packaging size is a concern. It has been well documented that the second gene following some commonly used IRESs are expressed at a lower level than the first gene (20–50% decrease in expression) [39].

Yet another method of achieving co-expression in gene therapy is by 2A “self-cleaving” peptides [40]. These small 18–22 amino acid sequences were identified first in the foot-and-mouth disease virus, and have now been found in multiple others including porcine teschovirus-1 and equine rhinitis A. The presence of a 2A sequence leads to “cleavage” of the flanking polypeptides during translation, although this terminology is misleading as the 2A mechanism actually entails ribosome skipping during translation, rather than a cleavage event post-translation [41, 42]. Successful 2A skipping during translation results in two separate proteins from the transcripts produced upstream and downstream of the 2A site, while failure to skip the 2A sequence causes a fusion protein.

Expression of the second gene following the 2A site can be reduced by 5–30% depending on the 2A sequence used and tissue type [43]. This technique therefore could allow for more robust expression of both genes in the vector as compared to an IRES approach. Additionally, 2A sequences are small in size, making them very useful in situations where genome size is a concern. An important aspect of the use of 2A sequences is that following cleavage several additional amino acids remain on the cleaved protein product, thus, any effect these additional amino acids have on protein function should be determined empirically.

3.1.2 Self-Complimentary rAAV Genomes

In rAAV-mediated gene therapy, one of the rate-limiting steps in vector expression is second strand synthesis. The naturally occurring AAV genome is packaged as single-stranded DNA, and following cellular entry either the compliment strand must be synthesized or multiple genomes (of sense and anti-sense polarity) come together in a process called strand annealing [44], before transcription can begin. In research and gene therapy applications, the time between viral delivery and gene expression may be too slow for a particular application. To mitigate this problem, self-complimentary AAV (scAAV) genomes have been developed. scAAV is designed as a single-stranded inverted repeat, such that the vector can fold back and anneal on itself to form a double-stranded DNA [45]. This bypasses the rate-limiting, second strand synthesis step in transduction, and expression can begin much more quickly following viral uncoating. Additionally, scAAV is 5–140-fold more efficient than single-stranded AAV in certain cell types [46]. The largest limitation of scAAV is packaging capacity. AAV already hosts one of the smallest recombinant genomes of typically used gene therapy vehicles and a scAAV vector reduces this packaging capacity by half. Because of this, scAAVs are most applicable for use with small genes or in gene silencing applications where short RNAi (e.g. shRNA) cassettes are used.

3.1.3 Cre Recombinase Dependent Genomes

In cases where viral transgene expression cannot be controlled with capsid tropism, cell-specific expression can be achieved using a Cre-lox recombination system, where transgenes are flanked by Lox sites and Cre recombinase is provided in *trans*. For example, the FLEx switch genomes are designed to allow gene expression only in Cre-expressing cells [47–49]. Within these genomes, the gene of interest, flanked by two Lox sites, is cloned downstream of the promoter in the antisense direction. When transduction occurs in a cell not expressing Cre recombinase, a nonsense transcript is made and subsequently degraded, so the desired protein is not expressed. In Cre-expressing cells, the Cre-recombinase initiates a recombination event in the genome which permanently flips the transgene into the sense direction, allowing for gene expression and protein production. This approach can be useful in situations where it is desired to target a single cell type within a heterogeneous population while avoiding gene expression in other cells. This Cre-dependent approach is only plausible in a research application, and therefore not relevant in clinical applications.

3.1.4 Oversized Genomes

As discussed above, the main limitation of rAAV vectors is their relatively small DNA carrying capacity. Two approaches currently exist to circumvent this issue, fragment AAV genome reassembly (fAAV) and split AAV vectors [50]. Large gene delivery by fAAV was developed based on the observation that DNA "fragments" of different sizes and both polarities are encapsidated when the transgenic cassette exceeds the rAAV carrying capacity [51, 52]. Following transduction with multiple fAAV particles in the same cell, single-stranded DNA fragments containing complementary sequences can anneal, followed by host-mediated DNA synthesis, generating the originally intended, "oversized", genome. The split AAV method takes advantage of the natural concatemerization of viral genomes that occurs following transduction. In this approach, the desired oversized cassette is "split" and packaged into two separate rAAV vectors. Three different strategies can be used to split the cassettes: (1) Overlapping; the transgenic cassette is split into two vectors such as they contain a region of overlapping sequence homology. (2) Trans-splicing; the transgenic cassette is split into two vectors to which splice donor and splice acceptor sites are added, respectively. (3) Hybrid trans-splicing; a combination of the two, utilizing both overlapping sequences and splicing sites [53, 54]. After transduction, large genome reconstruction occurs mediated by the host DNA repair machinery through homologous recombination and non-homologous end joining mechanisms.

3.2 Promoters

Promoters are an essential factor determining control of transgene expression, and special consideration must be given to selecting the promoter for your expression cassette with respect to the intended outcome. Based on their spatial pattern of activity, promoters can be classified as ubiquitous, cell/tissue specific, or activity dependent. Based on the temporal pattern of activity they can be classified as constitutive (always active) or regulatable. Finally, based on the eukaryotic RNA polymerase leading transcription, they can be classified as RNA Polymerase II (Pol II) or RNA Polymerase III (Pol III) promoters.

3.2.1 Ubiquitous Promoters

Ubiquitous promoters usually provide robust expression in virtually all tissues and cell types. Historically, viral promoters have been some of the most widely utilized, and a few of these, such as the simian virus 40 early promoter (SV40) and the cytomegalovirus immediate-early promoter (CMV) remain popular to date. These promoters are typically smaller, making them easy to accommodate into gene therapy vectors, and they also tend to be quite strong, mediating high levels of transcription. However, the main caveat with viral promoters is that they are prone to epigenetic silencing [55], which may not be surprising considering eukaryotic cells have evolved mechanisms to detect and stop expression of viral genes and are therefore unable to sustain transgene expression in vivo.

Several ubiquitous eukaryotic promoters have been adapted over the years to achieve strong, long term transgene expression in vivo. Some of the most broadly used are human ubiquitin C promoter (UBC), human elongation factor 1α promoter (EF1A), mouse phosphoglycerate kinase 1 promoter (PGK), and chicken β-Actin promoter (CBA) [56]. Finally, synthetic promoters with enhanced performance have been created by combining elements from several different sources. The most classical example being the CAG promoter (a.k.a. CAGGS, pCBA). A hybrid promoter made by combining the CMV enhancer element, the CBA promoter, the CBA first exon and first intron, and the splice acceptor of the rabbit beta-globin gene [57].

3.2.2 Tissue/Cell Specific Promoters

The use of tissue specific promoters offers significant advantages to the majority of in vivo gene therapy applications. Overall, tissue specific promoter can increase safety and efficacy by minimizing off-target expression and associated toxic effects. In addition, tissue specific promoters can facilitate sustained transgene expression, in turn reducing the vector load necessary to obtain desired levels of expression. The list of available tissue specific or disease specific promoters has grown considerably over the years. To address particular needs, researchers have validated and engineered promoters for: (a) tissue-specific expression, such as albumin for liver [58] and myosin heavy chain for muscle [59]; cell-specific expression within a tissue, like human Synapsin for neurons and glial fibrillary acidic protein for astroglia in the brain [60]; (b) disease-specific, such as progression-elevated gene-3 for cancer [61], and microtubule associated protein 1A for Parkinson's disease [62]. A second class of condition-specific promoters incorporates endogenous response elements that are activated by different kinds of stimuli such as ionizing radiation, hypoxia [63], light [64], etc. In some cases they can self-regulate and adjust transgene expression in response to physiological changes, like the glucose-6-phosphate (G6Pase) promoter, which is both suppressed by insulin and induced by glucose [65]. Of note in the selection and design of cell specific promoters is the Pleiades Promoter Project [66]. The Pleiades Promoter Project is a large collaborative effort aimed at producing human "MiniPromoters", which constructed and vetted numerous novel promoters in vivo in the adult mouse for selective expression in the brain. Moreover, to facilitate promoter choice selection and rational de novo promoter design, several data repositories have been built that provide easy access to annotated sequences and other relevant information. They differ by the validation technique used and the kind of data they present [67]. One of the most prominent examples is the Eukaryotic Promoter Database (EPD), which is a collection of databases containing promoter information curated from published articles (EPD) and derived from high-throughput validation techniques (EPDnew) [68, 69].

3.2.3 Regulatable Promoters

The possibility to efficiently induce expression of the desired transgene within a temporally restricted manner provides an extra layer of control that can help minimize the metabolic burden and potential toxic effects of viral transduction, and to provide an external means to titrate gene expression. The creation of regulatable or inducible promoters provides this type of temporal control. In these inducible systems, promoters (ubiquitous or cell-specific) are coupled to transcription factor response elements that can modulate whether a gene is turned on or off, depending on the presence of a specific stimulus. The majority of inducible promoters engineered to date are activated by a pharmacological ligand. By far, the most commonly used is the tetracycline-controlled transcriptional activator (tTA). This system was developed over 20 years ago from the *E.coli* tetracycline—resistance operon [70]. The tTA provides a dose dependent control of gene expression in the presence (Tet-ON) or absence (Tet-OFF) of the antibiotic tetracycline or one of its derivatives (e.g. doxycycline) [71]. To date, a handful of regulatable systems have been developed which fits within the confines of a single viral vector, and which have been fully vetted for the absence of leaky expression in vivo [71, 72]. In addition to the classical TET-systems, newer regulatable systems are based on the incorporation of destabilizing domains into the coding sequence of the transgene, where an exogenous agent is applied to "shield" the protein from degradation [73, 74].

3.2.4 Pol II vs. Pol III Promoters

RNA polymerase II is responsible for transcribing most eukaryotic genes, synthesizing mRNA from genes with highly variable and complex promoters and several interacting factors. On the other hand, RNA polymerase III transcribes a handful of short non-coding RNAs such as tRNAs and 5S rRNA, among others. All Pol III promoters are similar, and have been evolutionarily optimized for precise and efficient transcription initiation [75–77]. While all promoters required for expression of coding RNAs belong to the Pol II family, Pol III promoters are used for certain gene therapy applications in which precise and consistent expression of short non-coding RNAs is needed. This includes shRNA utilized for gene silencing by RNAi, or small guide RNAs utilized for gene editing by CRISPR/Cas9 technology. The two most commonly used Pol III promoters are the human U6 small nuclear promoter (U6) and the human H1 promoter (H1) [78].

3.3 Altering or Monitoring Cellular Physiology

Understanding and ultimately altering the tightly regulated molecular processes occurring within cells is a critical challenge in biomedical research and has become of particular importance in the neurosciences. One of the more recent developments in gene therapy applications is the ability to deliver transgenes encoding proteins that can respond to changes in cellular physiology. The development of these "genetically encoded indicators" has drastically increased the ability to monitor cellular physiology and pathophysiology within living cells. Alternatively, another advance in gene therapy is the development of expression cassettes that encode for proteins that are able to directly modulate the activity of excitable cells. Specifically, optogenetics and chemogenetics have revolutionized the field of neuroscience, where investigators can now manipulate activity within specific neurons of interest.

3.3.1 Genetically Encoded Indicators

The application of genetically encoded indicators has been most extensively used in the field of neuroscience. The nervous system is a dynamic and complex collection of highly specialized cells that have been notoriously difficult to study in live animals. Genetically encoded indicators have addressed this issue, allowing investigators to longitudinally monitor many aspects of subcellular physiology in specific cell types. Currently, genetically encoded indicators exist to monitor changes in intracellular calcium, neurotransmitter concentrations, transmembrane voltage, vesicle trafficking and release, and even mitochondrial redox potential.

Fluorescent reporters of calcium flux have advanced our ability to monitor intracellular calcium changes. One of the most commonly used calcium reporters is GCaMP; a fusion protein combining a GFP molecule with calmodulin and the M13 peptide sequence from myosin light chain. In the presence of calcium, the GCaMP molecule undergoes a conformational change resulting in an increase in the fluorescence emitted from the GFP fluorophore [79]. GCaMP molecules can also be fused with proteins in order to monitor calcium dynamics in a specific subcellular region. For example, GCaMP has been successfully fused to the presynaptic protein synaptophysin to monitor changes in presynaptic calcium concentrations within a specific subset of neurons [80–82]. Alternatively, there are GCaMP-based indicators, including GCaMP3 [83], that can be used for the detection of postsynaptic calcium changes. To enable multicolor imaging in single cells, red genetically encoded calcium indicators have also been developed, including R-GECO1 [84] and RCaMP [85], as well as a presynaptically-targeted version [86].

In addition to monitoring changes in calcium, genetically encoded indicators exist to monitor changes in neuronal membrane potential or vesicular dynamics, which in turn have provided researchers a means to visualizing synaptic transmission. Genetically encoded voltage indicators can be used to visualize neuronal depolarization occurring before or after synaptic transmission, however the signals are not strong enough to resolve activity at the single-cell level [87–90]. In contrast, the development of a pH-sensitive variant of GFP (pHluorin) allows investigators to study the dynamics of individual synaptic vesicles within the axon terminal. This technology takes advantage of the acidic lumen of synaptic vesicles, which quenches the fluorescent signal of pHluorin, so that the GFP only emits signal when the vesicle membrane is disrupted during endo- or exocytosis [91]. Similarly, the fusion of pHluorin with vesicle associated SNARE molecules (e.g. synaptophysin) [92]) or neurotransmitter transporters (e.g. Vglut) [93]) provides further insight into vesicle trafficking within phenotypically identified neurons. In addition to pH-sensitive GFP indicators, there are also pH-sensitive variants of other fluorescent proteins such as pHTomato, which allow investigators to simultaneously study multiple facets of vesicle handling in a single synapse.

Lastly, genetically encoded redox sensors have also been developed. One such redox sensor, roGFP, was created by inserting two cysteine residues into a GFP molecule. The resultant fluorophore displays different fluorescence properties depending on the oxidation state of these cysteine residues. roGFPs have two fluorescence excitation maxima at approximately 400 and 490 nm, allowing for a ratiometric measurement of the redox state of the cellular environment [94]. Similar to the aforementioned genetically encoded indicators, roGFP has been successfully fused to other cellular proteins to increase its specificity. For example, fusion of roGFP with glutaredoxin enables specific visualization of the redox status of glutathione.

3.3.2 Modulating Cellular Activity

Beyond the ability to monitor changes in cellular physiology, advances in gene therapy have now also enabled researchers to directly modulate the activity of excitable cells. The two most common systems used to modulate neuronal activity are optogenetics and chemogenetics.

As the name would suggest, optogenetics is the combination of optics and genetics to modulate the activity of specific cells within living tissue. Optogenetics utilizes light, delivered to discrete anatomical areas, in order to activate microbial proteins called opsins, which are ectopically expressed in cells of interest. Opsins are light-sensitive proteins that can mediate the flow of ions across a cellular membrane in response to light. Most opsins require a chemical cofactor, all-*trans*-retinal, in order to respond to a light stimulus. However, most vertebrate cells naturally express all-*trans*-retinal, thus, optogenetics using opsins can be achieved with the viral delivery of a single opsin gene. Following exposure to light, and in the presence of all-*trans*-retinal, opsins undergo a conformational change, opening a membrane channel or pump, and facilitating the flow of ions across a membrane. Opsins from different microbes can be used to either depolarize or hyperpolarize neurons, leading to neuronal activation or inhibition, respectively. For example, photostimulation of the optogenetic protein channelrhodospin-2 (ChR2) facilitates the flow of cations through its channel, resulting in neuronal depolarization [95]. In contrast, photostimulation of the halorhodopsin optogenetic protein from *Natronomonas* facilitates the flow of chloride anions through its channel, resulting in neuronal hyperpolarization [96]. There are currently a variety of different opsins being utilized in optogenetic applications, each with a different activation spectrum, kinetics, and resultant cellular effects, allowing researchers to tailor a particular optogenetic protein to their specific experimental needs. For example, beyond ionotropic optogenetic proteins, researchers have now also fused opsins to specific G-protein receptors [97], resulting in a light sensitive protein that can directly alter second messenger concentrations.

Chemogenetics is similar to optogenetics, however in lieu of light, chemogenetics relies on synthetic, biologically inert compounds, to activate ectopically expressed receptors in order to modulate neuronal activity. The most widely used chemogenetic technique is designer receptors exclusively activated by designer drugs or DREADDs. As the name implies, the DREADD system is a two-part system composed of an artificial G-protein-coupled receptor (designer receptor) and a chemical ligand (designer drug), engineered to bind exclusively to each other. Importantly, because the chemical ligands used are completely synthetic molecules and not found in nature, they presumably have no off-target effects. Different varieties of DREADDs can be used to either activate or inhibit neurons, depending on the type of G-protein it is coupled to [98]. For example, the hM3Dq DREADD receptor is coupled to the Gq protein and activation of this pathway leads to increase in cytosolic calcium and cellular activation [99]. Alternatively, the hM4Di DREADD receptor is coupled to a Gi protein, and activation of this pathway inhibits adenylate cyclase activity, in turn decreasing cyclic adenosine monophosphate (cAMP) and ultimately resulting in membrane hyperpolarization and neuronal silencing [100].

Genetically encoded indicators and neuronal modulators have specific advantages and limitations that need to be taken into account when designing a new experiment. Genetically encoded indicators are superior to dyes and other chemical indicators as they are functional in cells over long periods of time, and can be ectopically expressed in model organisms. Optogenetics operate on the order of milliseconds, giving unprecedented temporal control. On the other hand, chemogenetics depends on second-messengers and are therefore slower to respond than optogenetics, however, this approach produces a longer controlled modulation of cellular activity.

4 Route of Administration

The final decision that must be made when designing a gene therapy application is how the virus will be delivered to the target tissue. As mentioned above this decision will be affected by the type of virus used as well as the expression cassette. From a broad perspective, delivery methods fall into three general categories: In vitro delivery, ex vivo delivery, and in vivo delivery.

4.1 In Vitro Gene Therapy

In vitro gene therapy is the simplest form of gene therapy. In this paradigm a viral vector will be delivered directly to the media of cultured cells. When performing in vitro gene therapy, special consideration should be paid to the choice of viral vector, as some types of cultured cells are refractory to transduction by certain types of virus. Moreover, transduction properties of viral vectors differ widely between in vitro and in vivo applications. LV is arguably the most common viral vectors used for transduction of cultured cells, largely due to the broad tropism of the commonly used VSV-G pseudotype. However with the continual discovery and engineering of novel AAV serotypes, this vector is also gaining popularity for in vitro use [101]. Consideration should also be paid to the type and source of sera used in the culture media, as some reports indicate that certain types of serum may inhibit transduction [102]. There are methods to increase the transduction efficiency of certain viral vectors in vitro. For example, the use of polycations, such as polybrene, can be used in conjugation with LV in order to boost transduction efficiency [102]. These positively charged molecules help reduce the electrostatic repulsion between the negatively charged membrane of the target cell and the LV envelope, increasing transduction. Finally, cellular toxicity caused by the viral vector or the use of transduction reagents such as polybrene should be closely monitored. In these cases, it may be beneficial to change the cellular media at a certain time point after the application of the viral vector in order to avoid toxicity. Ultimately, the ideal vector and transduction protocol should be empirically determined for every unique in vitro application.

4.2 Ex Vivo Gene Therapy

Ex vivo gene therapy refers to any gene therapy manipulation targeting cells or tissues that have been removed from an organism. For example, in most ex vivo applications, living cells or tissue will be removed from an organism, genetically modified, and then transplanted back into the host organism. Benefits of ex vivo gene therapy include the ability to transduce target cells in a more controlled environment, better access of the vector to target cells, and preventing the exposure of non-target tissue to genetic manipulation. Ex vivo gene therapy has been successfully used to accomplish a number of different gene therapy strategies. For example, ex vivo gene therapy can be used to express a therapeutic transgene to compensate for a mutated gene [103], to express trophic factors or other secrete able molecules that can be released by the modified cell [104], or to correct a genetic mutation using genome editing tools such as CRISPR/Cas [105]. Again, consideration should be paid to the type of viral vector selected for ex vivo gene therapy. For example, if short lived gene expression is desired, then vectors such as rAAV should be considered, however, the vast majority of ex vivo applications require stable long term expression, in which case LV is very commonly used. Other considerations are whether the researcher desires the viral payload to integrate or remain episomal, which is extremely important if the target cells will replicate follow transplantation to the host, as well as how genetic expression is regulated (i.e. constitutive versus inducible expression). Descriptions of ex vivo gene therapy to create genetically modified stem cells are detailed in Chapter 17.

4.3 In Vivo Gene Therapy

In vivo gene therapy describes the genetic manipulation of a target cell in situ within a living organism. In vivo gene therapy is arguably both the most commonly used and the most complicated form of gene therapy. This is largely due to the multitude of variables that can be manipulated and/or need to be controlled when genetically modifying cells within an organism. For example, all of the variables discussed above (e.g. expression cassette, viral vector) need to be taken into consideration when designing an in vivo gene therapy approach. In addition, the route of administration also needs to be considered. For example, is the goal to transduce one specific cell type or many cell types throughout the organism, and accordingly will the viral vector be delivered directly to the target tissue/organ or systemically administered? As mentioned above, choosing the appropriate expression cassette, viral vector, and delivery method are decisions that must be made in concert in order to design a successful in vivo gene therapy experiment. Moreover, in terms of systemic delivery, considerations such as off-target gene expression and potential immunogenicity must be taken in to consideration. A large portion of this book, Chapters 12–21, details specific routes of administration for in vivo-viral vector based genetic manipulation of various organs, including the many details that must be considered when designing in vivo gene therapy experiments. However, three overarching modes of in vivo delivery exist: (1) Systemic delivery: This is most commonly achieved via intravenous delivery, and depending on the vector/capsid and age of subject, will give rise to various patterns of transduction throughout the organism [106, 107]. In this approach, highly vascularized organs (e.g. liver) will receive a high vector load, and thus exhibit a significant degree of transgene expression. (2) Delivery to specific organs: In order to minimize vector spread, specialized approaches targeting an organ directly (e.g. portal vein delivery [108], aerosol-mediated delivery [109], intrathecal delivery, etc.) results in efficient transduction of the target organ. (3) Stereotaxic delivery to the CNS: Using precise coordinates, researchers can transduce specific circuits within the brain, without affecting others [110].

5 Conclusion

Viral vectors are arguably the most promising tools for the future of gene therapy. With constant advances in virology and biomedical engineering, viral vectors are able to accomplish an astounding repertoire of sophisticated cellular and molecular manipulations. Making oneself aware of the potential of viral vectors is the first step in utilizing these powerful tools. Within subsequent chapters of this book many of these techniques will be discussed in more detail, providing in depth protocols that can be used by any scientist to execute a successful gene therapy experiment.

References

1. Samulski RJ, Berns KI, Tan M et al (1982) Cloning of adeno-associated virus into pBR322: rescue of intact virus from the recombinant plasmid in human cells. Proc Natl Acad Sci U S A 79:2077–2081
2. Samaranch L, Salegio EA, San Sebastian W et al (2012) Adeno-associated virus serotype 9 transduction in the central nervous system of nonhuman primates. Hum Gene Ther 23:382–389
3. Castle MJ, Turunen HT, Vandenberghe LH et al (2016) Controlling AAV tropism in the nervous system with natural and engineered capsids. Methods Mol Biol 1382:133–149
4. Kaplitt MG, Leone P, Samulski RJ et al (1994) Long-term gene expression and phenotypic correction using adeno-associated virus vectors in the mammalian brain. Nat Genet 8:148–154
5. Boutin S, Monteilhet V, Veron P et al (2010) Prevalence of serum IgG and neutralizing factors against adeno-associated virus (AAV) types 1, 2, 5, 6, 8, and 9 in the healthy population: implications for gene therapy using AAV vectors. Hum Gene Ther 21:704–712
6. Peden CS, Burger C, Muzyczka N et al (2004) Circulating anti-wild-type adeno-associated virus type 2 (AAV2) antibodies inhibit recombinant AAV2 (rAAV2)-mediated, but not rAAV5-mediated, gene transfer in the brain. J Virol 78:6344–6359
7. Peden CS, Manfredsson FP, Reimsnider SK et al (2009) Striatal readministration of rAAV vectors reveals an immune response against AAV2 capsids that can be circumvented. Mol Ther 17:524–537
8. Alba R, Bosch A, Chillon M (2005) Gutless adenovirus: last-generation adenovirus for gene therapy. Gene Ther 12(Suppl 1): S18–S27
9. Gallo P, Dharmapuri S, Cipriani B et al (2005) Adenovirus as vehicle for anticancer genetic immunotherapy. Gene Ther 12(Suppl 1): S84–S91
10. Benskey MJ, Manfredsson FP (2016) Lentivirus production and purification. Methods Mol Biol 1382:107–114
11. Kobayashi K, Kato S, Inoue K et al (2016) Altering entry site preference of lentiviral vectors into neuronal cells by pseudotyping with envelope glycoproteins. Methods Mol Biol 1382:175–186
12. Cannon JR, Sew T, Montero L et al (2011) Pseudotype-dependent lentiviral transduction of astrocytes or neurons in the rat substantia nigra. Exp Neurol 228:41–52
13. Marconi P, Argnani R, Epstein AL et al (2009) HSV as a vector in vaccine development and gene therapy. Adv Exp Med Biol 655:118–144

14. Antinone SE, Smith GA (2010) Retrograde axon transport of herpes simplex virus and pseudorabies virus: a live-cell comparative analysis. J Virol 84:1504–1512
15. Schmeisser F, Weir JP (2007) Incorporation of a lambda phage recombination system and EGFP detection to simplify mutagenesis of Herpes simplex virus bacterial artificial chromosomes. BMC Biotechnol 7:22
16. Manfredsson FP, Mandel RJ (2011) The development of flexible lentiviral vectors for gene transfer in the CNS. Exp Neurol 229:201–206
17. Kato S, Inoue K, Kobayashi K et al (2007) Efficient gene transfer via retrograde transport in rodent and primate brains using a human immunodeficiency virus type 1-based vector pseudotyped with rabies virus glycoprotein. Hum Gene Ther 18:1141–1151
18. Sinn PL, Hickey MA, Staber PD et al (2003) Lentivirus vectors pseudotyped with filoviral envelope glycoproteins transduce airway epithelia from the apical surface independently of folate receptor alpha. J Virol 77:5902–5910
19. Cronin J, Zhang XY, Reiser J (2005) Altering the tropism of lentiviral vectors through pseudotyping. Curr Gene Ther 5:387–398
20. Gao G, Vandenberghe LH, Alvira MR et al (2004) Clades of Adeno-associated viruses are widely disseminated in human tissues. J Virol 78:6381–6388
21. Marsic D, Zolotukhin S (2016) Altering tropism of rAAV by directed evolution. Methods Mol Biol 1382:151–173
22. Packer MS, Liu DR (2015) Methods for the directed evolution of proteins. Nat Rev Genet 16:379–394
23. Kienle E, Senis E, Borner K et al (2012) Engineering and evolution of synthetic adeno-associated virus (AAV) gene therapy vectors via DNA family shuffling. J Vis Exp 62:3819
24. Muller OJ, Kaul F, Weitzman MD et al (2003) Random peptide libraries displayed on adeno-associated virus to select for targeted gene therapy vectors. Nat Biotechnol 21:1040–1046
25. Marsic D, Govindasamy L, Currlin S et al (2014) Vector design Tour de Force: integrating combinatorial and rational approaches to derive novel adeno-associated virus variants. Mol Ther 22:1900–1909
26. Bartel M, Schaffer D, Buning H (2011) Enhancing the clinical potential of AAV vectors by capsid engineering to evade pre-existing immunity. Front Microbiol 2:204
27. Horowitz ED, Weinberg MS, Asokan A (2011) Glycated AAV vectors: chemical redirection of viral tissue tropism. Bioconjug Chem 22:529–532
28. Zhong L, Li B, Jayandharan G et al (2008) Tyrosine-phosphorylation of AAV2 vectors and its consequences on viral intracellular trafficking and transgene expression. Virology 381:194–202
29. Kanaan NM, Sellnow RC, Boye SL et al (2017) Rationally engineered AAV capsids improve transduction and volumetric spread in the CNS. Mol Ther Nucleic Acids 8:184–197
30. Boye SL, Bennett A, Scalabrino ML et al (2016) Impact of heparan sulfate binding on transduction of retina by recombinant adeno-associated virus vectors. J Virol 90:4215–4231
31. Gulbransen BD (2017) Emerging tools to study enteric neuromuscular function. Am J Physiol 312:G420–G426
32. Bjorklund T (2016) Expression of multiple functional RNAs or proteins from one viral vector. Methods Mol Biol 1382:41–56
33. Fagoe ND, Eggers R, Verhaagen J et al (2014) A compact dual promoter adeno-associated viral vector for efficient delivery of two genes to dorsal root ganglion neurons. Gene Ther 21:242–252
34. Amendola M, Venneri MA, Biffi A et al (2005) Coordinate dual-gene transgenesis by lentiviral vectors carrying synthetic bidirectional promoters. Nat Biotechnol 23:108–116
35. Benskey MJ, Sellnow RC, Sandoval IM et al (2018) Silencing alpha synuclein in mature nigral neurons results in rapid neuroinflammation and subsequent toxicity. Front Mol Neurosci 11:36
36. Gray SJ, Foti SB, Schwartz JW et al (2011) Optimizing promoters for recombinant adeno-associated virus-mediated gene expression in the peripheral and central nervous system using self-complementary vectors. Hum Gene Ther 22:1143–1153
37. de Leeuw CN, Dyka FM, Boye SL et al (2014) Targeted CNS delivery using human minipromoters and demonstrated compatibility with adeno-associated viral vectors. Mol Ther Methods Clin Dev 1:5
38. Ngoi SM, Chien AC, Lee CG (2004) Exploiting internal ribosome entry sites in gene therapy vector design. Curr Gene Ther 4:15–31
39. Mizuguchi H, Xu Z, Ishii-Watabe A et al (2000) IRES-dependent second gene expression is significantly lower than cap-dependent

first gene expression in a bicistronic vector. Mol Ther 1:376–382

40. Kim JH, Lee SR, Li LH et al (2011) High cleavage efficiency of a 2A peptide derived from porcine teschovirus-1 in human cell lines, zebrafish and mice. PLoS One 6:e18556
41. Donnelly ML, Luke G, Mehrotra A et al (2001) Analysis of the aphthovirus 2A/2B polyprotein 'cleavage' mechanism indicates not a proteolytic reaction, but a novel translational effect: a putative ribosomal 'skip. J Gen Virol 82:1013–1025
42. Donnelly ML, Hughes LE, Luke G et al (2001) The 'cleavage' activities of foot-and-mouth disease virus 2A site-directed mutants and naturally occurring '2A-like' sequences. J Gen Virol 82:1027–1041
43. Liu Z, Chen O, Wall JBJ et al (2017) Systematic comparison of 2A peptides for cloning multi-genes in a polycistronic vector. Sci Rep 7:2193
44. McCarty DM (2008) Self-complementary AAV vectors; advances and applications. *Molecular therapy : the journal of the American Society of.* Gene Ther 16:1648–1656
45. Raj D, Davidoff AM, Nathwani AC (2011) Self-complementary adeno-associated viral vectors for gene therapy of hemophilia B: progress and challenges. Expert Rev Hematol 4:539–549
46. McCarty DM, Monahan PE, Samulski RJ (2001) Self-complementary recombinant adeno-associated virus (scAAV) vectors promote efficient transduction independently of DNA synthesis. Gene Ther 8:1248–1254
47. Atasoy D, Aponte Y, Su HH et al (2008) A FLEX switch targets Channelrhodopsin-2 to multiple cell types for imaging and long-range circuit mapping. J Neurosci 28:7025–7030
48. Saunders A, Johnson CA, Sabatini BL (2012) Novel recombinant adeno-associated viruses for Cre activated and inactivated transgene expression in neurons. Front Neural Circuits 6:47
49. Schnutgen F, Doerflinger N, Calleja C et al (2003) A directional strategy for monitoring Cre-mediated recombination at the cellular level in the mouse. Nat Biotechnol 21:562–565
50. Hirsch ML, Wolf SJ, Samulski RJ (2016) Delivering transgenic DNA exceeding the carrying capacity of AAV vectors. Methods Mol Biol 1382:21–39
51. Dong B, Nakai H, Xiao W (2010) Characterization of genome integrity for oversized recombinant AAV vector. Mol Ther 18:87–92
52. Wu Z, Yang H, Colosi P (2010) Effect of genome size on AAV vector packaging. Mol Ther 18:80–86
53. Duan D, Yue Y, Engelhardt JF (2001) Expanding AAV packaging capacity with trans-splicing or overlapping vectors: a quantitative comparison. Mol Ther 4:383–391
54. Hirsch ML, Agbandje-McKenna M, Samulski RJ (2010) Little vector, big gene transduction: fragmented genome reassembly of adeno-associated virus. Mol Ther 18:6–8
55. Hsu CC, Li HP, Hung YH et al (2010) Targeted methylation of CMV and E1A viral promoters. Biochem Biophys Res Commun 402:228–234
56. Papadakis ED, Nicklin SA, Baker AH et al (2004) Promoters and control elements: designing expression cassettes for gene therapy. Curr Gene Ther 4:89–113
57. Niwa H, Yamamura K, Miyazaki J (1991) Efficient selection for high-expression transfectants with a novel eukaryotic vector. Gene 108:193–199
58. Gorski K, Carneiro M, Schibler U (1986) Tissue-specific in vitro transcription from the mouse albumin promoter. Cell 47:767–776
59. Rindt H, Gulick J, Knotts S et al (1993) In vivo analysis of the murine beta-myosin heavy chain gene promoter. J Biol Chem 268:5332–5338
60. Kugler S (2016) Tissue-specific promoters in the CNS. Methods Mol Biol 1382:81–91
61. Su ZZ, Sarkar D, Emdad L et al (2005) Targeting gene expression selectively in cancer cells by using the progression-elevated gene-3 promoter. Proc Natl Acad Sci U S A 102:1059–1064
62. Wettergren EE, Gussing F, Quintino L et al (2012) Novel disease-specific promoters for use in gene therapy for Parkinson's disease. Neurosci Lett 530:29–34
63. Greco O, Marples B, Dachs GU et al (2002) Novel chimeric gene promoters responsive to hypoxia and ionizing radiation. Gene Ther 9:1403–1411
64. Shimizu-Sato S, Huq E, Tepperman JM et al (2002) A light-switchable gene promoter system. Nat Biotechnol 20:1041–1044
65. Chen R, Meseck ML, Woo SL (2001) Auto-regulated hepatic insulin gene expression in type 1 diabetic rats. Mol Ther 3:584–590
66. Portales-Casamar E, Swanson DJ, Liu L et al (2010) A regulatory toolbox of MiniPromoters to drive selective expression in the brain. Proc Natl Acad Sci U S A 107:16589–16594

67. Majewska M, Wysokinska H, Kuzma L et al (2018) Eukaryotic and prokaryotic promoter databases as valuable tools in exploring the regulation of gene transcription: a comprehensive overview. Gene 644:38–48
68. Dreos R, Ambrosini G, Groux R et al (2017) The eukaryotic promoter database in its 30th year: focus on non-vertebrate organisms. Nucleic Acids Res 45:D51–D55
69. Dreos R, Ambrosini G, Perier RC et al (2015) The Eukaryotic Promoter Database: expansion of EPDnew and new promoter analysis tools. Nucleic Acids Res 43:D92–D96
70. Gossen M, Bujard H (1992) Tight control of gene expression in mammalian cells by tetracycline-responsive promoters. Proc Natl Acad Sci U S A 89:5547–5551
71. Manfredsson FP, Burger C, Rising AC et al (2009) Tight Long-term dynamic doxycycline responsive nigrostriatal GDNF using a single rAAV vector. Mol Ther 17:1857–1867
72. Chtarto A, Yang X, Bockstael O et al (2007) Controlled delivery of glial cell line-derived neurotrophic factor by a single tetracycline-inducible AAV vector. Exp Neurol 204:387–399
73. Quintino L, Manfre G, Wettergren EE et al (2013) Functional neuroprotection and efficient regulation of GDNF using destabilizing domains in a rodent model of Parkinson's disease. Mol ther 21:2169–2180
74. Breger L, Wettergren EE, Quintino L et al (2016) Regulated gene therapy. Methods Mol Biol 1382:57–66
75. Arimbasseri AG, Rijal K, Maraia RJ (2014) Comparative overview of RNA polymerase II and III transcription cycles, with focus on RNA polymerase III termination and reinitiation. Transcription 5:e27639
76. Paule MR, White RJ (2000) Survey and summary: transcription by RNA polymerases I and III. Nucleic Acids Res 28:1283–1298
77. Butler JE, Kadonaga JT (2002) The RNA polymerase II core promoter: a key component in the regulation of gene expression. Genes Dev 16:2583–2592
78. Ma H, Wu Y, Dang Y et al (2014) Pol III promoters to express small RNAs: delineation of transcription initiation. Mol Ther Nucleic Acids 3:e161
79. Nakai J, Ohkura M, Imoto K (2001) A high signal-to-noise Ca(2+) probe composed of a single green fluorescent protein. Nat Biotechnol 19:137–141
80. Dreosti E, Odermatt B, Dorostkar MM et al (2009) A genetically encoded reporter of synaptic activity in vivo. Nat Methods 6:883–889
81. Li H, Foss SM, Dobryy YL et al (2011) Concurrent imaging of synaptic vesicle recycling and calcium dynamics. Front Mol Neurosci 4:34
82. Nikolaou N, Lowe AS, Walker AS et al (2012) Parametric functional maps of visual inputs to the tectum. Neuron 76:317–324
83. Tian L, Hires SA, Mao T et al (2009) Imaging neural activity in worms, flies and mice with improved GCaMP calcium indicators. Nat Methods 6:875–881
84. Zhao Y, Araki S, Wu J et al (2011) An expanded palette of genetically encoded Ca (2)(+) indicators. Science 333:1888–1891
85. Akerboom J, Carreras Calderon N, Tian L et al (2013) Genetically encoded calcium indicators for multi-color neural activity imaging and combination with optogenetics. Front Mol Neurosci 6:2
86. Walker AS, Burrone J, Meyer MP (2013) Functional imaging in the zebrafish retinotectal system using RGECO. Front Neural Circuits 7:34
87. Siegel MS, Isacoff EY (1997) A genetically encoded optical probe of membrane voltage. Neuron 19:735–741
88. Sakai R, Repunte-Canonigo V, Raj CD et al (2001) Design and characterization of a DNA-encoded, voltage-sensitive fluorescent protein. Eur J Neurosci 13:2314–2318
89. Ataka K, Pieribone VA (2002) A genetically targetable fluorescent probe of channel gating with rapid kinetics. Biophys J 82:509–516
90. Perron A, Mutoh H, Launey T et al (2009) Red-shifted voltage-sensitive fluorescent proteins. Chem Biol 16:1268–1277
91. Miesenbock G, De Angelis DA, Rothman JE (1998) Visualizing secretion and synaptic transmission with pH-sensitive green fluorescent proteins. Nature 394:192–195
92. Granseth B, Odermatt B, Royle SJ et al (2006) Clathrin-mediated endocytosis is the dominant mechanism of vesicle retrieval at hippocampal synapses. Neuron 51:773–786
93. Balaji J, Ryan TA (2007) Single-vesicle imaging reveals that synaptic vesicle exocytosis and endocytosis are coupled by a single stochastic mode. Proc Natl Acad Sci U S A 104:20576–20581
94. Hanson GT, Aggeler R, Oglesbee D et al (2004) Investigating mitochondrial redox potential with redox-sensitive green fluorescent protein indicators. J Biol Chem 279:13044–13053
95. Boyden ES, Zhang F, Bamberg E et al (2005) Millisecond-timescale, genetically targeted

optical control of neural activity. Nat Neurosci 8:1263–1268
96. Zhang F, Wang LP, Brauner M et al (2007) Multimodal fast optical interrogation of neural circuitry. Nature 446:633–639
97. Spoida K, Eickelbeck D, Karapinar R et al (2016) Melanopsin variants as intrinsic optogenetic on and off switches for transient versus sustained activation of G protein pathways. Curr Biol 26:1206–1212
98. Farrell MS, Roth BL (2013) Pharmacosynthetics: reimagining the pharmacogenetic approach. Brain Res 1511:6–20
99. Alexander GM, Rogan SC, Abbas AI et al (2009) Remote control of neuronal activity in transgenic mice expressing evolved G protein-coupled receptors. Neuron 63:27–39
100. Armbruster BN, Li X, Pausch MH et al (2007) Evolving the lock to fit the key to create a family of G protein-coupled receptors potently activated by an inert ligand. Proc Natl Acad Sci U S A 104:5163–5168
101. Ellis BL, Hirsch ML, Barker JC et al (2013) A survey of ex vivo/in vitro transduction efficiency of mammalian primary cells and cell lines with Nine natural adeno-associated virus (AAV1-9) and one engineered adeno-associated virus serotype. Virol J 10:74
102. Denning W, Das S, Guo S et al (2013) Optimization of the transductional efficiency of lentiviral vectors: effect of sera and polycations. Mol Biotechnol 53:308–314
103. Hacein-Bey-Abina S, Le Deist F, Carlier F et al (2002) Sustained correction of X-linked severe combined immunodeficiency by ex vivo gene therapy. N Engl J Med 346:1185–1193
104. Tuszynski MH, Thal L, Pay M et al (2005) A phase 1 clinical trial of nerve growth factor gene therapy for Alzheimer disease. Nat Med 11:551–555
105. Li HL, Fujimoto N, Sasakawa N et al (2015) Precise correction of the dystrophin gene in duchenne muscular dystrophy patient induced pluripotent stem cells by TALEN and CRISPR-Cas9. Stem Cell Reports 4:143–154
106. Duan D (2016) Systemic delivery of adeno-associated viral vectors. Curr Opin Virol 21:16–25
107. Zincarelli C, Soltys S, Rengo G et al (2008) Analysis of AAV serotypes 1-9 mediated gene expression and tropism in mice after systemic injection. Mol Ther 16:1073–1080
108. Nakai H, Herzog RW, Hagstrom JN et al (1998) Adeno-associated viral vector-mediated gene transfer of human blood coagulation factor IX into mouse liver. Blood 91:4600–4607
109. Gruntman AM, Mueller C, Flotte TR et al (2012) Gene transfer in the lung using recombinant adeno-associated virus. Curr Protoc Microbiol Chapter 14:Unit14D.12
110. Benskey MJ, Manfredsson FP (2016) Intraparenchymal Stereotaxic Delivery of rAAV and Special Considerations in Vector Handling. Methods Mol Biol 1382:199–215

Part II

Novel Modes of Gene Therapy (Going Beyond Overexpression and Knockdown)

Chapter 2

Design and Assembly of CRISPR/Cas9 Lentiviral and rAAV Vectors for Targeted Genome Editing

Ivette M. Sandoval, Timothy J. Collier, and Fredric P. Manfredsson

Abstract

Clustered regularly interspaced short palindromic repeat (CRISPR/Cas) system has emerged as an extremely useful tool for biological research and as a potential technology for gene therapy approaches. CRISPR/Cas mediated genome editing can be used to easily and efficiently modify endogenous genes in a large variety of cells and organisms. Furthermore, a modified version of the Cas9 nuclease has been developed that can be used for regulation of endogenous gene expression and labeling of genomic loci, among other applications. This chapter provides an introduction to the basis of the technology and a detail protocol for the most classic application: gene inactivation by CRISPR/Cas9 nuclease system from *Streptococcus pyogenes*. This workflow can be easily adapted for other CRISPR systems and applications.

Key words Gene editing, Gene inactivation, CRISPR/Cas9, Cas9 nuclease, LV, AAV, guideRNA

1 Introduction

The ability to introduce desired genomic modification at a specific DNA sequence holds enormous potential for research and biomedical applications. Great progress in targeted gene editing has been enabled by the development of programmable nucleases. These enzymes allow the introduction of a double strand break (DSB) at the genomic locus to be modified. Repair of the break by the error-prone non-homologous end joining (NHEJ) pathway introduce insertion/deletion (indels) mutations which can disrupt the reading frame causing early stop codons, hence blocking gene expression. Alternatively, in the presence of a homologous DNA template, homology-directed repair (HDR) can allow precise introduction of a specific DNA sequence. Three major kinds of designer nucleases have been developed thus far [1]: ZFNs (zinc-finger nucleases), TALENs (transcription activator-like effectors nucleases), and CRISPR/Cas (clustered regularly interspaced palindromic repeats and its CRISPR-associated proteins: Cas). ZFNs

Fredric P. Manfredsson and Matthew J. Benskey (eds.), *Viral Vectors for Gene Therapy: Methods and Protocols*, Methods in Molecular Biology, vol. 1937, https://doi.org/10.1007/978-1-4939-9065-8_2,

and TALENs technologies use a modular arrangement of DNA-binding domains for sequence recognition. The CRISPR-Cas technology, on the other hand, uses a small RNA moiety with complementarity to the target DNA; therefore, the system is considerably easier and cheaper to implement and has quickly become the gene-editing tool of choice.

CRISPR-Cas9 is an adaptive immune system mechanism used by a variety of bacteria to detect and eliminate foreign genetic elements. The most current classification encompasses two major classes, six types, and sixteen subtypes: Class 1 effectors, which utilize multi-protein complexes and include Type I, III, and IV systems, and Class 2 effectors which rely on a single component protein and include Type II, V, and VI systems [2, 3]. In the simplest and best characterized version, the Type II CRISPR system (Fig. 1a), bacteria incorporate short sequences from invading DNA (known as protospacer) into its genome in a distinctive array of repetitive elements. Each protospacer sequence is 20 nucleotides long and in the target DNA this sequence is immediately followed by a protospacer adjacent motif (PAM), a short stretch of DNA required for Cas9 binding and activity. The CRISPR array is transcribed and processed into CRISPR RNAs (crRNAs), which hybridize with another auxiliary RNA called trans-activating CRISPR RNA (tracrRNA), and together they associate with the protein Cas9 nuclease. The RNA/Cas9 complex then binds and cleaves DNA with homology to the protospacer sequences [4].

The Type II CRISPR system from *Streptococcus pyogenes* was recently adapted as a tool for gene editing [5–7]. The most commonly used version consists of two components: the nuclease SpCas9 and a guide RNA (gRNA), which is a fusion of a crRNA and a tracrRNA (Fig. 1b). In this manner the Cas9 nuclease can be directed to a specific genomic location simply by modifying the 20 nucleotide protospacer sequence within the gRNA. The protospacer sequence can be easily designed by following Watson-Crick base complementarity rules to target any genomic location preceded by a PAM sequence. For spCas9 the PAM sequence motif is 5′-NGG (Fig. 1c), but it varies for different Cas9 orthologs [8]. Further, mutant variants of SpCas9 with altered PAM specificities have been created to help expand the number of potential genomic target loci [9].

A major concern with any genome targeting technology is the potential introduction of unwanted mutations at off-target locations. It is known that Cas9, like other nucleases, can cleave the genome at off-target sites with varying frequencies. Considerable effort has been invested in understanding the basis of Cas9 specificity and to develop approaches to minimize off-target effects. One method is to use a “nickase” mutant version of SpCas9. The Cas9 enzyme contains two nuclease domains: the RuvC-like domain and the HNH domain, each responsible for cleavage of one strand of

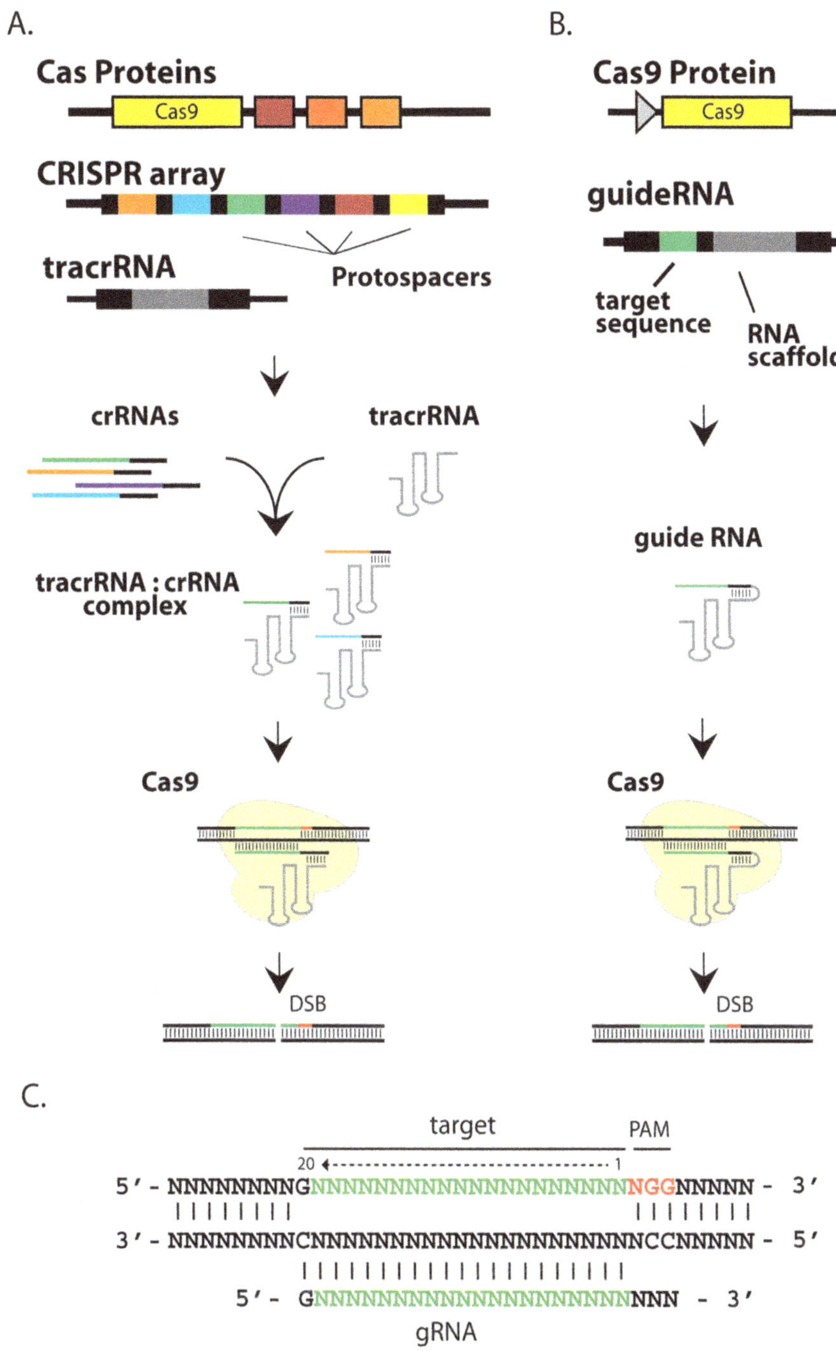

Fig. 1 Schematic of CRISPR Type II system and CRISPR/Cas9 gene-editing tool. (**a**) Naturally occurring CRISPR systems incorporate small fragment of foreign DNA (protospacers) into their genomes as CRISPR arrays. The array is processed into single crRNAs which hybridize with tracrRNAs. The pair associates with Cas9 nuclease. This complex can then bind and cleave target DNA bearing the protospacer sequence. (**b**) The most common CRISPR/Cas9 system uses a single guideRNA, a fusion of a crRNA with a portion of the tracrRNA sequence, to mediate cleavage by the Cas9 nuclease. (**c**) Example of guideRNA and target DNA sequences, 20 nucleotides of complementarity are shown in green and PAM sequence in red

the DNA duplex, together creating a blunt DSB. Point mutations that inactivate the catalytic activity of either one of the nuclease domains can turn the Cas9 nuclease into a "nickase" that creates a single strand break (SSB). A double nicking approach using a pair of gRNAs properly spaced can considerably improve specificity relative to the wild-type Cas9 nuclease [10–13]. Alternatively, variants of SpCas9 with enhanced specificity can be used. Researchers from the Zheng laboratory at the Broad Institute and the Joung laboratory at Harvard Medical School rationally mutated SpCas9 and created two novel variants: eSpCas9 [14] and SpCas9-HF1 [15], respectively. The mutations reduce the binding energy between the Cas9, the gRNA, and the target DNA resulting in decreased binding to off-target sites, yielding enhanced specificity. Other factors affecting off-target activity that should be taken into account and optimized to the greatest extent possible include the structure of the gRNA [16] and the concentration and time of exposure of the Cas9 enzyme.

Because of its simplicity, CRISPR/Cas9 technology has been quickly adopted by the research community. CRISPR/Cas9 has been successfully utilized for NHEJ-mediated gene knockout and HDR-mediated gene editing to generate cellular and animal models in many species, some intractable through traditional genetic engineering techniques. Also, the modular nature of the CRISPR array makes the system amenable for multiplexing, either simultaneously or in parallel. As such, it has been used for large-scale functional genomic screens using libraries [17, 18], and for targeting multiple genes together [6, 19]. Moreover, by tethering different effector molecules to a catalytically dead version of SpCas9 (dCas9), the CRISPR/Cas9 system has been used successfully for a number of targeted genome engineering applications beyond editing, including transcriptional activation [20, 21], transcriptional repression [22], DNA methylation [23], and DNA labeling [24], among others. Finally, while the vast majority of the current applications and research use CRISPR/Cas9 from *Streptococcus pyogenes,* there has been a recent focus to characterize orthologous CRISPR/Cas systems and adapt them as gene-editing tools [8, 25]. A few examples include: the Type II Cas9 system from *Staphylococcus aureus* (SaCas9) [26, 27], the Type V Cas12a system from *Francisella tularensis* (Cpf1) [28], and Type VI Cas13b system from *Leptotrichia wadei* (LwaCas13a) [29].

In order to use a CRISPR system, expression of both components, the Cas effector protein and the gRNA, is required in the target cells. The specific application and host cell/tissue type will determine the choice of method for delivery [30, 31]. Among the viral-mediated methods, recombinant adeno-associated virus (rAAV) is the vector of choice for most in vivo applications due to its selective tropism and low toxicity. However, due to its small carrying capacity (~4.7 Kb) and the large size of the SpCas9

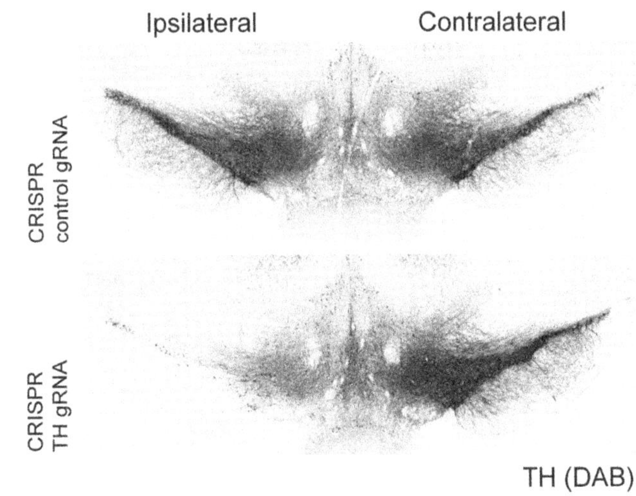

Fig. 2 AAV-CRISPR/Cas9-mediated targeting of the tyrosine hydroxylase (TH) gene in the rat *substantia nigra*. Representative images show robust reduction of TH immuno-reactivity only in the injected (bottom left) hemisphere of CRISPR-THgRNA-treated rats as compared to the intact hemisphere (right), or control gRNA-injected hemisphere (top)

protein, the applications of rAVV are limited. Nonetheless, two alternatives have been successfully implemented thus far: (1) a dual AAV vector system, with one AAV carrying the SpCas9 protein and a second AAV carrying the gRNA [32] and (2) The smaller Cas9 ortholog *Staphylococcus aureus:* SaCas9, nearly 1 Kb shorter than SpCas9, that can be accommodated into the AAV genome along with the gRNA [27]. Lentiviruses (LVs) on the other hand, are preferred for most ex vivo approaches. These vectors have a much larger carrying capacity than AAVs (7–9 Kb); thus, they offer more flexibility for design of expression cassettes. A single LV vector can usually accommodate the protein Cas9, the gRNA expression cassette and sometimes a reporter or selection marker. Further, LVs can efficiently and reproducibly transduce a large variety of cells, including difficult to transfect cell lines, primary cell cultures, and differentiated cells. In addition, because of the ability to integrate into the host genome LVs vectors offer the opportunity to easily create transgenic cell lines if desired.

In this chapter we describe a protocol for implementation of the most classic application: gene inactivation using CRISPR/SpCas9, including considerations on genomic target selection, gRNA design, construction of targeting constructs, as well as a cell-based assay for screening efficiency of genomic cleavage prior to packaging into LV or rAAV vectors. Using the same workflow we have succesfully knock-out the tyrosine hydroxylase gene from the midbrain of adult rats (Fig. 2). The protocol can be easily adapted to other desired applications or different CRISPR/Cas9 ortholog systems.

2 Materials

1. Backbone plasmids (*see* **Note 1**): pLentiCRISPRv2 (Addgene plasmid #52961, a gift from Feng Zhang). gRNA Cloning Vector [5] (Addgene plasmid #41824, a gift from George Church).
2. Oligonucleotides. Standard de-salted synthesis is adequate for all cloning procedures described in this protocol. Upon arrival, resuspend each oligo in sterile water to a final concentration of 100 μM.
3. Restriction enzymes and reaction buffers (*see* **Note 2**): BsmBI (Esp3I) and AflII (BspTI).
4. Recombinant alkaline phosphatase.
5. T4 polynucleotide kinase: T4 PNK.
6. Ligase enzyme.
7. Gibson Assembly® Master Mix (New England Biolabs, Ipswich, MA, USA).
8. Bacterial competent cells: One Shot Stbl3™ Competent *E. coli* cells (Invitrogen, Waltham, MA, USA) or Sure2® Supercompetent cells (Agilent Technologies, Santa Clara, CA, USA) and bacterial growth reagents (culture broth and agar plates, antibiotic) (*see* **Note 3**).
9. DNA Gel extraction kit and PCR purification kit.
10. Plasmid DNA purification miniprep and maxiprep kits.
11. Genomic DNA extraction kit.
12. Standard gel DNA electrophoresis equipment and reagents: agarose; 1× TBE (Tris-Borate-EDTA) running buffer: 89 mM Tris, pH 7.6, 89 mM boric acid, 2 mM EDTA; Ethidium bromide, DNA ladder marker.
13. SURVEYOR® Mutation Detection Kit (Transgenomics LTD, Omaha, NE, USA) (*see* **Note 4**).
14. High-fidelity DNA polymerase.
15. High-definition Agarose.
16. Cell line selected for mutation detection screen.
17. Cell culture reagents: Growth medium, dissociation reagent, and culture plates.

3 Methods

3.1 Select Target Sequence and Design gRNAs

1. Obtain the DNA sequence of the gene to be targeted, including annotations for introns, exons, and other relevant regulatory elements. This can be done using any available genome

database such as the National Center for Biotechnology Information (NCBI, https://www.ncbi.nlm.nih.gov) or Genome Browser from University of California Santa Cruz (https://genome.ucsc.edu).

2. Select the specific genomic region to be targeted based on the desired outcome. (a) For gene inactivation by NHEJ a general rule of thumb to follow is to design the gRNA to target the gene within 5–65% of the protein-coding regions. Avoiding genomic region too proximal to the N-terminus of the protein lessens the chance of the cell to utilize an alternative ATG start codon. Avoiding the genomic region close to the C-terminus of the protein mitigates the chances of creating a mutant truncated version of the protein. Targeting exons that code for known essential functional domains is also a good strategy to achieve gene inactivation. (b) For gene editing through HDR the target sequence for the gRNA needs to be close to the location where the edit is desired (*see* **Note 5**). (c) For gene activation or gene repression the gRNA needs to be targeted to the promoter region.
3. The specificity of the Cas9 nuclease protein is determined by a 20 nucleotide sequence in the gRNA with complementarity to the gene of interest, followed in the 3′-end by the PAM motif. For SpCas9 the PAM sequence is 5′-NGG (*see* **Note 6**). Multiple online tools are currently available that can identify potential target sites within a given sequence (*see* **Note 7**), providing a list of gRNAs design ranked based on their predicted on-target and off-target activity. E-CRISP (www.e-crisp.org/E-CRISP/) [33] and CRISPR Design tool (crispr.mit.edu) both provide functional and user-friendly platforms for gRNA design.
4. For in vivo gene therapy applications, it is recommended that 3–5 different gRNAs are designed for each desired target, and their efficiency of cleavage be empirically tested in cell culture in order to select the best candidates prior to packaging into viral vectors (*see* **Note 8**).

3.2 Synthesize gRNA and Build Plasmid Vectors

Once gRNAs have been designed they need to be cloned into a plasmid backbone. The exact cloning strategy depends on the specific plasmid vector that was chosen based on the delivery method and the tissue target. In most cases, plasmids have been built to allow insertion of targeting gRNA oligonucleotides by standard restriction-ligation cloning or by Gibson Assembly [34, 35]. As an example, we will describe in detail how to clone designed gRNAs into two plasmids backbones (*see* **Notes 1** and **9**): (1) pLentiCRISPRv2 [36] and (2) gRNA_Cloning Vector [5].

3.2.1 Cloning of gRNAs Target Sequence into pLentiCRISPRv2 Plasmid by Restriction-Ligation

pLentiCRISPRv2 is a popular all-in-one plasmid for the generation of lentivirus vectors expressing SpCas9 from ubiquitous EFS promoter and a cassette for easy cloning of single gRNAs. The expression cassette also contains puromycin-resistant gene for selection of positive transformants. Multiple versions of this plasmid currently exist that contain different reporter genes or selection markers (i.e., green fluorescent protein; mCherry fluorescent protein; blasticidin resistance).

1. Order oligonucleotides with appropriate adaptor sequences. For pLentiCRISPRv2:

   Oligo F: 5′-CACCG$(N)_{20}$-3′.

   Oligo R: 5′-AAAC$(N)_{20}$C-3′.

2. Anneal and phosphorylate oligo pairs: Prepare mixture as indicated in Table 1 (*see* **Note 10**) and incubate in thermocycler: (37 °C × 30 min); (95 °C × 5 min, cool down to 25 °C at 5 °C/min rate). Dilute annealed oligonucleotides 1:200 in sterile water.
3. Prepare pLentiCRISPRv2 entry plasmid for ligation by digesting with BsmBI restriction enzyme (*see* **Note 11**): Mix components as indicated in Table 2 (*see* **Note 12**) and incubate at 37 °C for 30 min. Run entire reaction in a 1% agarose gel and

Table 1
Oligo annealing reaction

Oligo Sense (100 μM)	1 μL
Oligo Antisense (100 μM)	1 μL
T4 ligation buffer (10×)	1 μL
T4 PNK	0.5 μL
ddH_2O	6.5 μL
Total	10 μL

Table 2
Entry vector digestion reaction

Plasmid DNA	XX μL (4 μg)
Restriction enzyme Buffer (10×)	5 μL
Alkaline phosphatase	2.5 μL
Restriction enzyme	2.5 μL
ddH_2O	XX μL
Total	50 μL

Table 3
Ligation reaction

Linearized entry plasmid	XX μL (50 ng)
Insert (oligo duplex)	1 μL
Ligase buffer (2×)	5 μL
ddH_2O	XX μL
Ligase	0.5 μL
Total	10 μL

purify the larger band (~12.8 Kb) using a DNA gel extraction kit. Elute in 30 μL of sterile water. The pLentiCRISPRv2 vector contains a 2 Kb filler sequence between the BsmBI restriction sites.

4. Set up ligation reaction: Mix components as indicated in Table 3 (*see* **Note 13**), adding ligase enzyme last. Mix gently by pipetting up and down, spin down briefly and incubate at room temperature for 10 min. Transfer mix to ice for transformation or store at −20 °C for future use.
5. Transform ligation mixture into One Shot Stbl3™ Competent *E. coli* cells by heat shock according to manufacturer's instructions (*see* **Note 3**).
6. Next day pick 5–6 single colonies for screening: grow overnight individually in 5 mL of culture broth and extract plasmid DNA using a miniprep DNA purification kit. Confirm insertion of the gRNA oligo by Sanger sequencing using LKO1 primer: 5′-GACTATCATATGCTTACCGT-3′.

3.2.2 Cloning of gRNAs Target Sequence into gRNA_Cloning Vector by Gibson Assembly

gRNA_Cloning Vector is a simple plasmid with an empty gRNA expression cassette for cloning of specific single gRNAs. The gRNA expression cassette can then be easily transferred by standard cloning techniques into any suitable plasmid backbone for generation of LV or rAAV. A second cassette for expression of the Cas9 is required for gene editing.

1. Order oligonucleotides with appropriate adaptor sequences. For gRNA_Cloning Vector:

   Oligo F: 5′-TTTCTTGGCTTTATATATCTTGTGGAAAGG ACGAAACACCG$(N)_{19}$-3′.

   Oligo R: 5′-AAAC$(N)_{19}$C-3′.
2. Anneal oligonucleotide pairs and extend to create a 100 bp double-stranded DNA fragment using a polymerase reaction: Mix components as indicated in Table 4 and incubate in thermocycler using the following program:

Table 4
Oligo annealing by PCR

Oligo Sense (25 μM)	1.25 μL
Oligo Antisense (25 μM)	1.25 μL
Reaction buffer (5×)	5 μL
High-fidelity polymerase	0.25 μL
10 mM dNTPs	0.5 μL
ddH_2O	16.75 μL
Total	25 μL

Table 5
Gibson Assembly reaction

pTR-gRNA-GFP plasmid	XX μL (50 ng)
Insert (oligo duplex)	1 μL
Master Mix (NEB) (2×)	5 μL
ddH_2O	XX μL
Total	20 μL

(98 °C × 30 s) × 1 cycle; (98 °C × 30 s, 62 °C × 15 s, 72 °C × 15 s) × 4 cycles.

(72 °C × 5 min) × 1 cycle. Product can be purified or used directly for Gibson Assembly.

3. Prepare gRNA-Cloning vector for ligation by digesting with AflII restriction enzyme: Mix components as indicated in Table 2 (*see* **Note 12**) and incubate at 37 °C for 30 min. Purify digested plasmid using a PCR purification kit and elute in 30 μL of sterile water.
4. Set up Gibson Assembly reaction: Mix reagents as indicated in Table 5 and incubate at 50 °C for 15 min.
5. Use 2 μL of mix to transform bacterial competent cells according to manufacturer's instructions (*see* **Note 3**).
6. Extract plasmid DNA using a plasmid purification kit and confirm insertion of the gRNA by sequencing using U6 promoter primer: 5′-GACTATCATATGCTTACCGT-3′.

3.3 Screening of gRNAs by On-Target Mutation Analysis

The SURVEYOR® mutation detection assay is a quick and easy way to assess gene targeting efficiency by quantifying the amount of indels at the expected genomic location. It is based on the ability of the Surveyor enzyme of the CEL family of endonucleases to detect

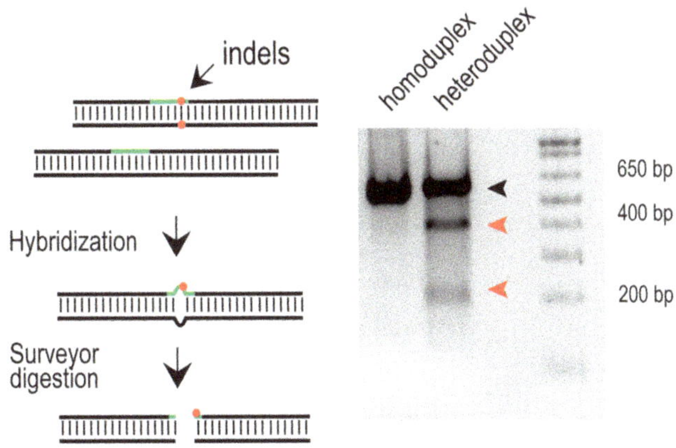

Fig. 3 Example of SURVEYOR® Mutation detection assay. Two DNA fragments containing a single mismatch were amplified by PCR and annealed alone (homoduplex control), or combined in equal amounts (heteroduplex test sample). Hybridized DNA was treated with Surveyor Nuclease. Digested DNA fragments were analyzed in 2% high-definition agarose in 1× TBE buffer. Red and black arrowheads indicate cleaved and parental DNA fragments, respectively

and cleave mismatches in DNA heteroduplexes [37, 38] (*see* **Note 14**; Fig. 3).

1. Select a cell line that contains the gene of interest in its genome (*see* **Note 15**). If the host cell line is amenable for efficient transfection, the plasmids constructs containing the CRISPR components can be directly delivered by lipofection or electroporation. Alternatively, cells can be transduced using LVs.
2. Seed cells in a 96-well plate. Next day, transfect plasmids or transduce with LV, using standard methodology (*see* **Note 16**). Incubate for 48 h and collect cells for analysis. Biological replicates are recommended.
3. Extract genomic DNA from cells. This can be done by standard phenol:chloroform extraction protocol or using a kit.
4. Amplify genomic region of interest by PCR from test (mutated) and reference (wild-type) DNA (*see* **Notes 17** and **18**). Mix reagents as indicated in Table 6 and incubate in thermocycler using the following program:

   (98 °C × 30 s) × 1 cycle; (98 °C × 30 s, 62 °C × 30 s, 72 °C × 30 s) × 30 cycles.

   (72 °C × 5 min) × 1 cycle. Purify PCR products and elute in sterile water.
5. Prepare DNA heteroduplexes sample by mixing equal amounts of test and reference DNA. Also prepare a homoduplex background control, using only reference DNA. Mix all reagents as

Table 6
Example of PCR reaction using high-fidelity polymerase

Component	Volume	Final concentration
Reaction buffer (10×)	10 μL	1×
Primer oligo sense (10 μM)	2.5 μL	0.5 μM
Primer oligo antisense (10 μM)	2.5 μL	0.5 μM
dNTP mix (10 mM)	1 μL	200 μM
DMSO	1.5 μL	3%
Polymerase	0.5 μL	1 unit
Template genomic DNA	XX μL	100 ng
ddH_2O	XX μL	
Total	50 μL	

Table 7
Heteroduplex formation reaction

200–400 ng of purified PCR product	XX μL
Hybridization buffer (10×)	1 μL
ddH_2O	XX μL
Total	10 μL

indicated in Table 7. Hybridization buffer (10×): 100 mM Tris–HCl pH 8.8, 15 mM $MgCl_2$, 50 mM KCl (*see* **Note 19**). Hybridize by denaturing and reannealing PCR products using a thermocycler using the following program:

(94 °C × 5 min); (94 °C–85 °C ramp down at −2 °C/s); (85 °C–25 °C ramp down at −0.1 °C/s).

6. Digest with Surveyor nuclease enzyme as follows: Add 1 μL of Enhancer S solution, mix by pipetting. Add 1 μL of Nuclease S solution, mix by pipetting. Incubate at 42 °C for 60 min. Add 1 μL of Stop solution.
7. Analyze fragments by standard agarose gel electrophoresis (*see* **Note 20**). The percentage of indels can be calculated from the relative intensities of the DNA bands. Fraction cleaved: (sum of cleaved fragments bands)/(sum of all bands, cleaved and intact fragments).

3.4 Screening of Off-Target Mutations

1. When possible, potential off-target sites should be scanned at a genome-wide scale. Otherwise, some high-risk sites should be screened by DNA sequencing or using Surveyor mutation detection assay (as described previously in Subheading 3.3). The majority of guide RNA design software provide a list of the most likely off-target sites with each guideRNA.

4 Notes

1. An ample variety of plasmids exist for the purpose of implementing CRISPR-based applications. Selection of the appropriate plasmid must be made based on the needs of each specific experiment, taking into account (choice of promoter, Cas9 variant, selection markers, convenient cloning, etc.). Plasmids can be acquired from Addgene plasmid repository or from commercial sources.
2. Required restriction enzymes depend on each specific cloning strategy.
3. The choice of bacterial competent cell must be made based on the characteristics of each cloning construct. LV transfer plasmids contain long terminal repeat sequences (LTR) and AAV transfer plasmids contain inverted terminal repeat sequences (ITR), both of which are prone to recombination [39, 40]. Therefore, recombination-deficient bacteria must be used for DNA amplification of LV or AAV plasmid constructs.
4. Other commercial options are available, such as GeneArt Genomic Cleavage Detection Kit (Life Technologies) or T7 endonuclease (NEB).
5. For HDR-mediated gene editing it is also necessary to supply a fragment of DNA to act as a repair template. The repair template requires certain amount of homology around the target sequence. Also, the repair template must be impervious to binding of the gRNA to avoid reintroduction of DSBs by Cas9. In general, HDR efficiency is much lower than NHEJ, and highly dependent on the host cell and stage of the cell cycle [41, 42]. Therefore, it is important to experimentally confirm the introduction of the desired genetic change by sequencing, and isolate individual clones when necessary.
6. Different Cas9 orthologs have different PAM sequence binding specificity.
7. CRISPR Software Matchmaker from Addgene compiled and compared features of several available software tools to help choosing the best one for the project needs.
8. Several, known and currently unknown, elements inherent to the gRNA can affect the cleavage efficiency by Cas9 nuclease.

For example, a gRNA with a G nucleotide located 1 bp upstream of the PAM sequence may be more efficacious than a gRNA with a C nucleotide, despite both having a perfect match to the target sequence.

9. We mentioned restriction enzymes specific for the examples provided here. However molecular biology reagents and protocols used have to be adapted according to the plasmid vector chosen and the cloning strategy.
10. PNK requires ATP for activity. Ensure buffer is fresh without repeated freeze-thaw cycles. Alternatively, supplement reaction with 1 mM ATP.
11. BsmBI restriction enzyme requires DTT for activity; adjust reaction to contain 10 mM final concentration. BsmBI can also be sold as its isoschizomer Esp3I (having the same recognition site and cleavage specificity).
12. Table 2 contains a general reference for preparing an entry plasmid for cloning by standard restriction-ligation. The plasmid must be linearized by digestion with a specific restriction enzyme(s). Dephosphorylation of the ends using a recombinant phosphatase (e.g., calf intestine or shrimp alkaline phosphatase) limits recircularization of the plasmid reducing background of nonrecombinant clones. Reagent concentrations and incubation times must be adjusted according to each specific enzyme manufacturer's instructions.
13. Reagent concentrations and reaction conditions vary depending on the specific ligase used. Adjust accordingly.
14. Other strategies can also be used to quantify the presence of mutations such as high-resolution melt-curve analysis [43], restriction fragment length polymorphism analysis, and heteroduplex mobility assay [44, 45].
15. If no cell lines with the correct genome are available, a common and easy to transfect cell line (such as human embryonic kidney HEK-293 cells) can be used. In this case the target sequence is co-transfected with the CRISPR/Cas components.
16. If not known, perform preliminary experiments in order to optimize transfection conditions. In the same manner, a range of conditions should be tested when transducing a cell line with LVs for the first time in order to determine the optimal multiplicity of infection (MOI) to obtain maximum transduction efficiency.
17. We recommend amplifying fragments in the 400 to 2000 bp size range, as they are more easily resolved from digested products in standard agarose gels. Design primers at a distance from the expected cleavage site so that: (1) cleaved products are at least 80 bp long, in order to be efficiently visualized in

agarose gels; (2) larger expected cleaved product is less than 85% of the total fragment size, so they can be effectively resolved from undigested product; (3) fragments of different sizes are produced after digestion, indicating the relative location of the mismatches.

18. The quality of the PCR amplicon is critical for a successful Surveyor nuclease assay. Using a high-fidelity polymerase is strongly recommended to avoid introduction of unwanted mutations that can confound the results. It is very important that there is a single species amplified, visible as a single band while analyzed by agarose gel electrophoresis. Optimize primers and PCR condition to this end. Gel purify PCR product if necessary. Concentration of the PCR product should be between 25 and 80 ng/μL, ideally ~50 ng/μL. Repeat PCR and/or concentrate DNA if needed.
19. Optimal Buffer composition is important for oligo duplex formation reaction and for digestion with the Surveyor enzyme. Some PCR buffers are compatible with the reaction; check the SURVEYOR® mutation detection assay user manual for details.
20. For larger DNA fragments (≥1000 bp) standard 1% agarose gels is suitable. For smaller fragments in the 200–1000 bp range it is better to use 2% high-definition agarose gels in 1× TBE buffer or 10% polyacrylamide gels. Also, if the equipment is available, fragments can be analyzed by electrofluidic electrophoresis using Agilent Bioanalyzer.

References

1. Gaj T, Gersbach CA, Barbas CF 3rd (2013) ZFN, TALEN, and CRISPR/Cas-based methods for genome engineering. Trends Biotechnol 31(7):397–405
2. Makarova KS et al (2015) An updated evolutionary classification of CRISPR-Cas systems. Nat Rev Microbiol 13(11):722–736
3. Koonin EV, Makarova KS, Zhang F (2017) Diversity, classification and evolution of CRISPR-Cas systems. Curr Opin Microbiol 37:67–78
4. Horvath P, Barrangou R (2010) CRISPR/Cas, the immune system of bacteria and archaea. Science 327(5962):167–170
5. Mali P et al (2013) RNA-guided human genome engineering via Cas9. Science 339 (6121):823–826
6. Cong L et al (2013) Multiplex genome engineering using CRISPR/Cas systems. Science 339(6121):819–823
7. Jinek M et al (2012) A programmable dual-RNA-guided DNA endonuclease in adaptive bacterial immunity. Science 337 (6096):816–821
8. Pyzocha NK, Chen S (2018) Diverse class 2 CRISPR-cas effector proteins for genome engineering applications. ACS Chem Biol 13 (2):347–356
9. Kleinstiver BP et al (2015) Engineered CRISPR-Cas9 nucleases with altered PAM specificities. Nature 523(7561):481–485
10. Mali P et al (2013) CAS9 transcriptional activators for target specificity screening and paired nickases for cooperative genome engineering. Nat Biotechnol 31(9):833–838
11. Kim E et al (2012) Precision genome engineering with programmable DNA-nicking enzymes. Genome Res 22(7):1327–1333
12. Ran FA et al (2013) Double nicking by RNA-guided CRISPR Cas9 for enhanced

genome editing specificity. Cell 154 (6):1380–1389
13. Shen B et al (2014) Efficient genome modification by CRISPR-Cas9 nickase with minimal off-target effects. Nat Methods 11 (4):399–402
14. Slaymaker IM et al (2016) Rationally engineered Cas9 nucleases with improved specificity. Science 351(6268):84–88
15. Kleinstiver BP et al (2016) High-fidelity CRISPR-Cas9 nucleases with no detectable genome-wide off-target effects. Nature 529 (7587):490–495
16. Kuscu C et al (2014) Genome-wide analysis reveals characteristics of off-target sites bound by the Cas9 endonuclease. Nat Biotechnol 32 (7):677–683
17. Wang T et al (2014) Genetic screens in human cells using the CRISPR-Cas9 system. Science 343(6166):80–84
18. Shalem O et al (2014) Genome-scale CRISPR-Cas9 knockout screening in human cells. Science 343(6166):84–87
19. Sakuma T et al (2014) Multiplex genome engineering in human cells using all-in-one CRISPR/Cas9 vector system. Sci Rep 4:5400
20. Konermann S et al (2015) Genome-scale transcriptional activation by an engineered CRISPR-Cas9 complex. Nature 517 (7536):583–588
21. Perez-Pinera P et al (2013) RNA-guided gene activation by CRISPR-Cas9-based transcription factors. Nat Methods 10(10):973–976
22. Larson MH et al (2013) CRISPR interference (CRISPRi) for sequence-specific control of gene expression. Nat Protoc 8(11):2180–2196
23. McDonald JI et al (2016) Reprogrammable CRISPR/Cas9-based system for inducing site-specific DNA methylation. Biol Open 5 (6):866–874
24. Tanenbaum ME et al (2014) A protein-tagging system for signal amplification in gene expression and fluorescence imaging. Cell 159 (3):635–646
25. Chaudhary K, Chattopadhyay A, Pratap D (2018) The evolution of CRISPR/Cas9 and their cousins: hope or hype? Biotechnol Lett 40(3):465–477
26. Friedland AE et al (2015) Characterization of Staphylococcus aureus Cas9: a smaller Cas9 for all-in-one adeno-associated virus delivery and paired nickase applications. Genome Biol 16:257
27. Ran FA et al (2015) In vivo genome editing using Staphylococcus aureus Cas9. Nature 520 (7546):186–191
28. Zetsche B et al (2015) Cpf1 is a single RNA-guided endonuclease of a class 2 CRISPR-Cas system. Cell 163(3):759–771
29. Abudayyeh OO et al (2017) RNA targeting with CRISPR-Cas13. Nature 550 (7675):280–284
30. DiCarlo JE, Deeconda A, Tsang SH (2017) Viral vectors, engineered cells and the CRISPR revolution. Adv Exp Med Biol 1016:3–27
31. Glass Z et al (2018) Engineering the delivery system for CRISPR-based genome editing. Trends Biotechnol 36(2):173–185
32. Swiech L et al (2015) In vivo interrogation of gene function in the mammalian brain using CRISPR-Cas9. Nat Biotechnol 33 (1):102–106
33. Heigwer F, Kerr G, Boutros M (2014) E-CRISP: fast CRISPR target site identification. Nat Methods 11(2):122–123
34. Gibson DG et al (2009) Enzymatic assembly of DNA molecules up to several hundred kilobases. Nat Methods 6(5):343–345
35. Gibson DG (2011) Enzymatic assembly of overlapping DNA fragments. Methods Enzymol 498:349–361
36. Sanjana NE, Shalem O, Zhang F (2014) Improved vectors and genome-wide libraries for CRISPR screening. Nat Methods 11 (8):783–784
37. Kulinski J et al (2000) CEL I enzymatic mutation detection assay. BioTechniques 29 (1):44–6, 48
38. Pimkin M et al (2007) Recombinant nucleases CEL I from celery and SP I from spinach for mutation detection. BMC Biotechnol 7:29
39. Choi VW, Samulski RJ, McCarty DM (2005) Effects of adeno-associated virus DNA hairpin structure on recombination. J Virol 79 (11):6801–6807
40. Chakiath CS, Esposito D (2007) Improved recombinational stability of lentiviral expression vectors using reduced-genome Escherichia coli. BioTechniques 43(4):466, 468, 470
41. Mao Z et al (2008) Comparison of nonhomologous end joining and homologous recombination in human cells. DNA Repair (Amst) 7 (10):1765–1771
42. Shrivastav M, De Haro LP, Nickoloff JA (2008) Regulation of DNA double-strand

break repair pathway choice. Cell Res 18 (1):134–147

43. Zaboikin M, Freter C, Srinivasakumar N (2018) Gaussian decomposition of high-resolution melt curve derivatives for measuring genome-editing efficiency. PLoS One 13(1): e0190192

44. Ota S et al (2013) Efficient identification of TALEN-mediated genome modifications using heteroduplex mobility assays. Genes Cells 18(6):450–458

45. Nakagawa Y et al (2014) Screening methods to identify TALEN-mediated knockout mice. Exp Anim 63(1):79–84

Chapter 3

Design, Construction, and Application of Transcription Activation-Like Effectors

Peter Deng, Sakereh Carter, and Kyle Fink

Abstract

Transcription activator-like effectors (TALEs) are modular proteins derived from the plant *Xanthomonas* sp. pathogen that can be designed to target unique DNA sequences following a simple cipher. Customized TALE proteins can be used in a variety of molecular applications that include gene editing and transcriptional modulation. Presently, we provide a brief primer on the design and construction of TALEs. TALE proteins can be fused to a variety of different effector domains that alter the function of the TALE upon binding. This flexibility of TALE design and downstream effect may offer therapeutic applications that are discussed in this section. Finally, we provide a future perspective on TALE technology and what challenges remain for successful translation of gene-editing strategies to the clinic.

Key words Gene editing, TALEs, Nucleases, Artificial transcription factor, Chromatin remodeling, DNA binding, Huntington's disease

1 Introduction

Transcription activator-like effectors (TALEs) are monomeric DNA-binding domains (DBDs) that can be readily designed to target unique genomic targets in a sequence-specific manner. Originally discovered as a secreted modulatory protein by the plant pathogen *Xanthomonas*, TALEs received their name due to their binding of gene promoters and alteration of gene expression patterns that favored *Xanthomonas* infection in plant cells. Since the deciphering of the TALE-binding code in 2009 [1], designer TALEs have provided a powerful new molecular tool for gene modification. Presently, we provide a brief discussion on the design, application, and future perspectives of TALEs as a molecular tool in mammalian models.

Fredric P. Manfredsson and Matthew J. Benskey (eds.), *Viral Vectors for Gene Therapy: Methods and Protocols*, Methods in Molecular Biology, vol. 1937, https://doi.org/10.1007/978-1-4939-9065-8_3,

2 Binding and Specificity of Tale Proteins

The specific DNA base-pair preference for TALEs has been well characterized, providing a simple cipher for interested researchers to use and engineer TALEs for gene modification. Currently, engineered TALES can be used for modeling disease in cells and animals. Recognition of individual DNA base pairs is mediated through a tandem of protein repeats. These repeats consist of 34 amino acids in length where single base-pair binding is dictated by the 12th and 13th amino acid, referred to as the Repeat Variable Di-residues (RVDs; Fig. 1). Tandem repeats consist of two alpha helices connected by a three-residue loop that consists of the RVDs that organize in a solenoid shape. The RVD "touches" the preferred DNA base pair via hydrogen bonding of the 13th amino acid with the major groove of the target base, while the 12th amino acid stabilizes the loop structure [2].

There are six widely used RVDs (NG, HD, NI, NK, NH, and NN) for the construction of TALEs that exhibit variable specificity

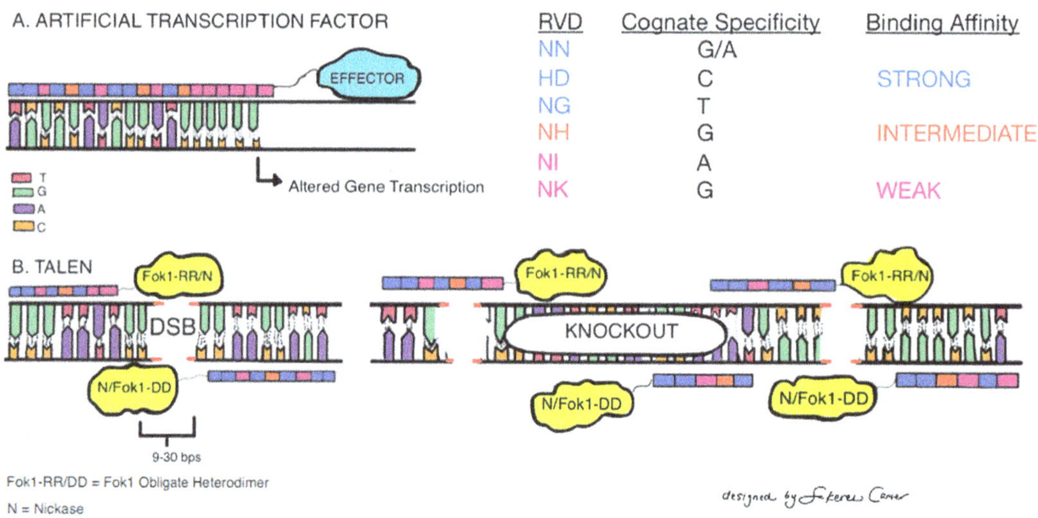

Fig. 1 Schematic of TALE Binding and Mechanisms of Action. TALEs can be used with a variety of different effector domains to depending on user preference. The TALE DBD can be fused with transcriptional repressors, activators, nucleases, or nickases. (**a**) TALEs fused with transcription activators or repressors (effector domains) can be targeted toward regulatory regions of target genes to elicit a change in endogenous expression via recruitment of additional transcription factors or chromatin remodeling. (**b**) TALEs can be fused with obligate nucleases such as variants of the Fok1 heterodimer (RR/DD) or nickases to create TALENs. (*Left Panel*) Pairs of TALENs may be used together to generate DSB when targeted near complementary strands of DNA. Fok1 nucleases cause DSB following dimerization of the Fok1 monomers of the two TALENs. Pairs of nickases can achieve DSB when targeted near of complementary strands of DNA. (*Right Panel*) Multiple pairs of TALENs can be used in conjunction with one another to excise larger regions of DNA from target loci. Both methodologies follow non-homologous end joining and homology-directed DNA repair pathways that can be directed by the user

and binding affinity (summarized in Fig. 1). “Strong” RVDs such as HD and NG almost exclusively bind to cytosine (C) and thymine (T) [3]. NN is also a “strong” RVD in its binding affinity for guanine (G); however, it has degenerate specificity in that it will also bind to adenine (A). “Weak” RVDs such as NI and NK will bind to A and G exclusively, however with low affinity [3]. NH has been shown to have exclusive binding toward G with intermediate affinity, favoring its use as a flanking RVD for strong RVDs. It is important to note that the methylation status of target DNA sites influences the binding affinity of TALEs. Methylated cytosine is not effectively bound by the canonical RVD HD; however, NG is capable of binding to methylated cytosine in part due to the highly structural similarity of methylated cytosine to thymine [4].

These tandem repeats can be tailored together in a modular fashion and confer sequence-specific binding of DNA based on a simple set of rules (*see* Subheading 3). The optimal target sequence length for TALE binding is between 17 and 20 BP, which is the common range observed in nature. Increased repeat array length does not yield greater specificity and may increase off-targeting effects due to an increased tolerance of mismatches from excess RVD length [5]. Additionally, naturally occurring TALE sites invariably have a 5′-Thymine in the “0-position” due to the presence of a conserved tryptophan in the N-terminal end of the TALE repeat that makes a van der Waal bond with the methyl group of thymine. The presence of this 5′-T is necessary for full target gene activation for naturally occurring TALEs [6]. Conversely, some TALEs have been shown to target 5′-Cytosines in the −0 position rather than thymine, and newer artificial TALE scaffolds do not appear to be constrained by the need for a 5′-T [7]. Crystallographic work has shown that TALE proteins bind to a single strand of DNA in a N-terminal to C-terminal manner in the 5′ to 3′ orientation where each RVD interacts with its corresponding unique nucleotide [2].

Interestingly, work by our colleagues demonstrated that TALE-DNA interaction exhibits a polarity effect. Base pair mismatches on the 5′ end of the target site demonstrate a greater disruptive effect of TALE binding as compared to mismatches on the 3′ end of the target sequence. Together, this suggests a greater impact of the N-terminal repeats to binding affinity as compared to the C-terminus [8]. This intolerance is not observed in naturally occurring TALEs such as the AvrBs3; however, naturally occurring TALEs are composed of “nonstandard” RVDs not used in the synthetic construction of TALEs. Subsequent nonstandard RVDs, such as NS, may take unusual RVD loop conformations that adjust structural constraints of TALE-DNA binding [8]. In practical consideration, synthetic TALE arrays are highly specific toward target genomic sites if prospective mismatches occur within the 5′ end (i.e., RVD position one through seven) rather than the 3′ end.

3 Design

The highly modular nature of TALEs and widely available in silico design tools have made TALE construction incredibly accessible for researchers. Many of these tools readily provide prospective TALE-binding sites as well as off-target sites and should be used to identify lead TALE target sites. Engineered TALEs follow simple design rules that have stayed relatively constant since their first description [9, 10].

1. Confirm that the target site is truly unique within the genome and is preceded by a 5′-Thymine in the "0–position" preceding the first targeted base pair.
2. If designing TALEs to modulate gene expression, ensure that the target falls within a regulatory domain of the gene of interest such as promoters or enhancers [11]. This can be visualized using online genomic browsers such as the UC Santa Cruz Genome Browser and reference human epigenomes from the NIH Roadmap Epigenomics Mapping Consortium (*see* **Note 1**).
3. TALE repeat length varies by target site [5]; however, TALE arrays consisting of 17–20 repeats are the most commonly occurring length within natural TALEs [3, 12].
4. Avoid clustering "strong" RVDs onto the N-termini as this results in greater off-target profile. Inversely, clustering "weak" RVDs in the C-terminus may also pose a risk for off-targets [8, 13].
5. Flank "weak" RVDs with "strong" RVDs with specific emphasis on "strong" RVDs on both termini of the TALE DBD [3, 13].
6. Avoid repeated stretches of identical RVDs. For example, three sequential NG repeats have been shown to result in abnormal folding of the ensuing protein [3].
7. Use the NH RVD to target G when discrimination between A & G is necessary (important in the context of discriminating between single base pair mismatches)[3]s. Use NI for specific targeting of A [3].
8. Use NG instead of HD when designing TALEs for regions with known CpG islands (*see* **Note 2**).
9. The C-terminal end of TALE fusion proteins serves as a linker between the DNA-binding domain of the TALE and the effector domain. Linker lengths vary depending on type of effector with shorter linkers (16 aa) for fused nucleases and longer linkers for transcriptional domains (65 aa) [9, 10, 14].

Currently, there are several commercially available kits and services to assemble TALEs. Both the Voytas [9] and Zhang [10] groups have designed a single cut-ligation GoldenGate Assembly method of cloning TALEs with corresponding plasmids encoding tandem repeats for any targeting position up to 31 BP [10]. Turnaround time for the construction of these TALEs typically range within 1 week from plasmid preparation to sequence verification, with an additional week for in vitro validation of TALE binding. Numerous companies will provide synthesis of TALE with comparable turnaround times.

4 Use of Tales for Gene Modification

Prior to the discovery of TALEs (as well as the CRISPR/Cas9 system) engineered zinc fingers (ZF) were commonly used for gene modification. ZFs, like TALEs, are monomeric DBD proteins that could be constructed in a modular way. These involved fusions with obligate nucleases such as Fokl and transcription effector domains such as VP64 and KRAB (Fig. 1a). ZF technology has provided great flexibility for researchers interested in modifying DNA. These fusions have subsequently been adapted for use with TALEs. The properties of TALEs, such as the readily synthesizable nature of repeat arrays and highly specific binding affinity of DBDs, have allowed for TALEs to be applied for generation disease models and therapeutic development.

4.1 TALENs

Transcription activator-like effector nucleases (TALENs) were constructed by fusing the TALE DBD to an obligate-heterodimeric version of the Fokl nuclease (Fig. 1b). To increase specificity of double-strand break events, TALENs were fused with monomers of the Fokl nuclease domain and targeted to adjacent regions of DNA. For a single DSB to occur, the Fokl nuclease domains must dimerize within 9–30 BP of one another in order to achieve nuclease activity—improving target specificity by expanding out the overall target sequence through two TALENs (Fig. 1b, Left Panel). Two pairs of TALENs can be used in concert to excise larger target sites where DNA repair is mediated through both non-homology-directed repair and homology-directed repair (Fig. 1b, Right Panel). TALEN technology has been successfully used to generate yeast [15], zebrafish [16], mammalian cell [14], rodent [17, 18], nonhuman primate [19], and livestock [20] models. Additionally, TALE DBDs have been fused with nickases that result in single-stranded breaks instead of DSBs with a single nickase or paired with two TALE-fused nickases to induce a DSB. Homology-directed and base excision repair is favored with nickases over NHEJ, reducing the potential genetic footprint that other nucleases may elicit [21]. This approach has been used to generate tuberculosis-resistant transgenic cattle [22]. Both the obligate nuclease and nickase approach may offer greater specificity for target DNA cutting when compared to the monomer nucleases such as CRISPR/Cas9 [23, 24].

Therapeutic approaches for TALENs have involved the ex vivo editing of genes in cell transplant approaches. TALENs have been used to knockout T-cell receptor alpha and CD52 genes in nonhuman leukocyte antigen-matched T cells [25]. These modified T cells successfully engrafted in infant leukemia patients who relapsed following lymphodepleting therapies and served as a scaffold until allogeneic stem cell transplantation therapies could be performed when the subjects were older.

4.2 Artificial Transcription Factors

Xanthamonas are plant pathogens that secrete effector proteins following infection into host plant cells. These proteins include TALE family proteins that include the TALE DBD fused to the AvrBs3-family effectors. These naturally occurring proteins act as eukaryotic-like transcription factors that induce the expression of specific host genes that facilitate Xanthamonas infection and propagation. Pioneering work in the early 2010s identified that TALE effectively binds to DNA of other eukaryotic organisms and these C-terminal AvsBs3 effectors could be readily exchanged with other effectors such as VP64 (tetrameric effector of VP16 form herpes simplex virus) or KRAB (kruppel-associated box) to create gene modifying DBDs [12–14, 26]. These artificial transcription factors (ATFs) modulate endogenous gene expression when targeted to regulatory regions of the genome (Fig. 1a).

Modification of endogenous mammalian genes with TALE activators have been shown with KLF4 and SOX2, but not the methylated c-Myc and Oct4 genes when targeted to gene promoters in 293 FT cells [10]. Activation of the methylated Oct4 gene has been shown by using TALE-VP64 in conjunction with small molecule drugs such as histone deactylase and DNA methyltransferase inhibitors [27]. Targeting of TALE activators to known enhancers for Oct4 and Nanog in mouse embryonic fibroblasts and epiblast stem cells have been shown to effectively reprogram these cells to induced pluripotent stem cells, respectively [28]. Thus, gene modification with TALEs appears to be highly dependent on the epigenetic landscape of the targeted genes. Increased expression using either TALE-VP64 or -p65 as effector domains targeted toward DHS of both protein-coding and non-coding DNA elements, such as microRNAs, have been shown in human cells. Additionally, multiple TALE activators targeted toward DHS of the same gene can drastically increase expression (<100 fold) as compared to single TALEs not targeted to DHS [26, 29].

The targeting of TALEs toward the coding strand of DNA has been shown to hinder gene expression, likely due to the inhibition of native transcription factors binding [30]. Inclusion of repressive effector domains such as the kruppel-associated box (KRAB) with TALEs targeted toward regulatory elements of protein-coding and non-coding elements reduce transcription expression in a transient fashion [28, 31]. Synergistic targeting of genomic regions with TALEs fused with DNA methyltransferases DNMT3a, DNMT3a, and KRAB sustained gene silencing across cell generations [32].

TALE-based ATFs provide valuable tools in interrogating molecular pathways in basic biology and disease from cells in a dish to animal modeling. Traditional approaches for gene therapy involved replacement of a deficient gene using viral-directed approaches (lentiviral reprogramming of ADA-SCID, adeno-associated virus episomal gene expression) or RNA interference strategies using siRNAs and antisense oligonucleotides. Target-specific TALE arrays provide direct control of endogenous gene expression. Monogenetic diseases provide valuable insight into the application of TALEs as a potential gene therapy.

Huntington's disease is an autosomal dominant neurodegenerative disorder characterized by a CAG trinucleotide-repeat expansion on exon1 of the Huntingtin gene in short arm of chromosome 4 that affects 4 in 100,000 individuals worldwide. Involuntary ballistic movements of the distal limbs termed chorea is observed in Huntington's and correlated with degeneration of the striatum. The causative disease agent in Huntington's disease is the formation of a large, misfolded huntingtin protein. TALENs targeted toward the CAG expansion may provide a therapeutically relevant gene-editing approach toward treating Huntington's disease; however, this method is predicated on successful in vivo gene correction. In practice, this approach is difficult to achieve due to the need to delineate between wild-type and mutant CAG expansion. There are approximately eight other genes (SCA variants, SBMA, DRPLA) that carry CAG expansions, posing potential off-target sites. Furthermore, such an approach would necessitate a seamless collapse of the CAG expansion to prevent unwanted insertion/deletion events. Current strategies for seamless gene correction require homology-directed repair, which in the current state of research is a low-efficiency approach.

ATFs provide an alternative approach toward treating Huntingtin's disease. TALEs fused with repressive factors can be designed to selectively silence the mutant Huntingtin gene using single nucleotide polymorphisms to distinguish the healthy allele from the mutant allele [33]. SNP-specific rational design utilizes the high on-target specificity of DBDs toward regulatory regions of the mutant Huntingtin gene and allow for a remodeling of the surrounding chromatin structure. Our group has shown the use of TALEs fused with a transcriptional repressor targeting multiple SNPs in the promoter region of mutant Huntingtin allele demonstrate high allele-specific knockdown of the muHtt. Furthermore, our group has demonstrated that gene knockdown is highly correlated with the presence of the TALE in the cell, suggesting that a future limitation in ATF therapies will be developing a robust delivery vehicle and not the efficacy of the ATF itself.

5 Challenges for Therapeutic Application

Both TALENs and TALEs offer considerable promise in treating previously undruggable diseases; however, there are an equal number of unmet challenges. In vivo delivery of both approaches remains difficult. TALEs are considerably large proteins, approximately ~2.3 KB in size for a repeat array targeting an 18 BP genomic site. The inclusion of the necessary promoter, transgene, and transcription elements bolsters the size of the TALE to 4.4 kb, approximately the upper limit of an adeno-associated virus. Dual TALENs, for DSB or gene knockout would require two separate AAV vectors. Lentiviral vectors have a greater capacity for transgene size; however, the highly repetitive repeat arrays result in unstable rearrangement of RVDs [34]. However, new TALE cloning methods have made this less of a concern [35]. Purified TALE protein may provide an avenue for direct in vivo injection; however, TALEs are bacterial in origin and the risk of an immune response in mammalian systems reduces the feasibility of such an approach. In vivo correction remains a challenge using nuclease approaches and may be ineffective in non-dividing tissue types such as neurons. Furthermore, despite the robust initial effects of ATFs like TALEs, sustained gene modification appears to degrade following loss of therapeutic protein [36–38] in actively proliferating cells. Rational design in gene therapy is a continually developing field however and there are lessons to be learned from traditional pharmacological approaches toward diseases such as cancer. Multi-pronged strategies using a variety of effectors appears to result in sustained gene silencing even in proliferating cell lines [32] and continued optimization of DNA vectors may functionally "extend" the half-life of the encoded therapeutic protein by prolonging transgene expression [37, 39–41].

Off-target binding remains a key concern for DBD-based therapeutic approaches. Appropriate vetting of proposed target sites in silico and in vivo remains paramount for successful gene-editing approaches. When designing TALEs, special consideration should be paid toward the 5′ seed region of the TALE as mentioned in Subheading 3. Prospective TALEs will better distinguish mismatches on the 5′ end, with weaker discrimination for mismatches on the 3′ end. All prospective sites should be vetted via BLAST on

NCBI and available online TALE design databases such as CHOP-CHOP [42]. On- and off-target binding efficiency should be measured empirically in interested cell lines using chromatin immunoprecipitation sequencing.

6 Future Perspectives

Beginning with the ability to synthetically design zinc fingers, the field of genetic modification and gene editing has undergone a rapid expansion. The use of TALE and CRISPR/Cas9 allowed for more diverse and higher throughput methods to identify and screen alterations on the genomic level. As the technology continues to evolve, the inherent strengths and weaknesses of each DBD (ZF, TALE, and Cas9, respectively) become more apparent. With the discovery of Cas9 the speed in which genomic alterations can be made has enhanced the ability to screen multiple genes at once. However, the use of Cas9 is limited by the necessity of an adjacent PAM site in the target sequence. As covered in above, TALEs are not limited in the target sequence outside of a 5′ T. This flexibility, along with the high specificity of RVD in the construction of TALE, allow for single base pair discrimination, allowing the possibility of allele-specific approaches with limited "off-target" binding.

While much of the ground-breaking work has been established in vitro, the ability to make genomic alterations in vivo still needs to be elucidated. Our group focuses on the in vivo application of ATFs for therapeutic interventions in central nervous system (CNS) disorders. As discussed above, the ability to package and express a DBD with an appropriate effector domain in viral vectors are limited by the total size of the plasmid. While work by our group and others focuses on the efficient packaging and delivery vehicle for these powerful genomic modifiers, this field is still very much in infancy. When deciding on a delivery modality there are several key factors to take into consideration. One must consider the inherent invasiveness to access the brain for delivery of any drug. The blood brain barrier remains a major hurdle for the delivery of ATFs as current approaches necessitate either direct intracranial injection or intrathecal infusion into the cerebral spinal fluid for CNS penetration. A second key factor to take into consideration is the transient nature of the ATF when selecting the effector domain to be used. As discussed above, the duration of effect of artificial epigenetic regulation in a post-mitotic neuron is unknown; however, optimization of the expression vectors may allow for longer expression and sustained therapeutic benefit. Optimization of tissue specific promoters and removal of non-essential components expression vectors such as bacterial selection markers to mitigate epigenetic silencing of the expression vectors may extend the therapeutic window of such an approach. Additionally, appropriate selection of an effector domain may increase the duration of effect. The Lombardo group has recently demonstrated a "cocktail" of effector domains using ATFs that result in inherited methylation status and gene silencing between cell generations, potentially providing an avenue for sustained silencing of targeted genes without the need for knocking out a gene with nucleases. These factors should play a role in the selection of delivery vehicles and a route of administration. A more invasive delivery, such as intracranial injection, may be feasible if the genomic alterations prove to be long-lasting. However less-invasive measures, such as systemic injection, intrathecal, or intranasal, may be required if repeated administration is necessary to confer long-lasting therapeutic benefit.

The field of gene editing or epigenetic modification has exploded in recent years. However, prior to widespread use in vivo several challenges must be addressed such as duration of effect, route of administration, immune tolerance, and distribution to many cells or organ system. Our group along with our collaborators and other laboratories are addressing these issues to advance the knowledge and use of these ATF for therapeutic interventions.

7 Notes

1. The effectiveness of TALE binding and activity is highly context-dependent on the physical characteristics of the target DNA. Chromatin structure, methylation status, and cell cycle state may act as steric barriers for TALE binding. Many of these features are well annotated on available genome browsers. Genomic sites with high DNAse I hypersensitivity sites (DHS) are suggestive of open chromatin where DBDs such as TALEs may readily bind. Histone marks for promoters or enhancer elements such as H3K4me1 and H3K27ac can be used for designing TALEs toward sites that modify gene expression. These characteristics differ between tissue types and should be considered prior to the construction of putative TALEs to best serve the researchers' interest.
2. It has been recommended by Deng et al. [4] that two TALEs should be designed when targeting CpG regions particularly due to dynamic nature of methylation status of DNA using the NG or HD RVD configuration.

References

1. Boch J, Scholze H, Schornack S et al (2009) Breaking the Code of DNA Binding Specificity of TAL-Type III Effectors. Science 326 (80):1509–1512. https://doi.org/10.1126/science.1178811
2. Mak AN-S, Bradley P, Cernadas RA et al (2012) The crystal structure of TAL effector PthXo1 bound to its DNA target. Science 335:716–719. https://doi.org/10.1126/science.1216211
3. Streubel J, Blücher C, Landgraf A, Boch J (2012) TAL effector RVD specificities and efficiencies. Nat Biotechnol 30:593–595. https://doi.org/10.1038/nbt.2304
4. Deng D, Yin P, Yan C et al (2012) Recognition of methylated DNA by TAL effectors. Cell Res 22:1502–1504. https://doi.org/10.1038/cr.2012.127
5. Guilinger JP, Pattanayak V, Reyon D et al (2014) Broad specificity profiling of TALENs results in engineered nucleases with improved DNA-cleavage specificity. Nat Methods 11:429–435. https://doi.org/10.1038/nmeth.2845
6. Römer P, Recht S, Strauß T et al (2010) Promoter elements of rice susceptibility genes are bound and activated by specific TAL effectors from the bacterial blight pathogen, Xanthomonas oryzae pv. oryzae. New Phytol 187:1048–1057. https://doi.org/10.1111/j.1469-8137.2010.03217.x
7. Lamb BM, Mercer AC, Barbas CF (2013) Directed evolution of the TALE N-terminal domain for recognition of all 5′ bases. Nucleic Acids Res 41:9779–9785. https://doi.org/10.1093/nar/gkt754
8. Meckler JF, Bhakta MS, Kim M-S et al (2013) Quantitative analysis of TALE-DNA interactions suggests polarity effects. Nucleic Acids Res 41:4118–4128. https://doi.org/10.1093/nar/gkt085
9. Cermak T, Doyle EL, Christian M et al (2011) Efficient design and assembly of custom TALEN and other TAL effector-based constructs for DNA targeting. Nucleic Acids Res 39:e82. https://doi.org/10.1093/nar/gkr218
10. Zhang F, Cong L, Lodato S et al (2011) Efficient construction of sequence-specific TAL effectors for modulating mammalian transcription. Nat Biotechnol 29:149–153. https://doi.org/10.1038/nbt.1775
11. Gaj T, Gersbach CA, Barbas CF III (2013) ZFN, TALEN, and CRISPR/Cas-based methods for genome engineering. Trends Biotechnol 31:397–405. https://doi.org/10.1016/j.tibtech.2013.04.004
12. Joung JK, Sander JD (2013) TALENs: a widely applicable technology for targeted genome editing. Nat Rev Mol Cell Biol 14:49–55. https://doi.org/10.1038/nrm3486
13. Cong L, Zhou R, Kuo Y et al (2012) Comprehensive interrogation of natural TALE DNA-binding modules and transcriptional repressor domains. Nat Commun 3:968. https://doi.org/10.1038/ncomms1962
14. Miller JC, Tan S, Qiao G et al (2011) A TALE nuclease architecture for efficient genome editing. Nat Biotechnol 29:143–148. https://doi.org/10.1038/nbt.1755
15. Li T, Huang S, Jiang WZ et al (2011) TAL nucleases (TALNs): hybrid proteins composed of TAL effectors and FokI DNA-cleavage domain. Nucleic Acids Res 39:359–372. https://doi.org/10.1093/nar/gkq704
16. Wood AJ, Lo T-W, Zeitler B et al (2011) Targeted genome editing across species using ZFNs and TALENs. Science 333 (80):307–307. https://doi.org/10.1126/science.1207773
17. Tesson L, Usal C, Ménoret S et al (2011) Knockout rats generated by embryo microinjection of TALENs. Nat Biotechnol 29:695–696. https://doi.org/10.1038/nbt.1940
18. Wefers B, Meyer M, Ortiz O et al (2013) Direct production of mouse disease models by embryo microinjection of TALENs and oligodeoxynucleotides. Proc Natl Acad Sci U S A 110:3782–3787. https://doi.org/10.1073/pnas.1218721110
19. Sato K, Oiwa R, Kumita W et al (2016) Generation of a nonhuman primate model of severe combined immunodeficiency using highly efficient genome editing. Cell Stem Cell 19:127–138. https://doi.org/10.1016/j.stem.2016.06.003
20. Carlson DF, Tan W, Lillico SG et al (2012) Efficient TALEN-mediated gene knockout in livestock. Proc Natl Acad Sci 109:17382–17387. https://doi.org/10.1073/pnas.1211446109
21. Kim E, Kim S, Kim DH et al (2012) Precision genome engineering with programmable DNA-nicking enzymes. Genome Res 22:1327–1333. https://doi.org/10.1101/gr.138792.112
22. Wu H, Wang Y, Zhang Y et al (2015) TALE nickase-mediated SP110 knockin endows cattle with increased resistance to tuberculosis.

Proc Natl Acad Sci U S A 112:E1530–E1539. https://doi.org/10.1073/pnas.1421587112

23. Wang X, Wang Y, Wu X et al (2015) Unbiased detection of off-target cleavage by CRISPR-Cas9 and TALENs using integrase-defective lentiviral vectors. Nat Biotechnol 33:175–178. https://doi.org/10.1038/nbt.3127
24. Ran FA, Hsu PD, Lin C-Y et al (2013) Double nicking by RNA-guided CRISPR Cas9 for enhanced genome editing specificity. Cell 154:1380–1389. https://doi.org/10.1016/j.cell.2013.08.021
25. Qasim W, Thrasher AJ (2014) Progress and prospects for engineered T cell therapies. Br J Haematol 166:818–829. https://doi.org/10.1111/bjh.12981
26. Maeder ML, Linder SJ, Reyon D et al (2013) Robust, synergistic regulation of human gene expression using TALE activators. Nat Methods 10:243–245. https://doi.org/10.1038/nmeth.2366
27. Bultmann S, Morbitzer R, Schmidt CS et al (2012) Targeted transcriptional activation of silent oct4 pluripotency gene by combining designer TALEs and inhibition of epigenetic modifiers. Nucleic Acids Res 40:5368–5377. https://doi.org/10.1093/nar/gks199
28. Gao X, Tsang JCH, Gaba F et al (2014) Comparison of TALE designer transcription factors and the CRISPR/dCas9 in regulation of gene expression by targeting enhancers. Nucleic Acids Res 42:e155. https://doi.org/10.1093/nar/gku836
29. Perez-Pinera P, Ousterout DG, Brunger JM et al (2013) Synergistic and tunable human gene activation by combinations of synthetic transcription factors. Nat Methods 10:239–242. https://doi.org/10.1038/nmeth.2361
30. Uhde-Stone C, Cheung E, Lu B (2014) TALE activators regulate gene expression in a position- and strand-dependent manner in mammalian cells. Biochem Biophys Res Commun 443:1189–1194. https://doi.org/10.1016/j.bbrc.2013.12.111
31. Zhang Z, Xiang D, Heriyanto F et al (2013) Dissecting the roles of miR-302/367 cluster in cellular reprogramming using TALE-based repressor and TALEN. Stem Cell Reports 1:218–225. https://doi.org/10.1016/j.stemcr.2013.07.002
32. Amabile A, Migliara A, Capasso P et al (2016) Inheritable silencing of endogenous genes by hit-and-run targeted epigenetic editing. Cell 167:219–232.e14. https://doi.org/10.1016/j.cell.2016.09.006
33. Fink KD, Deng P, Gutierrez J et al (2016) Allele-Specific reduction of the mutant huntingtin allele using transcription activator-like effectors in human huntington's disease fibroblasts. Cell Transplant 25:677–686. https://doi.org/10.3727/096368916X690863
34. Holkers M, Maggio I, Liu J et al (2013) Differential integrity of TALE nuclease genes following adenoviral and lentiviral vector gene transfer into human cells. Nucleic Acids Res 41:e63–e63. https://doi.org/10.1093/nar/gks1446
35. Yang L, Guell M, Byrne S et al (2013) Optimization of scarless human stem cell genome editing. Nucleic Acids Res 41:9049–9061. https://doi.org/10.1093/nar/gkt555
36. Hathaway NA, Bell O, Hodges C et al (2012) Dynamics and memory of heterochromatin in living cells. Cell 149:1447–1460. https://doi.org/10.1016/j.cell.2012.03.052
37. Munye MM, Tagalakis AD, Barnes JL et al (2016) Minicircle DNA provides enhanced and prolonged transgene expression following airway gene transfer. Sci Rep 6:23125. https://doi.org/10.1038/srep23125
38. Lewis O, Woolley M, Johnson D et al (2016) Chronic, intermittent convection-enhanced delivery devices. J Neurosci Methods 259:47–56. https://doi.org/10.1016/j.jneumeth.2015.11.008
39. Kay MA, He C-Y, Chen Z-Y (2010) A robust system for production of minicircle DNA vectors. Nat Biotechnol 28:1287–1289. https://doi.org/10.1038/nbt.1708
40. Lu J, Zhang F, Kay MA (2013) A mini-intronic plasmid (MIP): a novel robust transgene expression vector in vivo and in vitro. Mol Ther 21:954–963. https://doi.org/10.1038/mt.2013.33
41. Lu J, Williams JA, Luke J et al (2017) A 5′ noncoding exon containing engineered intron enhances transgene expression from recombinant AAV vectors *in vivo*. Hum Gene Ther 28:125–134. https://doi.org/10.1089/hum.2016.140
42. Montague TG, Cruz JM, Gagnon JA et al (2014) CHOPCHOP: a CRISPR/Cas9 and TALEN web tool for genome editing. Nucleic Acids Res 42:W401–W407. https://doi.org/10.1093/nar/gku410

Chapter 4

Practical Considerations for the Use of DREADD and Other Chemogenetic Receptors to Regulate Neuronal Activity in the Mammalian Brain

Patrick Aldrin-Kirk and Tomas Björklund

Abstract

Chemogenetics is the process of genetically expressing a macromolecule receptor capable of modulating the activity of the cell in response to selective chemical ligand. This chapter will cover the chemogenetic technologies that are available to date, focusing on the commonly available engineered or otherwise modified ligand-gated ion channels and G-protein-coupled receptors in the context of neuromodulation. First, we will give a brief overview of each chemogenetic approach as well as in vitro/in vivo applications, then we will list their strengths and weaknesses. Finally, we will provide tips for ligand application in each case.

Each technology has specific limitations that make them more or less suitable for different applications in neuroscience although we will focus mainly on the most commonly used and versatile family named designer receptors exclusively activated by designer drugs or DREADDs. We here describe the most common cases where these can be implemented and provide tips on how and where these technologies can be applied in the field of neuroscience.

Key words Chemogenetics, Ligand-gated ion channel, G-Protein-coupled receptor, DREADD, TRPV1, GluCl, PSAM/PSEM, Allatostatin, Ligand, Engineered receptors, Neuronal modulation, Gene therapy

1 Introduction to Chemogenetics

Chemogenetics is the process by which macromolecule proteins, usually ionotropic or metabotropic receptors, are engineered to interact with an otherwise biologically inactive exogenous small molecule chemical actuator ligand [1–3]. This method shares many applications with the well-known optogenetics approach. However, while optogenetics uses light and light-sensitive ion channels (known as opsins), chemogenetics takes advantage of chemical ligands to modulate genetically engineered receptors. Two common approaches to generate novel macromolecule-

Fredric P. Manfredsson and Matthew J. Benskey (eds.), *Viral Vectors for Gene Therapy: Methods and Protocols*, Methods in Molecular Biology, vol. 1937, https://doi.org/10.1007/978-1-4939-9065-8_4,

chemical actuator pairs have been to hijack protein–actuator ligand pairs from other tissues and species, or to reengineer endogenous receptors to introduce affinity and selectivity to novel ligands. For a broad overview of available technologies that are used to achieve cellular modulatory control, a comprehensive flowchart is available (*see* Fig. 1).

Chemogenetics was originally used to describe the effects of mutations on chalcone isomerase activity on substrate specificities in the flowers of *Dianthus caryophyllus* [1]. The main use of chemogenetics today, however, is as a toolkit designed to modulate the activity of specific populations of cells in order to study their function, cellular signaling, or in future therapeutic potential. Most of these chemogenetic techniques allow for unidirectional excitation or inhibition of neuronal activity, although some allow for bidirectional modulation.

Several important considerations must be taken into account when using various chemogenetic approaches however; these include baseline activity of the engineered protein and actuator ligand, specificity of expression, toxicity, ligand application, and kinetics. For example, it is important for the receptor-actuator pair to not interact with the organism being studied in the absence of the other component, i.e., have very low or preferably nonexistent baseline activity. Toward these aims, a large number of chemogenetic families have emerged over the past two decades with different approaches to address these considerations, each generation generally improving on the previous. These various iterations of chemogenetic platforms, engineered from different macromolecules, have now been used in several biological fields and include kinases [4–9], non-kinase enzymes [10–12], ligand-gated ion channels, [13–16], and G protein-coupled receptors (GPCRs) [17–19]. GPCRs are by far the most widely adopted type of chemogenetic approach. Therefore, this chapter will introduce the reader to the different chemogenetic approaches but focus on current GPCR-based chemogenetic technology.

2 Materials

The materials and other components covered in this chapter are mainly conceptual, binding together decades of published molecular engineering and other methods used when developing chemogenetic technologies. Most characterized chemogenetic approaches described in this chapter have been made available on Addgene as finished expression plasmids. However, molecular cloning may be required in order to optimize each technology for new applications. In addition, some chemogenetic technologies described in this chapter may need further characterization in terms of pharmacokinetics and receptor expression in certain neuronal subtypes. It is

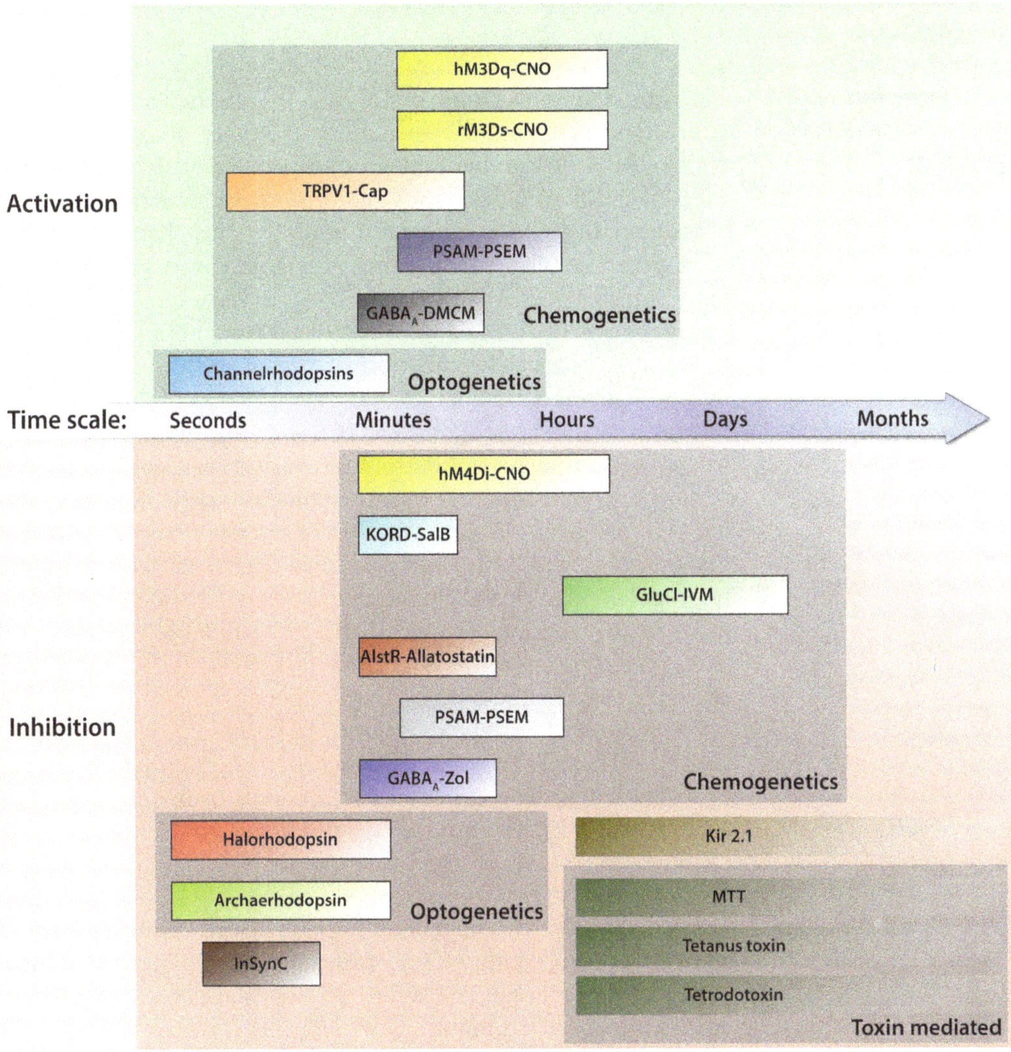

Fig. 1 Chart of available methods for modulating neuronal activity-based time resolution from in vivo data. Methods for increasing neuronal excitability through treatment with clozapine-N-oxide (CNO) include hM3Dq and hM3Ds DREADDs. Increase in neuronal excitability can also be achieved with TRPV1 receptors treated with capsaicin (cap), PSAM/PSEM pairs, GABAA receptor treated with methyl-6,7-dimethoxy-4-ethyl-beta-carboline-3-carboxylate (DMCM) or through non chemogenetic approaches such as optogenetic channel rhodopsins. Chemogenetic methods for inhibiting neuronal excitability include hM4Di-mediated inhibition by CNO, KORD via treatment of salvinorin B (SalB), GluCl activation through treatment of ivermectin (IVM), allatostatin interacting with the allatostatin receptor (AlstR), PSAM/PSEM pairs and GABA$_A$ receptors treated with zolpidem (Zol). Other approaches for neuronal silencing include optogenetic approaches utilizing halorhodopsin or archaerhodopsin or through toxin-mediated silencing through membrane tethered toxins (MTT), tetanus toxin or tetrodotoxin

therefore highly recommended that researchers new to chemogenetic techniques characterize these when applying them in specific contexts not previously described in the literature.

3 Methods

3.1 Overview of Ligand-Gated Ion Channels

Ligand-gated ion channels (LGICs) are particularly effective in controlling and modulating neuronal activity due to their direct control of electrical membrane potential. This is facilitated by the LGICs' ability to quickly move anions or cations across the cellular membrane in response to ligand binding, thus changing the electrical potential of the membrane. The rapid control of ion conductance makes LGICs well suited for precise temporal control of neuronal activity. LGICs were first used to manipulate membrane potential in neurons by genetically targeted expression of ion channels to change neuronal excitability. However, this resulted in neurotoxicity over longer time periods [20]. Pioneering studies using LGIC to achieve neuronal activation or silencing in animal behavior required the use of intracranial infusions of the actuator, e.g., glutamate [21]. However, this invasive approach limits the use in behavioral studies and also prevents selective targeting of specific neuronal populations as glutamate receptors are widely expressed in many neuronal populations. A different small molecule was desired that could interact with a selectively expressed receptor and preferably able to cross the blood barrier. Therefore, several LGICs have been engineered for ligand-specific pharmacological modulation of neurons. These LGICs can be categorized into either invertebrate LGICs, such as the ivermectin-gated GluCl, or mammalian LGICs such as the TRPV1 and PSAMs discussed further below in this chapter.

3.2 Ivermectin-Gated Chloride Channels

3.2.1 Overview

One of the first chemogenetic approaches to silencing neuronal activity was utilizing the ivermectin-gated chloride channel [22]. This chemogenetic approach was developed to allow for long-term chronic inhibition of neurons, peaking after 12–48 h and lasting several days. Nematode worms such as the C. elegans express a high conductance glutamate-gated chloride channel (GluCl). This is present in the cell membrane as heteropentamers consisting of three α and two β subunits [23]. Since the GluCl receptor channel is only expressed in worms, it has been targeted for anti-parasite drugs such as ivermectin. Binding of Ivermectin to the worm GluCl causes hyperpolarization. By genetic modification of the α and β subunits, researchers have been able to significantly reduce glutamate sensitivity while retaining ligand binding for ivermectin. This was achieved by introducing a single-point mutation (Y182F) in the glutamate-binding site of the β subunit [24]. The modified chloride channel caused hyperpolarization and a strong reduction in generated action potentials over long periods of time [24, 25], primarily through a drop in input resistance across the membrane [24] (Fig. 2g). The long-term silencing reported during in vivo studies is likely not only due to binding

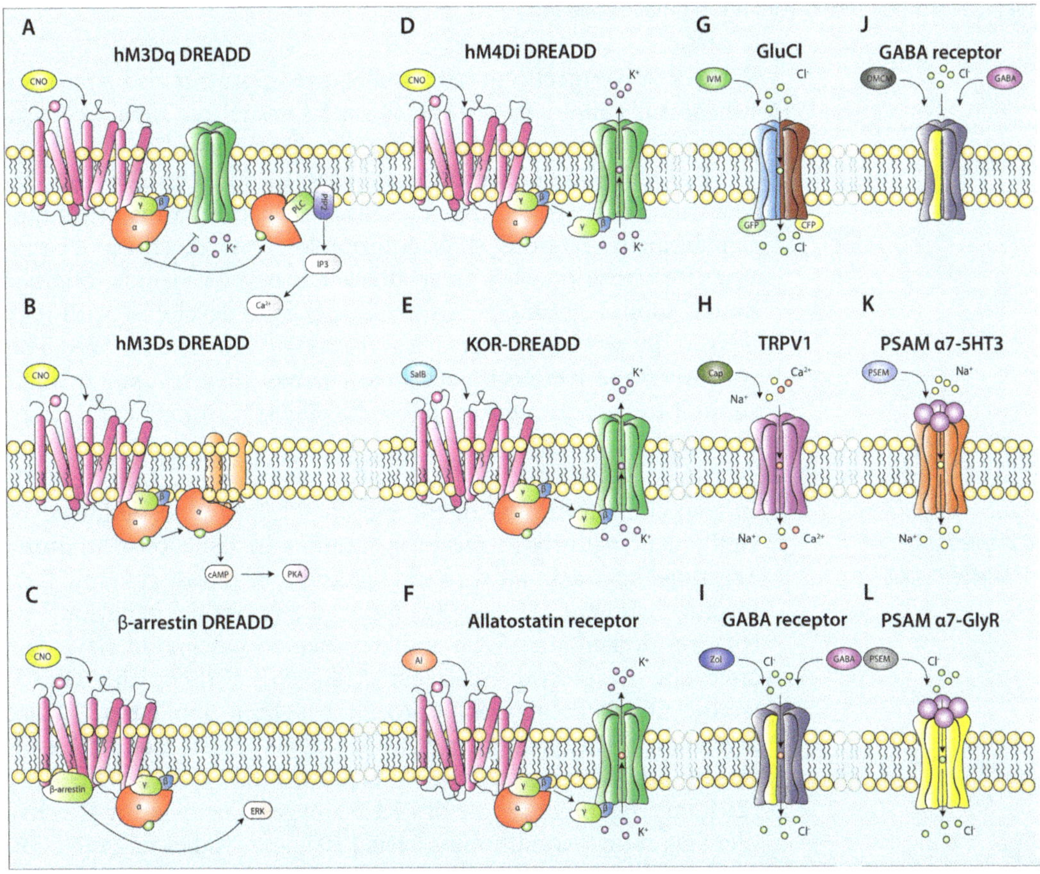

Fig. 2 Schematic illustrations of all chemogenetic receptors discussed in this chapter as well as their main mode of action. (**a–f**) Chemogenetic receptors based on various GPCRs. (**a**) the hM3Dq DREADD receptor is a GPCR that upon binding with CNO increases neuronal excitability. This is done through a $G\alpha$-mediated downstream pathway where the $G\alpha$ subunit activates PLC which in turn increases IP3. (**b**) The GPCR-based rM3Ds DREADD on also increases neuronal excitability following CNO binding. However, downstream signaling differs in that $G\alpha$-mediated signaling increases activity of adenylyl cyclase, thereby ramping up intracellular cAMP. (**c**) The β-arrestin DREADD modulates neuronal activity through a GPCR-β-arrestin dependent pathway following binding of CNO, recruiting downstream signaling molecules such as ERK. (**d**) The hM4Di DREADD on the other hand is a GPCR-based DREADD that inhibits neuronal activity following CNO binding. This is achieved through β/γ-mediated opening of GIRK membrane channels, leading to hyperpolarization of the cellular membrane. The other two GPCR-based inhibitory chemogenetic receptors, KORD (**e**) and Allatostatin receptor inhibit neuronal activity in a similar fashion following binding of SalB or Allatostatin, respectively. (**g–l**) Chemogenetic receptors based on LGICs. (**g**) The GluCl is an inhibitory modified LGIC made up of two subunits that upon binding with the ligand Ivermectin opens to Cl^- ions, thus hyperpolarizing the membrane potential. (**h**) The TRPV1 on the other hand is a LGIC that increases neuronal excitability by opening the ion channel to cations upon ligand binding of capsaicin, thus leading to membrane depolarization. (**i–j**) The $GABA_A$ LGIC receptor which in the context of chemogenetics has regained binding affinity to Zolpidem and DMCM. It can both activate or inhibit neuronal activity by modulating responsiveness to GABA. In the case of Zolpidem treatment (**i**) responsiveness to GABA is increased, thus opening the channel to Cl^- ions that hyperpolarizes the membrane. However, in the context of DMCM treatment the $GABA_A$ receptor becomes insensitive to GABA-mediated opening of the ion channel, thus blocking inhibition. (**k–l**) PSAM/PSEM can mediate neuronal activation or inhibition depending on the properties of the fusion channel. (**k**) α7-5HT3 LGIC opens to transport of Na^+ ions following ligand binding leading to increased neuronal excitability. Ligand binding to the α7-GlyR on the other hand opens the LGIC to Cl^- ions, resulting in hyperpolarization and neuronal inhibition

kinetics of ivermectin to the modified receptor but also a result of ivermectin's ability to accumulate in fat tissue that then acts as an ivermectin reservoir following the initial treatment [26].

Further modifications through rational engineering of the protein subunits led to improved localization to the plasma membrane of mammalian neurons, thus requiring lower ivermectin concentrations for activation [27]. In an alternative development, a modified human alpha1 glycine receptor has also been developed with high ivermectin sensitivity but a low sensitivity to glycine [28].This version of the ivermectin-sensitive channel requires only a single subunit to be expressed as opposed to the two subunits utilized by the original system.

3.2.2 In Vivo Characterization of Modified GluCl

In vivo characterization of the modified GluCl was conducted by virally expressing both receptor subunits in the mouse striatum. Treating these animals with amphetamine 4 h post systemic ivermectin treatment (10 mg/kg) induced a rotational behavior. The rotation peaked after 12 h and returned to baseline levels 4 days after the ivermectin treatment [14]. Since then, the GluCl-ivermectin chemogenetic approach has been applied to silence several different types of neurons in various animal models. Examples include studying the role of the pre-frontal cortex in narcolepsy [29], investigating the role of PKCδ amygdala neurons by selective silencing in fear conditioning tests [30], and studies of aggression and mating behavior by silencing of hypothalamic neurons [31].

3.2.3 Considerations and Limitations When Using the Modified GluCl Channel

One obvious limitation of the GluCl approach is the need to express two individual subunits in order to achieve full functional silencing in response to ivermectin. As both subunits are of substantial size, virally expressing both subunits with the common AAV vector requires at least two separate AAV constructs. This could potentially reduce efficacy and result in receptor expression variability.

The slow kinetics of the GluCl-Ivermectin system also requires careful consideration in the experimental design, as several days are required until full reversal of the neuronal inhibition. It is currently unknown how prolonged neuronal silencing may affect any given neuronal system. However, this particular aspect of the GluCl chemogenetic approach may be advantageous when studying the effects of chronic neuronal inhibition. Ivermectin's ability to readily cross the blood brain barrier has also come into question as only very low levels of ivermectin have been shown to accumulate in brain tissue following oral administration of ivermectin [32]. The low penetrance of ivermectin into the brain is thought to be due to the P-glycoprotein expressed in the endothelial cells making up the blood brain barrier. P-glycoprotein functions as an ATP-dependent membrane efflux pump with multiple known substrates, which may also include ivermectin. Indeed, the concentration of ivermectin in the brain tissue is increased by 50- to 100-fold in P-glycoprotein knockout mice [32]. However, utilizing such mice may not be the optimal solution as ivermectin may interact with endogenous $GABA_A$ and glycine receptors and cause neurotoxicity at high concentrations. Therefore, the iterations of the modified GluCl receptor with increased localization to the plasma membrane with increased ivermectin sensitivity may be the preferred choice.

3.2.4 Ivermectin Ligand Administration

Ivermectin has been used in a number of neuronal silencing studies both in vitro and in vivo. Electrophysiology recording of primary hippocampal neurons first dissolved the ivermectin in artificial cerebrospinal fluid containing 0.1% DMSO. Solutions that effectively silence the cultured neurons were reported to be in the 1–20 nM range [27].

For in vivo experiments, ivermectin a 1% stock solution has been diluted further in sterile propylene glycol and injected 5–10 mg/kg in mice [14]. Ivermectin has also been prepared for oral administration in mice. The researchers conducting the study diluted the ivermectin in 70% sesame oil and 30% propanediol at concentrations allowing dosing at a ratio of 10:1 per gram of mouse [32].

3.3 Transient Receptor Potential Channels

3.3.1 Overview

Transient receptor potential (TRP) channels have been used as chemogenetic approach allowing for neuronal activation with a seconds-to-minutes time resolution. The TRP channel was, together with modified GluCl channels, one of the first attempts to create chemogenetic receptors for driving neural activation. This approach was based on the mammalian cation ligand-gated ion channel subfamily V member 1 (TRPV1). TRPV1 is an ionotropic, nonselective cation channel that exists as a homotetramer, i.e., four associated identical subunits. The TPRV1 channel is mainly expressed in the PNS by nociceptive neurons and is opened in response to heat, low pH, and, importantly, binding by capsaicin (a small molecule found in hot chili peppers) [33]. Thus, ligand binding of capsaicin by TRPV1 in cultured neurons was first shown to cause membrane depolarization and induce action potentials [33, 34] (Fig. 2h).

3.3.2 In Vivo Characterization of TRPV1

Neuronal activation in an organism using the TRPV1-capsaicin paradigm was first achieved in the C. elegans nematode, where researchers transgenetically expressed the TRPV1 receptor in ASH (Amphid neurons single ciliated endings) nociceptive neurons [35]. These neurons normally do not express the TRPV1 receptor, nor do they interact with capsaicin. Exposure of capsaicin to the transgenic C. elegans induced a strong avoidance behavior through activation of the TRPV1 channel expressed in the ASH neurons. This chemogenetic approach was later also expanded to include the menthol-activated TRPM8 channel, a receptor in the same subfamily as TRPV1. Both TRP channels have since been shown to function in mammalian cultured hippocampal neurons, where photo release of caged ligands induced robust action potentials [13]. A Cre-inducible knock-in TRPV1 transgenic mouse line has also been created, allowing for cell type or region-specific expression of TRPV1 [15]. Intracerebral application of capsaicin into this mouse line, after crossing with a nestin-Cre driver mouse line, has been shown to induce potent action potentials in neurons expressing the TRPV1 receptor as well as evoking contralateral turning behavior lasting for several minutes following local capsaicin infusion into the striatum [15]. To avoid possible endogenous activation by capsaicin within the CNS, transgenic mouse lines have also been bred on a TRPV1 knockout background [36].

3.3.3 Considerations and Limitations When Using TRPV1-Capsaicin System

Although shown to induce action potentials and behavior alterations in several organisms, there are several important challenges to consider when using TRPV1-based chemogenetic approaches. Cell specificity in mammalian brains is one major concern as the TRPV1 channel is endogenously expressed in the CNS, although at a lower level compared to the PNS. This may cause capsaicin to interact and activate other neurons than those targeted to express TRPV1, thus complicating the readout. This concern would be alleviated by restricting this chemogenetic approach to animals with a TRPV1 null background. Ligand toxicity is another concern as high local concentrations of capsaicin may cause agonist-induced excitotoxicity [37]. Receptor baseline effects and activation by endogenous ligands have also been observed. For instance, resting membrane potential of hippocampal neurons expressing TRPV1 has been shown to be elevated [13] and TRPV1 can be activated by endogenous endocannabinoids.

3.3.4 Capsaicin Ligand Administration

Ligand administration of capsaicin is broadly considered to be limited to direct intracranial infusion. This is due to capsaicin's poor crossing of the blood brain barrier. However, it has been reported that capsaicin i.p. injections are able to activate TRPV1 receptors expressed on dopaminergic neurons in the CNS [36]. Unless working on TRPV1 knockout animals, systemic administration is however not advisable, as this would cause activation of nociceptive neurons within the PNS.

The TRPV1-capsaicin system has been used in a number of different applications and conditions. These include studies in TRPV1-expressing primary sensory neurons which display sensitivity to capsaicin in the range of 3 nM-1 μM (diluted in 100% DMSO) without apparent toxicity in vitro [38]. For in vivo studies in TRPV1-expressing C. elegans, capsaicin has been shown effective in modulation neuronal activation when diluted to 50 μM in M13 and 2% ethanol.

Capsaicin caged with 4,5-dimethoxy-2-nitrobenzyl chloroformate has been used to modulate the activity of hippocampal primary neurons after photo release and can be performed according to Zemelman et al. [13]. Caged capsaicin was in this set up dissolved at 100 mM in DMSO and diluted to 5 μM for extracellular recordings.

Whole-cell clamp recordings of layer 5 cortical pyramidal neurons have been performed with capsaicin diluted to 250 nM [15]. In the same study, acute infusion in animals was performed using a capsaicin concentration of (0.5–1 μM). This resulted in measurable and reproducible bursts of capsaicin-induced action potentials. For systemic i.p. injections of capsaicin in TRPV1 knockout mice, capsaicin was dissolved in 0.9% NaCl with 10% Tween-80 and sonicated until completely dissolved [36].

3.4 $GABA_A$ Receptors

3.4.1 Allosteric Modulation of the $GABA_A$ Receptor

Allosteric modulation of the $GABA_A$ receptor is a chemogenetic approach that allows for bidirectional modulation of physiological GABA-mediated neurotransmission in the same animal on a time scale of minutes.

This approach takes advantage of molecules binding to the benzodiazepine-binding site of the $GABA_A$ receptor. This binding site is distinct from the binding site of endogenous GABA and is located between an α subunit and the γ2 subunit. Ligands binding to this site can act as allosteric modulators that change the affinity of GABA and the ability of GABA binding to open the chloride channel. For example, the drugs diazepam and zolpidem can increase the efficiency of $GABA_A$ to mediate cellular chloride influx upon binding to GABA, inhibiting neuronal activity (Fig. 2i). On the other hand, allosteric modulation of the $GABA_A$ receptor mediated by binding of the β-carboline methyl-6,7-dimethoxy-4-ethyl-beta-carboline-3-carboxylate (DMCM) to the $GABA_A$ receptor decreases the ability of GABA to induce influx of chloride ions across the cellular membrane, thus enhancing neuronal activity [39] (Fig. 2j). Of note is that both DMCM and zolpidem can be acutely antagonized by flumazenil. Mutation I77F γ2 on the $GABA_A$ receptor abolishes DMCM and zolpidem binding to the receptor while retaining affinity for GABA [40, 41]. To take advantage of this, researchers first developed a mouse line with a I77F γ2 mutation on exon 4 which was also engineered to have flanking loxP sites. This made the mice insensitive to DMCM and zolpidem in the absence of Cre recombinase [42, 43]. Combining these mice with Cre-mediated substitution of γ2F77 for γ2I77 by crossing the mice with a Cre-expressing mouse line or virally infusing a Cre-expressing virus creates cell selective restoration for DMCM and zolpidem and makes bidirectional modulation of the selected neuronal population possible. Neuronal modulation using this type of approach is unlike most other chemogenetic approaches. It is achieved not through direct modulation of the neuron but by enhancing or inhibiting the efficacy endogenous GABA release, making the modulation context dependent.

3.4.2 In Vivo Characterization of Allosteric Modulation of the $GABA_A$ Receptor

Chemogenetic modulation using allosteric modulation of the $GABA_A$ receptor was first achieved in cerebellar Purkinje cells, where zolpidem treatment enhanced inhibitory post synaptic currents. This resulted in mice with an ataxia phenotype following i.p. injection of zolpidem [44]. A viral vector Cre approach was also able to restore DMCM and zolpidem sensitivity in cortical pyramidal neurons [45].

3.4.3 Considerations and Limitations When Using Allosteric Modulation of the $GABA_A$ Receptor

The main limitation of utilizing allosteric modulation of the $GABA_A$ receptor is the dependence on mice with a DMCM and zolpidem insensitive genetic background. This somewhat limits the potential applications for the system. It is also important to note that this approach does not allow for direct control of selected neurons but only modulates the effect of endogenous GABA release. This may lead to vastly different responsiveness in various neuronal populations. Researchers conducting experiments utilizing DMCM for chemogenetic control should note that DMCM induced convulsions in wild-type animals and may therefore not be suitable for studies where wild-type controls are required.

3.4.4 Zolpidem and DMCM Ligand Administration

For electrophysiological recordings, Zolpidem has been pre-dissolved in DMSO to a concentration of 50 mM and then diluted in ACSF to a final concentration of 1 μM for recordings [43, 44]. Zolpidem has been utilized in vivo in γ2I77 mice at concentrations in the range of 1–30 mg/kg i.p [43].

In vivo studies utilizing DMCM for chemogenetic control in γ2I77 mice have used concentrations ranging from 20 to 60 mg/kg. DMCM for in vivo studies have been dissolved in 30–60 μL of 0.1 M HCl and then diluted to final concentration using sterile saline before i.p injection [46].

3.5 Pharmacologically Selective Actuator Modules

3.5.1 Overview

PSAM (pharmacologically selective actuator modules) together with their specific paired agonists PSEM (pharmacologically selective effector molecules) are made up of various chimeric actuator domains fused to ion pores engineered for bimodal modulation of neuronal activity. Members of this chemogenetic family include the following:

- **$PSAM^{L141F,\ Y115F}$-5HT3 HC** is modulated by the ligand $PSEM^{89S}$ which is then activating the cell by allowing cations to cross the cell membrane into the cell (Fig. 2k).
- **$PSAM^{L141F,\ Y115F}$-GlyR** is modulated by the ligand $PSEM^{89S}$ which then silence neurons (Fig. 2l).
- **$PSAM^{Q79G,\ L141S}$-nAChR V13** is modulated by the ligand $PSEM^{9S}$ and works by enhancing calcium signaling.

At its core, this method utilizes the ligand-binding domain of the Cys-loop α7 nicotinic acetylcholine receptor (nAChR) which acts as an independent actuator [16]. The ligand-binding domain of the nAChR was first mutated to lose the affinity to the endogenous ligand acetylcholine and gain ligand-binding affinity to the selected synthetic PSEM ligands [16]. Altered ligand recognition was achieved by implementing the bump-hole molecular engineering strategy. This approach was used to create a family of chemogenetic PSAMs using different ion pore combinations [16, 47, 48] with specific ion conductance properties, operating through distinct pathways. Upon PSEM ligand binding they are therefore capable of either neuronal activation or inhibition. Different conductance properties can be studied depending on the fused ion pore domain, which also allows for fully orthogonal studies while utilizing different PSAM/PSEM pairs.

Ion pore domains selective for calcium, cations, and chloride have all been utilized in this approach. One such PSEM combines a mutated nAChR ligand-binding domain with the ion pore domain of the 5-HT3 serotonergic receptor. This allows for cation entry across the cellular membrane and thus induce membrane depolarization and neuronal activation [16]. Similarly, fusion of the mutated nAChR actuator domain with the glycine receptor ion pore domain allows for chloride intake across the cellular membrane and thus induces hyperpolarization and neuronal inactivation [16]. Other domains from $GABA_C$ and nicotinic acetylcholine receptors have also been utilized in this manner. This highlights the clear advantage of utilizing this method, as different ionic conductance can be modulated and investigated in various cellular contexts.

In slice recordings, addition of PSEM ligand causes pronounced neuronal modulation within seconds to minutes. The glycine receptor PSAM was shown to be extremely potent in this context, effectively silencing neurons even when hundreds of picoamps of depolarizing current was applied. This property makes it suitable for neuronal silencing even in contexts of strong excitatory input. Proof-of-concept for this potency was demonstrated while dissecting the role of interneuron subtypes in the CA1 region of the hippocampus where strong synaptic drive into the CA1, by ontogenetically activated Schaeffer collateral axon projections, was effectively silenced [49].

3.5.2 *In Vivo Characterization of PSAM/PSEM*

In vivo characterization studies have showed that robust behavioral impacts mediated by the glycine PSAM are observable around 30 min post i.p. $PSEM^{89S}$ injection [16]. The glycine PSAM-induced silencing was demonstrated when expressed together with channelrhodopsin-2 in AgRP hypothalamic neurons where it was able to abolish light-induced feeding behavior following i.p. injection of $PSEM^{89S}$ [16]. Further studies utilizing

PSAM^{L141F,Y115F}-GlyR in vivo neuronal silencing include suppression of contextual fear learning when silencing hippocampal neurons [50] and silencing of neurons of the ventral medullary reticular formation in skilled reaching task tests [51].

The orthogonal potential of utilizing multiple PSAM/PSEM combinations to achieve bidirectional control of hippocampal interneurons was demonstrated when silencing PSAM^{L141F,Y115F}-5HT3-mediated neuronal activation by simultaneous treatment with PSEM^{308S} acting on the inhibitory PSAM^{L141F,Y115F}-GlyR receptor [52]. This study highlights the flexibility of bidirectional orthogonal neuronal modulation using the PSAM/PSEM approach.

3.5.3 Considerations and Limitations When Using PSAM/PSEM

The largest consideration utilizing the different PSAM/PSEM combinations is the limited characterization of on/off kinetics with all known combinations in different neuronal cell populations. Although not limiting to the technology, the field would benefit from further characterization.

Although the neuronal silencing or activation achieved through LGIC is fast and potent, the system lacks the ability to study more complex intracellular pathways compared to those of GPCR-based DREADDs. One limitation compared to DREADDs is also the lack of developed transgenic animal models.

3.5.4 PSEM Ligand Administration

PSEM ligands have been used to activate the various PSAM receptors that are available. For electrophysiological recordings PSEM^{89S} and PSEM^{22S} have been used in AGRP hypothalamic neurons at 10–30 μM baths [16]. Bath application of the second generation PSEM^{308S} has been utilized potently at 3 μM [49]. PSEM ligands have, for electrophysiological recordings, been dissolved in DMSO and then diluted in ACSF to the final concentration.

In vivo data available for PSEM^{89S} and PSEM^{22S} suggest that following i.p. injections, PSEM^{89S} concentrations rise rapidly in serum and brain tissue and is cleared from both compartments after about 1 h. PSEM^{22S} appears to have slower pharmacokinetics and persists in the brain following clearance from serum [16].

PSEM^{89S} and PSEM^{22S} have been dissolved in sterile saline (12 μL/g) and injected at 30 mg/kg [16]. The second generation PSEM^{308S} has however been successfully used at 5 mg/kg [51].

3.6 G Protein-Coupled Receptors and DREADDs

3.6.1 Overview

Another class of receptors that have been extensively utilized for modulation of neuronal activity are the G-Protein-coupled receptors (GPCRs). This is the largest class of chemogenetic receptors [53] and are membrane-bound proteins consisting of seven transmembrane domains that upon extracellular ligand binding promotes intracellular signal cascades. Ligand binding to GPCRs may inhibit or excite neuronal firing through three major intracellular signaling cascade pathways. Which pathway is activated depend on the class of G protein associated with the receptor. These pathways include the Gq, Gs, and Gi pathways that change the cell state through distinct intracellular pathways. The Gq pathway activates phospholipase C (PLC), thus elevating levels of diacylglycerol (DAG) and inositol triphosphate (IP3) [54]. The Gs pathway, on the other hand, activates adenylyl cyclase (AC) and thus increases levels of cyclic adenosine monophosphate (cAMP). Finally, activation of the Gi pathway decreases levels of cAMP via CA inhibition. Thus, Gq and Gs pathways drive neuronal activation while the Gi GPCRs inhibit neuronal activity [55]. Temporal resolution and strength of GPCR-induced modulation may differ between different cell types and mainly depends on the availability of intracellular second messenger proteins.

The first development of utilizing GPCRs as modulators of cellular activity came in the early 90s, where researchers used site-directed mutagenesis to increase affinity to novel ligands. In the first study, a mutant β2-adrenergic receptor was designed to have low affinity to the endogenous ligand adrenaline. This novel GPCR had however an increased affinity for binding a synthetic ligand, 1-(3′,4′-dihydroxyphenyl)-3-methyl-L-butanone (L-185,870) [56]. This first iteration of chemogenetic GPCRs suffered from the low potency of the engineered receptor to induce cellular modulation. Nevertheless, it was an important first step of showing how GPCRs could be utilized for modulation of cellular activity.

Further advances in engineering of GPCRs for selective neuronal control resulted in the next iteration of chemogenetic GPCRs known as receptor activated solely by synthetic ligand (RASSLs). The first RASSL was derived from the Gi-coupled k-opioid (KOR) receptor which was engineered to be insensitive to the endogenous ligand while retaining affinity for the KOR agonist spiradoline [57]. Since then, several RASSLs were created to form a family of chemogenetic receptors [58]. However, the utility of RASSLs in the field of neuroscience have been limited due to the off-target pharmacological interactions of the ligands as well as several of the RASSLs displaying high levels of constitutive activity in the absence of ligand [59, 60]. To remedy these shortcomings, a new iteration of chemogenetic receptors known as designer receptor exclusively activated by designer drugs (DREADDs) was developed and these will be discussed in this chapter further below.

3.6.2 Allatostatin Receptor

Another early attempt to develop a GPCR that could inhibit neuronal activity came in the form of the allatostatin receptor (AlstR). AlstR is a GPCR originating from the *Drosophila* where it is specifically activated by the insect peptide allatostatin. Transplanting the allatostatin receptor to mammalian neurons induces neuronal silencing upon allatostatin ligand binding lasting from minutes to hours [61]. Ligand binding to the receptor effectively induces the Gi coupled inwardly rectifying GIRK potassium channel, thus hyperpolarizing the cell membrane [62] (Fig. 2f). Local allatostatin administration has been reported to efficiently silence AlstR-expressing neurons within minutes in slice preparations from a variety of mammalian CNS tissues. These include the mice spinal cord, hippocampus, cortex, amygdala, and rat brainstem [62–66]. As a whole, this chemogenetic method has been successfully employed both in vitro in single neurons and in vivo in various neuronal subtypes.

3.6.3 In Vivo Characterization of Allatostatin Receptor

Local CNS infusion of allatostatin have been shown to induce neuronal silencing in neuronal populations involved in locomotion, respiration, and fear in mice [61, 63, 64, 66, 67] as well as reversible neuronal silencing in nonhuman primates [68]. Silencing time resolution in these applications has been in the order of minutes to hours.

3.6.4 Considerations and Limitations When Using the Allatostatin Receptor

The major limitation of using the allatostatin chemogenetic approach is the inability of the allatostatin ligand to cross the blood brain barrier. Therefore, studies that have utilized this system have been relegated to use local infusion within the CNS. Combined with limited tissue diffusion of the allatostatin ligand, this may lead to silencing variability and limited efficacy and may complicate the experimental design.

3.6.5 DREADDs

The most recently developed and currently most commonly used group of engineered GPCRs are the designer receptors exclusively activated by designer drugs (DREADDs). With the exception of the newly developed kappa opioid receptor (KOR) DREADD, DREADDs are based on engineered muscarinic acetylcholine receptors that have lost their affinity to the endogenous ligand acetylcholine and gained affinity to the designer ligand Clozapine-n-oxide (CNO), a metabolite of Clozapine. To date, several DREADDs have been developed where the receptor is coupled to either Gq (Fig. 2a), Gs/Golf (Fig. 2b), Gi (Fig. 2d), or β-arrestin (Fig. 2c) GPCRs. Most of these have now entered the realm of mainstream neuroscience and have been successfully utilized in a large number of studies using both viral delivery [69] and transgenic DREADD mouse lines [19].

hM3Dq DREADD Overview

The hM3Dq DREADD is an engineered GPCR evoking action potentials in neurons through membrane depolarization, caused by Gq-coupled closure of KCNQ channels dependent on phospholipase C. The human (h)M3Dq DREADD was developed by directed molecular evolution through random mutagenesis in genetically engineered yeast [70] grown in CNO containing media. Yeast grown under these screening conditions could only survive in selective media when expressing a mutant M3 receptor that displayed CNO agonist activity [18, 71, 72]. Mutant receptors that displayed CNO-binding activity, but also constitutive activity, were identified and discarded. Following multiple rounds of mutagenesis and selection, a M3 receptor with only two-point mutations (Y149C and A239G) was selected. The selected hM3Dq receptor displayed nano-molar activation by CNO, insensitivity to the native ligand acetylcholine, and undetectable levels of constitutive activity. As the mutated regions are highly conserved between muscarinic receptors, these were also used to create M1- and M5-based Gq DREADDs [18]; however, the hM3-based Gq DREADD is to this day the most commonly used Gq DREADD.

The hM3Dq DREADD has by now been utilized in innumerable in vitro systems including transfected HEK cells [18], neurons [19], hepatocytes [73], and pancreatic β-cells [74]. Importantly, activation of the hM3Dq receptor by CNO ligand binding effectively imitate downstream signaling of native M3 muscarinic receptor binding by acetylcholine [75, 76]. One very important feature of the hM3Dq receptor is that overexpression does not enhance endogenous baseline activity in multiple different contexts including neurons, astrocytes, hepatocytes, and pancreatic β-cells.

In Vivo Characterization of Gq DREADDs

The hM3Dq DREADD receptor has been successfully utilized in a large number of cell types in various behavioral contexts, most prominently in various neuronal subtypes. For example, the hM3Dq DREADD has been used successfully to modulate neuronal activity through c-fos driven expression [77], AGRP and orexin neurons of the hypothalamus [78, 79], VTA neurons [80], induced neurons in parkinsonian rats [81], grafted neurons [82], neurons of the locus ceruleus [83], neurons in the insular cortex [84], cholinergic neurons of the nucleus tegmenti pedunculopontine and striatum [85, 86] and of glutamatergic neurons in the basolateral amygdala [87] among many others. The large number of cell types that have been modulated by the hM3Dq DREADD show the versatility and flexibility of the system in vivo.

rM3Ds DREADDs Overview

The rM3Ds is a DREADD receptor that activates the Gs pathway upon CNO binding. This mediates neuronal activation which lasts several hours after systemic CNO administration. The rM3Gs receptor was developed in an effort to expand the DREADD toolbox to include alternate activation pathways to the hM3Dq DREADD. The rM3Ds DREADD is a chimeric receptor that was created by swapping the intracellular regions of the rat (r)M3Dq receptor to that of the turkey erythrocyte β adrenergic Gs-coupled receptor [88]. The rM3Gs receptor has been shown to mediate efficient activity through both the Gs [88] and Golf [89] pathways in response to CNO ligand binding. Activation of the Gs pathway results in an increase in activity of adenylyl cyclase, a protein that catalyzes the conversion of ATP into cAMP, increasing cellular activity [89] (Fig. 2b). In the initial study, the rM3Ds receptor was used to modulate pancreatic β-cells in an effort to deduce the intracellular signaling essential for insulin secretion [88]. However, the rM3Ds DREADD has since then been shown to increase cAMP-mediated neuronal signaling upon binding of the CNO receptor ligand. Though not as broadly utilized as the hM3Dq receptor DREADD, the rM3Ds has been useful for dissecting cAMP-specific signaling pathways in a variety of cell types and behavioral contexts [82, 90].

In Vivo Characterization of rM3Ds DREADDs

An early description of cAMP-mediated activation in the neuronal context was described by researchers studying striatal neurons. In this study, activation of striatal medium spiny neurons led to increased phosphorylation of DARPP-32, which is regulated by cAMP [89]. The rM3Ds DREADD receptor has since been utilized in a variety of in vivo contexts, including but not limited to Suprachiasmatic neurons [91], amygdala CRF neurons [92], A2A striatal neurons [89], striatal cholinergic interneurons [86] transplanted fetal dopamine neurons [82] and pancreatic beta cells [88, 91].

hM4Gi DREADDs Overview

The hM4Di DREADD is an engineered inhibitory GPCR based on the human muscarinic M4 receptor that causes hyperpolarization in neurons through the Gi pathway. In vivo, the hM4Di receptor displays similar kinetics to the hM3Dq receptor in response to CNO with the inhibitory effect lasting several hours [69, 93].

Regions within the acetylcholine-binding pocket of the muscarinic receptor are highly conserved between receptor subtypes and across species [72]. Using this knowledge and utilizing the experience gained from the directed molecular evolution approach of the hM3Dq receptor, researchers introduced two point mutations (Y113C and A203G) to the hM4 muscarinic receptor. These mutations rendered the receptor insensitive to the endogenous acetylcholine ligand and induced a nanomolar sensitivity to CNO [18]. The same point mutations could also be used to create the inhibitory hM2Di DREADD based on the M2 receptor [18] but the hM4Di receptor has been the most widely utilized of these inhibitory receptors. In the first study developing the hM4Di receptor, researchers found that receptor overexpression induced hyperpolarization and silencing of spontaneous and depolarization-evoked firing in pyramidal hippocampal neurons in response to CNO treatment [18]. As with the endogenous M4 receptor, the hM4Di activates Gi-mediated signaling and induces hyperpolarization mainly through inwardly rectifying potassium channels (GIRKs) [18] (Fig. 2d). However, the hM4Di also induces silencing through other intracellular pathways including the ERK/-MAPK pathway and inhibition of direct neurotransmitter release [94].

In Vivo Characterization of hM4Di DREADDs

The first studies to demonstrate that the hM4Di receptor could be utilized to effectively silence neurons in vivo came from researchers investigating the direct and indirect pathways of the basal ganglia in the context of drug addiction and reinforcement behaviors [69]. Since then, the hM4Di has been utilized in a broad range of studies in different cell types including but not limited to ARN hypothalamic neurons [78], thalamic neurons [95, 96], amygdala neurons [82, 96], parvalbumin neurons [97], grafted dopaminergic fetal neurons [82], and GABAergic neurons [82, 98]. It should be noted that the hM4Di receptor, along with other DREADDs, has also successfully been combined with optogenetics, allowing for bimodal modulation, utilizing fully orthogonal inducers. In addition, the decay kinetics of the hM4Di allows for PET imaging, permitting analysis of whole-brain effects due to silencing of specific neurons [99].

On these accolades, the hM4Di receptor is now in broad use both in vivo and in vitro in a wide range of cellular contexts where noninvasive cellular silencing is required.

β-arrestin DREADD Overview

The β-arrestin receptor is a CNO-activated DREADD with preferential downstream signaling through β-arrestin. GPCRs do not only provide intracellular signaling through G-proteins but also through β-arrestin (Fig. 2c). In order to study this specific aspect of GPCR signaling, the β-arrestin DREADD Rq(R165L) was developed [100]. This was done by mutating an amino acid part of the hM3Dq DREADD that is required for G-protein-mediated intracellular signaling. Arrestin pathways have been associated with immune and inflammatory response [101] as well as the actions of several psychoactive drugs, such as lithium [102], opioids [103], and antipsychotics [104], which potentially could benefit from further studies in various cellular contexts with the utilization of the β-arrestin DREADD.

Considerations and Limitations When Using CNO-Mediated DREADDs

Even though DREADDs have been successfully utilized in a large number of cellular contexts, it is important to note that as with all tools there are limitations. One of the major concerns about utilizing DREADDs was highlighted in a recent study, questioning the use of the CNO as a ligand [105]. In this study, researchers using ligand-binding assays and autoradiography found that CNO does not cross the blood brain barrier in mice to any large extent. Maybe even more troubling, they also found that DREADD-mediated neuronal modulation is primarily mediated by in vivo back metabolism of CNO into clozapine which crosses the blood brain barrier with high efficiency. Back metabolism of CNO into clozapine has previously been reported in humans [106], monkeys [107], and guinea pigs [108] but this paper was the first to confirm this event in mice. This complicates the CNO-DREADD approach as clozapine is a potent antipsychotic drug used in schizophrenia with numerous targets inside and outside of the CNS. This in turn could lead to dubious readouts of cellular activity and behavior. It is also important to note that CNO may not be inert in higher doses even though no off-target effects have been observed in previous studies [109–111]. Researchers have reported that doses at 5 mg/kg of CNO attenuates amphetamine-induced hyperlocomotion [112]. Although this could be due to back metabolism of CNO, it may be prudent not to exceed this concentration in future animal studies. These limitations of the technology may be circumvented in a number of ways. For instance, researchers could use subthreshold concentrations of clozapine or perlapine [113] as the DREADD agonist instead of high doses of CNO. Although this would require very stringent controls as both clozapine and perlapine may have off-target effects. Another possibility is to utilize the novel DREADD ligand compound 21 which has been shown to be a very potent hM3Dq agonist that does not interact with the endogenous M3 receptor [113].

Although the temporal resolution of DREADDs is adequate for in vivo studies, limitations of the DREADDs and other chemogenetic technologies may arise in applications that need neuronal modulation with millisecond temporal resolution. For applications requiring such temporal control, optogenetic technology may be preferable. However, development of photo-releasable, caged CNO may alleviate some of these concerns. It is however difficult to deliver specific and reversible amounts of DREADD ligand to a given neuronal cell population and different cell types may respond differently to various concentrations of DREADD ligand, requiring precise titration for each specific context. While optogenetic control may provide an easier approach for precise control of stimuli, this would be limiting in studies focusing on GPCR downstream effects.

Researchers that opt to use DREADD technology should also be aware of the possibility that overexpressing the DREADD receptor may impact the expression of the endogenous receptor, thus acting as a dominant negative receptor. However, there is currently no data available on the subject. The DREADD receptor expression in itself may be of some concern in certain contexts. For instance, it has been reported that ligand-independent changes in ion channel activity and second messenger signaling occur when expressing the hM4Di DREADD in PNS neurons [114].

Also, although the family of DREADD receptors were developed with low or nonexistent levels of constitutive activity in mind, the rM3Ds DREADD has been shown to have low levels of constitutive activity in certain cellular contexts [82, 88]. It is therefore important to utilize stringent controls when modulating cellular activity using the Gs DREADD.

The main limitation of the β-arrestin Rq(R165L) DREADD is that this approach has only been shown to activate β-arrestin pathways in vitro. Moreover, this activation appears to require a relatively high concentration of CNO due to low potency of the receptor. It is possible that the high concentration of CNO required may limit the potential use of the β-arrestin DREADD in vivo.

Given the limitations of the technology presented in this chapter, especially since the recent reports regarding CNO back conversion, it is important for researchers using chemogenetic approaches to carefully consider all available options and approaches available for neuronal modulatory control.

CNO Ligand Administration

A large number of in vitro and in vivo studies have utilized the CNO-DREADD system in various cellular contexts. Below are a few examples on how to use CNO in some of these circumstances.

Electrophysiological recording in HEK cells have utilized 10 mM CNO when applied directly onto the cells [18]. Patch clamp recordings of CA1 pyramidal neurons in hippocampal slices have used CNO with a concentration of 500 nM to depolarize neurons in hM3Dq mice [19]. Calcium flux assays in primary hippocampal neurons have also been utilized where CNO was added at a concentration of 1 μM [19].

For most systemic in vivo applications, CNO is pre-diluted in 0.5–10% DMSO and vortexed until a clear yellow solution appears without any visible solids remaining in the solution. The DMSO stock solution has then been diluted in sterile saline solution to the appropriate concentration. CNO concentrations have been utilized in vivo in a range from 0.1 mg/kg [19] up to 10 mg/kg in rats [82] and mice [83, 93]. Most in vivo studies have however utilized 3 mg/kg [112].

In vivo chronic activation of DREADDs has been achieved through adding CNO to the animals drinking water. For example, CNO has been pre-diluted in 0.025% DMSO and then diluted in the drinking water to 5 mg/kg/day assuming an average body weight of 22 g with a water intake of 3.5 mL daily [115]. When using CNO in the animal's drinking water, the water bottle should be light protected, e.g., by wrapping the bottle in aluminum foil. The water should also be prepared fresh daily and saccharine can be added to promote drinking and mask the bitter taste of CNO.

Recently researchers showed that CNO dissolved in 0.9% saline solution could activate neurons in the brain through eye drop delivery [116]. This approach of delivery could be important when stress levels of the animal are needed to be kept to a minimum.

Kappa Opioid DREADD Overview

The kappa opioid receptor DREADD (KORD) is the latest receptor in the DREADD toolbox. The KORD receptor is an inhibitory DREADD that reaches the highest potency of neuronal inhibition minutes after systemic injection and the ligand is quickly cleared from the CNS [117]. Unlike previous DREADD receptors, The KORD receptor has been developed to utilize the ligand Salvinorin B (SalB) in the place of CNO. As previous DREADDs have all utilized CNO as their ligand, this has limited bidirectional control of selected neuronal populations and thus a novel DREADD was needed. With the advent of the KORD receptor, the DREADD toolbox now allows for bidirectional control using fully orthogonal ligands.

The KORD receptor was developed through a rational design approach based on crystallographic and molecular modeling studies of hKOR [117, 118]. This cumbersome approach taken as the initial attempts of directed molecular evolution through random mutagenesis [18, 71] failed due high levels of constitutive activity [18, 71]. The rational design studies identified a single mutation (D138N) which abolishes binding of all known KOR agonists while retaining the binding affinity for Salvinorin [117, 119]. The resulting KORD receptor appears to inhibit neuronal activity through the same downstream pathways as the hM4Di receptor, mainly through hyperpolarization of the cellular membrane through activation of $G\beta/\gamma$-mediated G protein GIRKs [117] (Fig. 2e). The other important mechanism of neuronal silencing is believed to be mediated by inhibition of presynaptic neurotransmitter release [94, 117]. Importantly, SalB has been shown not to have any apparent activity at any other tested molecular targets including GPCRs, ion channels, transporters, and enzymes [117, 120]. The pharmacokinetic aspects of the SalB ligand provides flexibility in comparison to CNO ligand DREADDs, as neuronal silencing occurs within minutes of systemic SalB ligand administration and is fully reversed within an hour. This means that neuronal silencing can be established within minutes but also extended to several hours if needed by repeated injections.

In Vivo Characterization of KORD

The mutant KOR receptor effectively silenced neuronal firing upon ligand binding with SalB in several cellular contexts to date. The KORD receptor has been shown to silence midbrain GABAergic neurons in VGAT-Cre mice, paraventricular hypothalamic neurons in SIM1-Cre mice, AGRP neurons in AGRP-Cre mice [117, 121], and midbrain dopaminergic neurons [122]. In addition, the KORD receptor has been utilized in conjunction with other DREADDs for orthogonal multiplexed neuronal modulation [82, 86, 123].

Considerations and Limitations When Using KORD

An important aspect when using the KORD chemogenetic approach is to note that the SalB ligand does have a low affinity to endogenous KOR albeit at high concentrations [124]. It is therefore prudent of any research utilizing the KORD to use the lowest possible dose of SalB as well as implementing stringent controls. The fast-acting kinetics of the KORD mean that the peak of silencing last only minutes with all behavior effects completely abolished within an hour. This may make the KORD unsuitable for long-term neuronal silencing when repeated systemic injections are undesirable. However, this is highly dependent on the nature of the behavior test in question. Also, it is yet to be shown if SalB can be administered through drinking water. If possible, it could be utilized to achieve chronic inhibition using the KORD-SalB approach.

SalB Ligand Administration

Due to the novelty of the KORD receptor, far less studies characterized this DREADD approach compared to the CNO DREADDs. However, a number of studies have utilized the KORD DREADD receptor. For electrophysiology recordings 100 nM SalB has been used and reportedly leads to potent neuronal silencing [117].

In vivo studies have mainly utilized pure DMSO in order to dilute SalB due to the compounds low solubility. SalB has therefore been diluted at fairly high concentrations (25–30) mg/mL and vortexed until dissolved before injecting at a low volume s.c. [82, 86, 117]. This is due to the irritable effect of DMSO when injected into animals. Repeated DMSO administrations may also result in skin lesions for the animal. One study has however pre-diluted SalB in DMSO and then further diluted the compound using sunflower oil [123]. SalB has been utilized at concentrations ranging from 1.0 to 17.0 mg/kg s.c. [82, 86, 117, 123], though 10 mg/kg appears to be the most commonly used dose. SalB has also been successfully diluted in tween 80; ethanol; sterile water (1:1:8) [125], although this may be more practical for recordings than in vivo use as the mixture contains ethanol.

DREADD Delivery and Expression In Vivo

The technologies available today gives researchers great amount of flexibility and control when utilizing DREADD technology to modulate specific neuronal populations. Two strategies have mainly been implemented in order to achieve this specificity of DREADD expression in vivo. These include targeted intracranial viral infusions, mainly utilizing AAVs and transgenic mouse lines engineered to express the DREADDs in specific neuronal subpopulations. Cell-specific DREADD expression using viral vectors has mainly utilized cell-specific promoters or recombinase-driven technology such as the Cre/Lox system or FLP recombinase.

A large selection of transgenic mouse lines is available for cell-specific expression of various DREADDs. For example, Cre-inducible DREADD mice for hM3Dq, rM3Ds, and hM4Di DREADDs as well as A2A D2-specific rM3Ds transgenic mice are available from Jax. Other examples include tet-on and tet-off based DREADD expression for hM3Dq [19] and hM4Di [126] DREADDs.

In addition, an FLP-induced hM4Di DREADDs has also been created [93] for specific expression of different cellular populations. An interesting approach has utilized a fos-based approach to express DREADDs in neurons that are activated in specific sensory contexts [77]. Transgenic DREADD approaches have the advantage of displaying very similar expression of the receptor in question across all animals for multiple generations, thus reducing between-mouse variability.

Transgenic lines do however require extensive breeding times, and maintenance costs may therefore be high. In addition, as each transgenic line is only specific for a certain cell type, creating new lines of transgenic mice if required, may be time consuming and costly. Therefore, utilizing Cre-dependent mice lines may be of preference.

The alternative to transgenic mice lines are viral vector-based approaches to induce targeted expression in specific cell types. Infusion of viral vectors containing chemogenetic constructs is a flexible approach to induce DREADDs or indeed any other chemogenetic receptor into a targeted cell population. Viral vectors are usually infused into the target region of the brain using stereotactic coordinates relative the bregma of the animal's skull which can easily be determined from a brain atlas.

The ability of the viral vector to diffuse in the brain and transduce the targeted cell population is determined by several factors such as viral family, serotype, viral titer, volume, and method of delivery. Several different viruses have been utilized by researchers for in vivo gene transfer in the CNS, including lenti, adeno, CAV, and rabies virus strains. However, the most broadly utilized viral vector is the adeno-associated virus (AAV). The low immunogenicity, tropism in the CNS, lack of insertional mutagenesis, and the ease of production and versatility means that the AAV has many advantages for in vivo gene transfer to the CNS over other available vectors. There are a number of different approaches to targeting DREADD expression to a specific cell population using viral vectors. First off, the physical targeting of the viral infusion into the brain gives spatial specificity but low cell type selectivity. Cell type specificity can be increased by tailoring the plasmid constructs delivered to the cell via AAVs. For instance, greater specificity can be achieved by driving the expression of the chemogenetic receptor in a specific cell type by utilizing a tissue-specific promoter system. However, promoter-driven systems may not always be tight, and leakiness to undesired cell populations may be an issue. Another approach is to utilize Cre/loxP or FLP/FRT recombination systems. By flanking the chemogenetic DNA construct with a set of loxP sites, only specific neurons expressing Cre recombinase will induce chemogenetic expression. Therefore, combining

Cre-expressing transgenic organisms with AAVs containing a Cre-inducible chemogenetic construct is a powerful tool for expressing the receptors in specific cell population. This approach usually targets the cell soma; however, utilizing retrogradely transported viruses containing Cre recombinase allows for Cre-targeted expression in neurons projecting to a specific anatomical region. This approach has successfully been implemented for specific DREADD expression based on projection localization using retrogradely transported herpes simplex viruses [69] and the canine adeno virus [127]. Recently, modified AAVs capable of efficient retrograde transport have also been developed [128].

Utilizing a viral approach for chemogenetic expression has a number of advantages compared to the use of transgenic mouse lines. Firstly, viral approaches offer flexibility and do not require the laborious effort of creating novel transgenic organisms for each new application. Viral approaches are also applicable on complex organisms for which no transgenic animals are available such as nonhuman primates [68, 129, 130]. However, viral approaches also do have some disadvantages compared to transgenic animals. For example, viral infusion and transduction of cells does not give any control of copy number and therefore expression levels can vary between transduced cells. In addition, due to the virus diffusing from the injection site, cells closest to the injection site will more readily be transduced by the viruses while peripheral cells may not be as readily transduced. This may result in an uneven distribution of the chemogenetic receptor within the tissue. Combined with the biological variability between animals and the subsequent variations between injection sites as well infusion rates may induce significant variability. The difference in transduction between distinct viral serotypes should also be taken into consideration. The most commonly utilized AAVs within the field of neuroscience are AAV1, 2, 5, 6, 8, and 9. The different AAV serotypes may have different affinity and transduction to different subtypes of neurons. For example, transducing striatal cholinergic interneurons, the most effective serotype is AAV8 [86] while cortical neurons are effectively transduced utilizing AAV6 [131]. It may therefore be prudent to test common serotypes in pilot studies when conducting experiments in novel cell populations. The different AAV serotypes may also have distinct spread within the CNS. In general, AAV serotypes 1–4 display limited spread, while serotypes 5–9 diffuse more readily to throughout larger brain structures. Modified AAV serotypes such as the AAV-DJ [132] or php.B [133] also display increased anatomical spread within the CNS. AAVs with large anatomical spread may be desirable when transducing large brain structures; however, they may also be detrimental if a small population of neurons in close proximity are targeted, thus making the choice of correct AAV for the right application essential.

3.7 Concluding Remarks

Chemogenetics are powerful and flexible approaches that enable elucidation of complex circuitry and function by neuronal activation or inhibition throughout the CNS. As such, chemogenetics and especially DREADDs have been utilized in ever-broadening research fields. The main strength of chemogenetics compared to other modulatory approaches, such as optogenetics, is the ability to target widely distributed neuronal circuits in a comparatively noninvasive way. However, optogenetic approaches have the advantage of millisecond temporal control which chemogenetics generally lack. Chemogenetics also allows for long-term reversible studies of neuronal function which enables complex behavior to be analyzed in a manner that is more functionally and physiologically relevant compared to optogenetic and toxin-induced neuronal modulation approaches. This is especially true for GPCR-based chemogenetic receptors such as DREADDs, as these act as endogenous GPCRs using the same downstream signaling pathways.

It is important to note that each chemogenetic approach comes with specific advantages and limitations, as discussed above, that researchers utilizing chemogenetic approaches should be aware of. It is imperative that the right chemogenetic approach is selected for the correct application, especially when working in the complex setting of the CNS. It is therefore important for researchers to have clear research questions in order to take full advantage of the chemogenetic approaches available. For instance, the choice of chemogenetic approach utilized for a given experimental setting should consider if neuronal activation or inhibition is enough or if a fully orthogonal approach utilizing different chemogenetic approaches for bimodal modulation is required. Other important aspects of chemogenetics researchers should also consider include, duration of neuronal modulation for a certain behavior experiment, ligand properties such as half-life, penetrance of the blood brain barrier, and possible off-target effects. When taking these aspects of chemogenetics into consideration and effectively utilizing the strengths, chemogenetics provide means for elucidation of complex research questions through noninvasive, often physiologically relevant and orthogonal modulatory control of neurons and wider neuronal circuity.

Acknowledgments

This work was supported by grants from the Swedish Research Council (K2014-79X-22510-01-1 and ÄR-MH-2016-01997); Michael J Fox foundation, Swedish Parkinson Foundation;

Crafoord foundation and the Bagadilico Linnaeus consortium. TB is supported by Ass. Senior lectureship from the Bente Rexed foundation.

References

1. Forkmann G, Dangelmayr B (1980) Genetic control of chalcone isomerase activity in flowers of Dianthus caryophyllus. Biochem Genet 18(5–6):519–527
2. Sternson SM, Roth BL (2014) Chemogenetic tools to interrogate brain functions. Annu Rev Neurosci 37:387–407
3. Strobel SA (1998) Ribozyme chemogenetics. Biopolymers 48(1):65–81
4. Bishop AC et al (1998) Design of allele-specific inhibitors to probe protein kinase signaling. Curr Biol 8(5):257–266
5. Bishop AC et al (2000) A chemical switch for inhibitor-sensitive alleles of any protein kinase. Nature 407(6802):395–401
6. Liu Y et al (1998) Engineering Src family protein kinases with unnatural nucleotide specificity. Chem Biol 5(2):91–101
7. Cohen MS et al (2005) Structural bioinformatics-based design of selective, irreversible kinase inhibitors. Science 308 (5726):1318–1321
8. Chen X et al (2005) A chemical-genetic approach to studying neurotrophin signaling. Neuron 46(1):13–21
9. Dar AC et al (2012) Chemical genetic discovery of targets and anti-targets for cancer polypharmacology. Nature 486(7401):80–84
10. Haring D, Distefano MD (2001) Enzymes by design: chemogenetic assembly of transamination active sites containing lysine residues for covalent catalysis. Bioconjug Chem 12 (3):385–390
11. Collot J et al (2003) Artificial metalloenzymes for enantioselective catalysis based on biotin-avidin. J Am Chem Soc 125(30):9030–9031
12. Klein G et al (2005) Tailoring the active site of chemzymes by using a chemogenetic-optimization procedure: towards substrate-specific artificial hydrogenases based on the biotin-avidin technology. Angew Chem Int Ed Engl 44(47):7764–7767
13. Zemelman BV et al (2003) Photochemical gating of heterologous ion channels: remote control over genetically designated populations of neurons. Proc Natl Acad Sci U S A 100(3):1352–1357
14. Lerchner W et al (2007) Reversible silencing of neuronal excitability in behaving mice by a genetically targeted, ivermectin-gated Cl-channel. Neuron 54(1):35–49
15. Arenkiel BR et al (2008) Genetic control of neuronal activity in mice conditionally expressing TRPV1. Nat Methods 5(4):299–302
16. Magnus CJ et al (2011) Chemical and genetic engineering of selective ion channel-ligand interactions. Science 333(6047):1292–1296
17. Redfern CH et al (1999) Conditional expression and signaling of a specifically designed Gi-coupled receptor in transgenic mice. Nat Biotechnol 17(2):165–169
18. Armbruster BN et al (2007) Evolving the lock to fit the key to create a family of G protein-coupled receptors potently activated by an inert ligand. Proc Natl Acad Sci U S A 104 (12):5163–5168
19. Alexander GM et al (2009) Remote control of neuronal activity in transgenic mice expressing evolved G protein-coupled receptors. Neuron 63(1):27–39
20. Nadeau H et al (2000) ROMK1 (Kir1.1) causes apoptosis and chronic silencing of hippocampal neurons. J Neurophysiol 84 (2):1062–1075
21. Stanley BG et al (1993) Lateral hypothalamic injections of glutamate, kainic acid, D, L-alpha-amino-3-hydroxy-5-methyl-isoxazole propionic acid or N-methyl-D-aspartic acid rapidly elicit intense transient eating in rats. Brain Res 613(1):88–95
22. Slimko EM et al (2002) Selective electrical silencing of mammalian neurons in vitro by the use of invertebrate ligand-gated chloride channels. J Neurosci 22(17):7373–7379
23. Degani-Katzav N et al (2016) Subunit stoichiometry and arrangement in a heteromeric glutamate-gated chloride channel. Proc Natl Acad Sci U S A 113(5):E644–E653
24. Li P, Slimko EM, Lester HA (2002) Selective elimination of glutamate activation and introduction of fluorescent proteins into a Caenorhabditis elegans chloride channel. FEBS Lett 528(1–3):77–82
25. Slimko EM, Lester HA (2003) Codon optimization of Caenorhabditis elegans GluCl ion channel genes for mammalian cells dramatically improves expression levels. J Neurosci Methods 124(1):75–81

26. McKellar QA et al (1992) Clinical and pharmacological properties of ivermectin in rabbits and guinea pigs. Vet Rec 130 (4):71–73
27. Frazier SJ, Cohen BN, Lester HA (2013) An engineered glutamate-gated chloride (GluCl) channel for sensitive, consistent neuronal silencing by ivermectin. J Biol Chem 288 (29):21029–21042
28. Lynagh T, Lynch JW (2010) An improved ivermectin-activated chloride channel receptor for inhibiting electrical activity in defined neuronal populations. J Biol Chem 285 (20):14890–14897
29. Oishi Y et al (2013) Role of the medial prefrontal cortex in cataplexy. J Neurosci 33 (23):9743–9751
30. Haubensak W et al (2010) Genetic dissection of an amygdala microcircuit that gates conditioned fear. Nature 468 (7321):270–276
31. Lin D et al (2011) Functional identification of an aggression locus in the mouse hypothalamus. Nature 470(7333):221–226
32. Schinkel AH et al (1994) Disruption of the mouse mdr1a P-glycoprotein gene leads to a deficiency in the blood-brain barrier and to increased sensitivity to drugs. Cell 77 (4):491–502
33. Caterina MJ et al (1997) The capsaicin receptor: a heat-activated ion channel in the pain pathway. Nature 389(6653):816–824
34. Tominaga M et al (1998) The cloned capsaicin receptor integrates multiple pain-producing stimuli. Neuron 21(3):531–543
35. Tobin DM et al (2002) Combinatorial expression of TRPV channel proteins defines their sensory functions and subcellular localization in *C. elegans* neurons. Neuron 35(2):307–318
36. Guler AD et al (2012) Transient activation of specific neurons in mice by selective expression of the capsaicin receptor. Nat Commun 3:746
37. Crawford DC et al (2009) Comparative effects of heterologous TRPV1 and TRPM8 expression in rat hippocampal neurons. PLoS One 4(12):e8166
38. Ahluwalia J et al (2003) Anandamide regulates neuropeptide release from capsaicin-sensitive primary sensory neurons by activating both the cannabinoid 1 receptor and the vanilloid receptor 1 in vitro. Eur J Neurosci 17(12):2611–2618
39. Campo-Soria C, Chang Y, Weiss DS (2006) Mechanism of action of benzodiazepines on GABAA receptors. Br J Pharmacol 148 (7):984–990
40. Sancar F et al (2007) Structural determinants for high-affinity zolpidem binding to GABA-A receptors. Mol Pharmacol 71(1):38–46
41. Buhr A, Baur R, Sigel E (1997) Subtle changes in residue 77 of the gamma subunit of alpha1beta2gamma2 GABAA receptors drastically alter the affinity for ligands of the benzodiazepine binding site. J Biol Chem 272 (18):11799–11804
42. Ogris W et al (2004) Affinity of various benzodiazepine site ligands in mice with a point mutation in the GABA(A) receptor gamma2 subunit. Biochem Pharmacol 68 (8):1621–1629
43. Cope DW et al (2004) Abolition of zolpidem sensitivity in mice with a point mutation in the GABAA receptor gamma2 subunit. Neuropharmacology 47(1):17–34
44. Wulff P et al (2007) From synapse to behavior: rapid modulation of defined neuronal types with engineered GABAA receptors. Nat Neurosci 10(7):923–929
45. Sumegi M et al (2012) Virus-mediated swapping of zolpidem-insensitive with zolpidem-sensitive GABA(A) receptors in cortical pyramidal cells. J Physiol 590(7):1517–1534
46. Leppa E et al (2005) Agonistic effects of the beta-carboline DMCM revealed in GABA (A) receptor gamma 2 subunit F77I point-mutated mice. Neuropharmacology 48 (4):469–478
47. Eisele JL et al (1993) Chimaeric nicotinic-serotonergic receptor combines distinct ligand binding and channel specificities. Nature 366(6454):479–483
48. Grutter T et al (2005) Molecular tuning of fast gating in pentameric ligand-gated ion channels. Proc Natl Acad Sci U S A 102 (50):18207–18212
49. Lovett-Barron M et al (2012) Regulation of neuronal input transformations by tunable dendritic inhibition. Nat Neurosci 15 (3):423–430, S1-3
50. Lovett-Barron M et al (2014) Dendritic inhibition in the hippocampus supports fear learning. Science 343(6173):857–863
51. Esposito MS, Capelli P, Arber S (2014) Brainstem nucleus MdV mediates skilled forelimb motor tasks. Nature 508(7496):351–356
52. Donato F, Rompani SB, Caroni P (2013) Parvalbumin-expressing basket-cell network plasticity induced by experience regulates adult learning. Nature 504(7479):272–276
53. Allen JA, Roth BL (2011) Strategies to discover unexpected targets for drugs active at G protein-coupled receptors. Annu Rev Pharmacol Toxicol 51:117–144

54. Masseck OA et al (2011) Light- and drug-activated G-protein-coupled receptors to control intracellular signalling. Exp Physiol 96 (1):51–56
55. Gainetdinov RR et al (2004) Desensitization of G protein-coupled receptors and neuronal functions. Annu Rev Neurosci 27:107–144
56. Strader CD et al (1991) Allele-specific activation of genetically engineered receptors. J Biol Chem 266(1):5–8
57. Coward P et al (1998) Controlling signaling with a specifically designed Gi-coupled receptor. Proc Natl Acad Sci U S A 95(1):352–357
58. Conklin BR et al (2008) Engineering GPCR signaling pathways with RASSLs. Nat Methods 5(8):673–678
59. Hsiao EC et al (2008) Osteoblast expression of an engineered Gs-coupled receptor dramatically increases bone mass. Proc Natl Acad Sci U S A 105(4):1209–1214
60. Sweger EJ et al (2007) Development of hydrocephalus in mice expressing the G(i)-coupled GPCR Ro1 RASSL receptor in astrocytes. J Neurosci 27(9):2309–2317
61. Tan EM et al (2006) Selective and quickly reversible inactivation of mammalian neurons in vivo using the Drosophila allatostatin receptor. Neuron 51(2):157–170
62. Lechner HA, Lein ES, Callaway EM (2002) A genetic method for selective and quickly reversible silencing of Mammalian neurons. J Neurosci 22(13):5287–5290
63. Marina N et al (2010) Essential role of Phox2b-expressing ventrolateral brainstem neurons in the chemosensory control of inspiration and expiration. J Neurosci 30 (37):12466–12473
64. Zhou Y et al (2009) CREB regulates excitability and the allocation of memory to subsets of neurons in the amygdala. Nat Neurosci 12(11):1438–1443
65. Wehr M et al (2009) Transgenic silencing of neurons in the mammalian brain by expression of the allatostatin receptor (AlstR). J Neurophysiol 102(4):2554–2562
66. Gosgnach S et al (2006) V1 spinal neurons regulate the speed of vertebrate locomotor outputs. Nature 440(7081):215–219
67. Zhang Y et al (2008) V3 spinal neurons establish a robust and balanced locomotor rhythm during walking. Neuron 60(1):84–96
68. Nielsen KJ, Callaway EM, Krauzlis RJ (2012) Viral vector-based reversible neuronal inactivation and behavioral manipulation in the macaque monkey. Front Syst Neurosci 6:48
69. Ferguson SM et al (2011) Transient neuronal inhibition reveals opposing roles of indirect and direct pathways in sensitization. Nat Neurosci 14(1):22–24
70. Schmidt C et al (2003) Random mutagenesis of the M3 muscarinic acetylcholine receptor expressed in yeast. Identification of point mutations that "silence" a constitutively active mutant M3 receptor and greatly impair receptor/G protein coupling. J Biol Chem 278(32):30248–30260
71. Dong S, Rogan SC, Roth BL (2010) Directed molecular evolution of DREADDs: a generic approach to creating next-generation RASSLs. Nat Protoc 5(3):561–573
72. Rogan SC, Roth BL (2011) Remote control of neuronal signaling. Pharmacol Rev 63 (2):291–315
73. Li JH et al (2013) A novel experimental strategy to assess the metabolic effects of selective activation of a G(q)-coupled receptor in hepatocytes in vivo. Endocrinology 154 (10):3539–3551
74. Jain S et al (2013) Chronic activation of a designer G(q)-coupled receptor improves beta cell function. J Clin Invest 123 (4):1750–1762
75. Alvarez-Curto E et al (2010) Ligand regulation of the quaternary organization of cell surface M3 muscarinic acetylcholine receptors analyzed by fluorescence resonance energy transfer (FRET) imaging and homogeneous time-resolved FRET. J Biol Chem 285 (30):23318–23330
76. Alvarez-Curto E et al (2011) Developing chemical genetic approaches to explore G protein-coupled receptor function: validation of the use of a receptor activated solely by synthetic ligand (RASSL). Mol Pharmacol 80(6):1033–1046
77. Garner AR et al (2012) Generation of a synthetic memory trace. Science 335 (6075):1513–1516
78. Krashes MJ et al (2011) Rapid, reversible activation of AgRP neurons drives feeding behavior in mice. J Clin Invest 121(4):1424–1428
79. Sasaki K et al (2011) Pharmacogenetic modulation of orexin neurons alters sleep/wakefulness states in mice. PLoS One 6(5):e20360
80. Boender AJ et al (2014) Combined use of the canine adenovirus-2 and DREADD-technology to activate specific neural pathways in vivo. PLoS One 9(4):e95392
81. Dell'Anno MT et al (2014) Remote control of induced dopaminergic neurons in parkinsonian rats. J Clin Invest 124(7):3215–3229

82. Aldrin-Kirk P et al (2016) DREADD Modulation of Transplanted DA Neurons Reveals a Novel Parkinsonian Dyskinesia Mechanism Mediated by the Serotonin 5-HT6 Receptor. Neuron 90(5):955–968
83. Vazey EM, Aston-Jones G (2014) Designer receptor manipulations reveal a role of the locus coeruleus noradrenergic system in isoflurane general anesthesia. Proc Natl Acad Sci U S A 111(10):3859–3864
84. Mizoguchi H et al (2015) Insular neural system controls decision-making in healthy and methamphetamine-treated rats. Proc Natl Acad Sci U S A 112(29):E3930–E3939
85. Pienaar IS et al (2015) Pharmacogenetic stimulation of cholinergic pedunculopontine neurons reverses motor deficits in a rat model of Parkinson's disease. Mol Neurodegener 10:47
86. Aldrin-Kirk P et al (2018) Chemogenetic modulation of cholinergic interneurons reveals their regulating role on the direct and indirect output pathways from the striatum. Neurobiol Dis 109(Pt A):148–162
87. Sengupta A et al (2016) Disrupted prediction error links excessive amygdala activation to excessive fear. J Neurosci 36(2):385–395
88. Guettier JM et al (2009) A chemical-genetic approach to study G protein regulation of beta cell function in vivo. Proc Natl Acad Sci U S A 106(45):19197–19202
89. Farrell MS et al (2013) A Galphas DREADD mouse for selective modulation of cAMP production in striatopallidal neurons. Neuropsychopharmacology 38(5):854–862
90. Alcacer C et al (2017) Chemogenetic stimulation of striatal projection neurons modulates responses to Parkinson's disease therapy. J Clin Invest 127(2):720–734
91. Brancaccio M et al (2013) A Gq-Ca2+ axis controls circuit-level encoding of circadian time in the suprachiasmatic nucleus. Neuron 78(4):714–728
92. Pleil KE et al (2015) NPY signaling inhibits extended amygdala CRF neurons to suppress binge alcohol drinking. Nat Neurosci 18 (4):545–552
93. Ray RS et al (2011) Impaired respiratory and body temperature control upon acute serotonergic neuron inhibition. Science 333 (6042):637–642
94. Stachniak TJ, Ghosh A, Sternson SM (2014) Chemogenetic synaptic silencing of neural circuits localizes a hypothalamus-->midbrain pathway for feeding behavior. Neuron 82 (4):797–808
95. Jurik A et al (2015) Roles of prefrontal cortex and paraventricular thalamus in affective and mechanical components of visceral nociception. Pain 156(12):2479–2491
96. Rei D et al (2015) Basolateral amygdala bidirectionally modulates stress-induced hippocampal learning and memory deficits through a p25/Cdk5-dependent pathway. Proc Natl Acad Sci U S A 112 (23):7291–7296
97. Ognjanovski N et al (2017) Parvalbumin-expressing interneurons coordinate hippocampal network dynamics required for memory consolidation. Nat Commun 8:15039
98. Weber F et al (2015) Control of REM sleep by ventral medulla GABAergic neurons. Nature 526(7573):435–438
99. Michaelides M et al (2013) Whole-brain circuit dissection in free-moving animals reveals cell-specific mesocorticolimbic networks. J Clin Invest 123(12):5342–5350
100. Nakajima K, Wess J (2012) Design and functional characterization of a novel, arrestin-biased designer G protein-coupled receptor. Mol Pharmacol 82(4):575–582
101. Sharma D, Parameswaran N (2015) Multifaceted role of beta-arrestins in inflammation and disease. Genes Immun 16(8):499–513
102. Beaulieu JM et al (2008) A beta-arrestin 2 signaling complex mediates lithium action on behavior. Cell 132(1):125–136
103. Bohn LM et al (1999) Enhanced morphine analgesia in mice lacking beta-arrestin 2. Science 286(5449):2495–2498
104. Allen JA et al (2011) Discovery of beta-arrestin-biased dopamine D2 ligands for probing signal transduction pathways essential for antipsychotic efficacy. Proc Natl Acad Sci U S A 108(45):18488–18493
105. Gomez JL et al (2017) Chemogenetics revealed: DREADD occupancy and activation via converted clozapine. Science 357 (6350):503–507
106. Chang WH et al (1998) Reversible metabolism of clozapine and clozapine N-oxide in schizophrenic patients. Prog Neuro-Psychopharmacol Biol Psychiatry 22 (5):723–739
107. Raper J et al (2017) Metabolism and distribution of clozapine-N-oxide: implications for nonhuman primate chemogenetics. ACS Chem Neurosci 8(7):1570–1576
108. Jann MW, Lam YW, Chang WH (1994) Rapid formation of clozapine in guinea-pigs and man following clozapine-N-oxide administration. Arch Int Pharmacodyn Ther 328 (2):243–250
109. Alves-Rodrigues A et al (1996) Binding of clozapine metabolites and analogues to the

histamine H3 receptor in rat brain cortex. Arch Pharm 329(8–9):413–416

110. Salmi P, Ahlenius S (1996) Further evidence for clozapine as a dopamine D1 receptor agonist. Eur J Pharmacol 307(1):27–31
111. Wong G et al (1996) Effects of clozapine metabolites and chronic clozapine treatment on rat brain GABAA receptors. Eur J Pharmacol 314(3):319–323
112. MacLaren DA et al (2016) Clozapine N-oxide administration produces behavioral effects in long-evans rats: implications for designing DREADD experiments. eNeuro 3:5
113. Chen X et al (2015) The first structure-activity relationship studies for designer receptors exclusively activated by designer drugs. ACS Chem Neurosci 6(3):476–484
114. Saloman JL et al (2016) Gi-DREADD expression in peripheral nerves produces ligand-dependent analgesia, as well as ligand-independent functional changes in sensory neurons. J Neurosci 36(42):10769–10781
115. Padilla SL et al (2017) AgRP to Kiss1 neuron signaling links nutritional state and fertility. Proc Natl Acad Sci U S A 114(9):2413–2418
116. Keenan WT et al (2017) Eye-drops for activation of DREADDs. Front Neural Circuits 11:93
117. Vardy E et al (2015) A new DREADD facilitates the multiplexed chemogenetic interrogation of behavior. Neuron 86(4):936–946
118. Wu H et al (2012) Structure of the human kappa-opioid receptor in complex with JDTic. Nature 485(7398):327–332
119. Kane BE et al (2006) A unique binding epitope for salvinorin A, a non-nitrogenous kappa opioid receptor agonist. FEBS J 273 (9):1966–1974
120. Chavkin C et al (2004) Salvinorin A, an active component of the hallucinogenic sage salvia divinorum is a highly efficacious kappa-opioid receptor agonist: structural and functional considerations. J Pharmacol Exp Ther 308 (3):1197–1203
121. Denis RG et al (2015) Palatability can drive feeding independent of AgRP neurons. Cell Metab 22(4):646–657
122. Marchant NJ et al (2016) Role of ventral subiculum in context-induced relapse to alcohol seeking after punishment-imposed abstinence. J Neurosci 36(11):3281–3294
123. Rapanelli M et al (2017) Histamine modulation of the basal ganglia circuitry in the development of pathological grooming. Proc Natl Acad Sci U S A 114(25):6599–6604
124. Roth BL (2016) DREADDs for neuroscientists. Neuron 89(4):683–694
125. Wang Y et al (2008) 2-Methoxymethyl-salvinorin B is a potent kappa opioid receptor agonist with longer lasting action in vivo than salvinorin A. J Pharmacol Exp Ther 324(3):1073–1083
126. Zhu H et al (2014) Chemogenetic inactivation of ventral hippocampal glutamatergic neurons disrupts consolidation of contextual fear memory. Neuropsychopharmacology 39 (8):1880–1892
127. Carter ME et al (2013) Genetic identification of a neural circuit that suppresses appetite. Nature 503(7474):111–114
128. Tervo DG et al (2016) A designer AAV variant permits efficient retrograde access to projection neurons. Neuron 92(2):372–382
129. Eldridge MA et al (2016) Chemogenetic disconnection of monkey orbitofrontal and rhinal cortex reversibly disrupts reward value. Nat Neurosci 19(1):37–39
130. Nagai Y et al (2016) PET imaging-guided chemogenetic silencing reveals a critical role of primate rostromedial caudate in reward evaluation. Nat Commun 7:13605
131. Aldrin-Kirk P et al (2014) Novel AAV-based rat model of forebrain synucleinopathy shows extensive pathologies and progressive loss of cholinergic interneurons. PLoS One 9(7): e100869
132. Grimm D et al (2008) In vitro and in vivo gene therapy vector evolution via multispecies interbreeding and retargeting of adeno-associated viruses. J Virol 82(12):5887–5911
133. Deverman BE et al (2016) Cre-dependent selection yields AAV variants for widespread gene transfer to the adult brain. Nat Biotechnol 34(2):204–209

Part III

Viral Vectors

Chapter 5

AAV Production Using Baculovirus Expression Vector System

Quentin Sandro, Karima Relizani, and Rachid Benchaouir

Abstract

Gene transfer and gene therapy are powerful approaches for many biological research applications and promising avenues for the treatment of many genetic or cancer diseases. The most efficient gene transfer tools are currently derived from viruses. Among them, the recombinant adeno-associated viruses (AAVs) are vectors of choice for many fundamental and therapeutic applications. The increasing number of clinical trials involving AAVs demonstrates the need to implement production and purification processes to meet the quantitative and qualitative demands of regulatory agencies for the use of these vectors in clinical trials. In this context, the rise of production levels on an industrial scale appeared essential. The introduction, in 2002, of an AAV process using a baculovirus expression vector system (BEVS) has circumvented this technological lock. The advantage of BEVS in expanding the AAV production in insect cells has been to switch the process to bioreactor systems, which are the ideal equipment for scaling up. We describe here a method for producing AAV vectors using the BEVS which can be easily used by research laboratories wishing to overcome the difficulties associated with the scaling up of production levels. The method provides sufficient quantities of AAV vectors to initiate preclinical projects in large animal models or for research projects where a single batch of vectors will consolidate the repeatability and reproducibility of in vitro and especially in vivo experimental approaches.

Key words AAV vectors, Baculovirus, Production, Purification, Upstream process, Downstream process

1 Introduction

AAV-derived viral vectors are effective tools for in vivo gene transfer and are currently used in many clinical trials. A proof of their interest for gene therapy approaches was brought by the launch, in 2012, of the first AAV marketing approval for the treatment of familial lipoprotein lipase deficiency [1]. In this context, it becomes necessary to implement vector production structures capable of responding favorably to the quantitative and qualitative requirements of other future clinical trials.

Fredric P. Manfredsson and Matthew J. Benskey (eds.), *Viral Vectors for Gene Therapy: Methods and Protocols*, Methods in Molecular Biology, vol. 1937, https://doi.org/10.1007/978-1-4939-9065-8_5,

At research scale, the classical method for producing AAV vectors uses a method of tri-transfection of the adherent human embryonic kidney 293 (HEK293) cell line [2–5]. This method uses the transfection of three distinct plasmid constructs. The first harbors the transgene of interest flanked by the viral inverted terminal repeat (ITR) sequences, which are essential elements for genome replication and packaging. The second plasmid harbors the serotype-dependant *rep* and *cap* gene sequences, deleted from the wild-type AAV genome between the ITR elements, and coding for the expression of the viral enzymes and capsid structural proteins, respectively. The third large plasmid contains adenoviral helper genes coding for the expression of proteins that facilitate the replication and the packaging of the recombinant AAV genome. Despite the effectiveness of this system in producing functional AAV vectors, the adherent nature of the cells makes it difficult to increase production toward industrial scales. Alternative methods have been proposed to circumvent this technological lock. Among them, we can mention the tri-transfection of HEK293 cells grown in suspension [6], the method of production through stable cell lines [7], or even the use of Herpes simplex virus as a vector for AAV production by mammalian cells [8]. However, the system described in 2002 by Dr. Robert Kotin's team using the baculovirus expression vector system (BEVS) to produce AAV vectors marked a real technological turning point in the field [9]. BEVS itself is not new since it has been used by the industrial world for decades to produce recombinant proteins. The challenge was to use this system for the production of a microorganism as complex as a virus with the variety of proteins needed for its efficient production and adequate structure.

Here we describe a BEVS-based AAV production protocol using baculovirus-infected insect cells (BIICs) as raw material [10]. This approach, described for the first time in 2009 for the production of a specific recombinant protein, was successfully implemented in 2011 for AAV vectors processed by BEVS [11]. BIICs have the advantage to better cryo-preserve recombinant baculoviruses (rBAC) and to simplify the insect cell line infection protocol. We must also note that the production of AAV vectors by the BEVS system does not require the use of helper functions as for the tri-transfection approach. Although the helper functions have not yet been clearly identified, the baculoviral genome itself can efficiently complement this deficit. The downstream process, here based on simple ultracentrifugation method, should be accessible to any research laboratory wishing to scale up their vector productions [12]. It yields sufficient quantities of high quality vector to perform a significant number of in vitro and more importantly in vivo experiments from a single vector batch.

2 Materials

2.1 Equipment

1. Incubator shaker (adjusted at 28 °C and 110 rpm).
2. Dry bath/or water-bath heated to 28 °C.
3. Peristatic pump.
4. Analog manometers.
5. Automatic cell counter with vital stain (e.g., Trypan).
6. Ultracentrifuge with rotor 70Ti (Beckman Coulter or equivalent).
7. Tube topper system (Beckman Coulter).
8. Fraction Recovery system (Beckman Coulter).

2.2 Upstream Process and Cell Preparation

1. Commercially available Sf9 insect cell line cryovial (ThermoFisher Scientific).
2. Bac-to-Bac baculovirus expression system kit (ThermoFisher Scientific).
3. pFastBac Dual plasmid (ThermoFisher Scientific).
4. Serum-free SFX medium (GE Healthcare).
5. Vented cap Erlenmeyer vessel of 1 L (Corning).
6. Dimethylsulfoxide (DMSO) solution.
7. 1 mL cryovials.
8. Cryoboxes.
9. Freezing medium: fresh SFX medium supplemented with 15% (v/v) of Dimethylsulfoxide (DMSO) solution.

2.3 Downstream Process

1. Triton X-100 detergent used at a 0.5% (v/v) final working concentration.
2. RNAse A (Sigma-Aldrich) used at a 5 μg/mL final working concentration.
3. 0.8–0.45 μm double layer polyethersulfone membrane filter of 17.3 cm^2 (Sartorius AG).
4. For optional TFF: 115 cm^2 polyethersulfone membrane hollow fiber unit of 500 kDa molecular weight cutoff.
5. Phosphate Buffered Saline (PBS), 1× and 10× (GIBCO).
6. Discontinuous iodixanol gradient: Prepare a 5× PBS-MK buffer (500 mL): 250 mL PBS 10× (GIBCO), 2.5 mL of 1 M $MgCl_2$, 6.25 mL of 1 M KCl, and 241.25 mL of ultrapure H_2O. Prepare, from the 60% iodixanol stock solution (Sigma-Aldrich), 50 mL of a 25% iodixanol solution diluted in 1× PBS-MK buffer (10 mL 5× PBS-MK buffer, 20.8 mL of iodixanol stock solution, 19.2 mL of ultrapure H_2O), and

50 mL of a 40% iodixanol solution diluted in 1× PBS-MK buffer (10 mL 5× PBS-MK buffer, 33.3 mL of iodixanol stock solution, 6.7 mL of ultrapure H_2O).

7. 25 × 89 mm Quick-Seal Polyallomer dome-top tubes (Beckman Coulter).
8. Glass Pasteur pipettes.
9. 22G needle.
10. Filtration unit Amicon Ultra-15 (Millipore) with a 100 kDa molecular weight cutoff.
11. 0.22 μm polyvinylidene difluoride membrane of 0.1 cm^2 (Millipore).

3 Methods

The protocol focuses mainly from the stage of BIIC production (*see* **Note 1** for the initial steps of molecular cloning and recombinant baculovirus production) to the final downstream process stage. The upstream process is described for up to 400 mL of production performed in erlenmeyer vessels. The typical infection cycle of a baculovirus in insect cells being stunting, increasing the mean cell size and rapid cell death (by lysis), all of these characteristics should be scrupulously followed during the production step. For a bioreactor system process, the protocol remains highly identical; however, the manipulators are expected to get closer to the equipment supplier for specific method instructions. The Sf9 insect cell line used for the upstream process must not exceed a number of passages greater than 20 compared to the original Master Cell Bank. The different stages of the downstream process are sequentially described for a better clarity.

3.1 Sf9 Master Cell Bank (MCB) Preparation

1. Prepare a Master Cell Bank from a commercially available Sf9 insect cell line cryovial according to the manufacturer instructions. We recommend using the serum-free SFX medium for Sf9 cells cultivation and expansion. Passage the cells between 5×10^5 and 8×10^5 cells/mL, two to three times per week; avoid diluting them at a concentration below than 3×10^5 cells/mL. Always keep the cell culture above 4 to 5×10^6 cells/mL and do not exceed 20 to 22 passages.

3.2 Plasmid Constructions and rBAC Production

1. The initial molecular cloning of the genes of interest has to be performed following instructions provided with the Bac-to-Bac baculovirus expression system kit (*see* **Note 1**).

3.3 Production and Cryo-Preservation of Baculovirus-Infected Insect Cells (BIICs)

1. At day 1, 400 mL of Sf9 cells are prepared at a final concentration of 8×10^5 living cells/mL in a 1 L vented cap erlenmeyer vessel in SFX medium pre-warmed at 28 °C. The cell culture is then incubated in a shaker warmed at 28 °C and rotating at 110 rpm.
2. At day 2, count the cell concentration and note the mean size of the living cells (*see* **Note 2**): living cell concentration must be between 1×10^6 and 2×10^6 living cells/mL (*see* **Note 3**).
3. If the cell concentration is correct, inoculate 1 mL of previously clarified batch of rBAC (produced from the previous stage of rBAC amplification; *see* **Note 1**) to infect the 400 mL Sf9 cells. The cell culture is then incubated at 28 °C/110 rpm.
4. At day 3, count the cell concentration and quantify the mean size of the living cells (*see* **Note 2**): cell viability must be greater than 90% and the diameter of living cells must have increased by at least 120% compared to the mean size of the cells prior to their rBAC infection at day 2. If the diameter of the living cells has not yet reached the minimum size, allow the culture to spread for another 6 h, 12 h, or even 24 h. Just carefully prevent cell viability from falling below 90%.
5. As soon as the expected criteria of size and viability are obtained, calculate the final volume Vf needed to resuspend the total living cells (pellet) to have a final concentration of 10^7 living cells/mL; then prepare a volume of freezing medium equivalent to half of the previously calculated volume Vf, and the number of 1 mL cryovials (annotated) needed according to the total number of living cells (*see* **Note 4**); in parallel, sufficient number of cryoboxes have to be ready to accommodate all the cryovials for the cell freezing (*see* **Note 5**); after all these preparation steps, stop the cell culture (BIICs) to initiate the freezing procedure.
6. Gently centrifuge the cell suspension at $500 \times g$ for 10 min at room temperature in adapted tubes; recover the conditioned medium in a sterile bottle and immediately use half of the previously calculated Vf to resuspend the cell pellet with part of this conditioned medium; then complete the other half with the pre-warmed (28 °C) freezing medium.
7. Gently mix the BIICs by slow pipetting and aliquot the BIICs per 1 mL fractions into the opened cryovials (under sterile hood equipment); immediately close and put all the cryovials in the cryoboxes, and place them in a freezer at –80 ° C (*see* **Note 6**).

3.4 Upstream Process (400 mL of Gross AAV Vector Production)

1. At day 1, 400 mL of Sf9 cells are prepared at a final concentration of 8×10^5 living cells/mL in a 1 L vented cap erlenmeyer vessel in SFX medium pre-warmed at 28 °C. The cell culture is then incubated in a shaker warmed at 28 °C and rotating at 110 rpm.
2. At day 2, count the cell concentration and note the mean size of the living cells (*see* **Note 2**): living cell concentration must be comprised between 1×10^6 and 2×10^6 living cells/mL.
3. Thaw a cryovial from each BIIC (BIIC-Rep/Cap and BIIC-ITR-Transgene-ITR) in a dry bath or a water-bath heated to 28 ° C (*see* **Note 7**).
4. If the cell concentration is correct, inoculate 40 μL of each pre-warmed BIIC to infect the 400 mL Sf9 cells (respect the ratio BIIC versus Sf9 cells of $1–10^4$ v/v). The cell culture is then incubated at 28 °C/110 rpm for 72 h.
5. 72 h post-BIIC inoculation, count the cell concentration and note the mean size of the living cells (*see* **Note 2**): cell viability should have dropped below 60% and insect cell size diameter should be at least equal to 130% of the cell mean size observed before baculovirus infection.
6. If all of the characteristics of infection described above are correct, proceed with the downstream process; if it is not the case, allow the culture to spread for another 6 h, 12 h, or even 24 h.

3.5 Downstream Process (for 400 mL Bulk Production)

1. The clarification step is initiated by chemical cell lysis through the addition of Triton X-100 detergent into the erlenmeyer vessel: use 2 mL of Triton X-100 to be at 0.5% final concentration.
2. RNAse A is consecutively added at a final concentration of 5 μg/mL and the Sf9 cell culture is maintained at 37 °C for 1 h under shaking at 110 rpm.
3. Centrifuge the resulting cell suspension at 3500 × *g* for 20 min at 16 °C.
4. Proceed to filtration of the supernatant through 0.8–0.45 μm double layer polyethersulfone membrane filter of 17.3 cm^2 using a system with a peristaltic pump.
5. ***Optional step***: concentration of the clarified product by using a tangential flow filtration (TFF) process. For volumes of 400 mL of cultures, we recommend a step of concentration by TFF through a 115 cm^2 polyethersulfone membrane hollow fiber unit of 500 kDa molecular weight cutoff. Instead, cassette TFF system can also be used while maintaining the aforementioned surface and membrane characteristics (*see* **Note 8**). A 25× concentration factor can be reached leading to a final

volume of approximately 16 mL; perform a wash of the system with 15 mL of PBS 1×; a total volume of concentrate equal to 31 mL is now ready to be used for purification step.

6. Prepare a discontinuous iodixanol gradient (*see* Subheading 2.3 for preparation of the different percent solutions of iodixanol). Fill a 25 × 89 mm Quick-Seal Polyallomer dome-top tube with 30 mL of viral suspension through a glass Pasteur pipette whose tip is well placed at the bottom of the tube. This suspension is raised up by successive addition of 3 mL of the 25% iodixanol solution, 4 mL of the 40% iodixanol solution, and 2 mL of the 60% iodixanol stock solution.
7. Ultracentrifuge tubes are heat-sealed with the Tube toppersystem, placed in Ti70 rotor (or equivalent) with their corresponding caps, and ultracentrifuged at 500,000 × *g* for 20 min in an adapted ultracentrifuge system.
8. The Fraction Recovery system is used to puncture the tube from the bottom side while a small air hole is made using a small needle (22G) on the upper part of the ultracentrifuge tube to allow liquid flow from the bottom; the first 4.5 mL, containing AAV vectors, are collected.
9. Diafiltration is then performed through centrifugal concentration and filtration unit Amicon Ultra-15 with a 100 kDa molecular weight cutoff membrane, according to the manufacturer instructions. The diafiltration is performed three times with PBS 1×, leading to a 450× dilution factor of the initial collected fraction.
10. A 0.22 μm polyvinylidene difluoride membrane of 0.1 cm^2 is used for sterilizing filtration of the previous diafiltrated solution.
11. AAV vectors are ready to be used for titration and evaluation of their transduction efficiency.

4 Notes

1. Briefly, the most optimal process uses two distinct rBAC for AAV vector production. The rBAC-Rep/Cap, coding for the *rep* and *cap* genes, is generated after cloning of these two genes separately (depending of the desired serotype) into the pFast-Bac Dual plasmid (not provided with the kit). The *rep* and the *cap* genes have to be cloned under the control of the pPolh (polyhedrin) and the p10 baculoviral promoters respectively following the instructions of the supplier. It is highly recommended to codon optimize the *rep* and the *cap* genes for a better transcription in insect cell environment as previously described [12]. Constructions harboring the transgene

cDNAs, inserted between the two AAV2 inverted terminal repeats (ITR), and controlled by their specific promoter, can be cloned in pFast-Bac 1 plasmid (provided with the kit) to produce the pFB-ITR-Transgene-ITR plasmid. These previous molecular constructions are each transformed in DH10Bac Escherichia coli competent cells (provided with the kit) to produce, by a transposition mechanism, the pBAC-Rep/Cap or pBAC-ITR-Transgene-ITR, the two recombinant baculoviral genomes. As recommended by the manufacturer, the pBAC genomes are transfected with Cellfectin II transfection reagent (provided with the kit) into Sf9 insect cell line to produce rBAC-Rep/Cap or rBAC-ITR-Transgene-ITR recombinant baculovirus vectors. Sf9 cells are then grown in adherent conditions in SFX medium without any serum supplementation. Five rBAC clones of each original construction are picked up from viral plaque assay and amplified in suspension conditions through two consecutive infectious passages in Sf9 cells. 24 h after the last infectious process performed in up to 400 mL of Sf9 cell suspension, baculovirus-infected insect cells (BIIC) are generated (*see* then Subheading 3.3).

2. Automatic cell counter with Trypan blue or equivalent solution (vital stain) is preferred for cell counting to better quantify living and dead cells and to simply follow other important insect cell characteristics (size diameter, percentage of viability, etc.).
3. Infection of Sf9 cells by baculoviral vectors is optimal during their exponential phase of growth when the cell concentration is generally comprised between 10^6 and 2×10^6 cells/mL.
4. For instance, if you count 10^8 total living cells in your 400 mL cell suspension, then 10 cryovials will be necessary, each with a final volume of 1 mL of BIIC suspension.
5. The cryoboxes must be chosen so that they can guarantee progressive freezing of the BIICs at a rate of −1 ° C/min.
6. The stability of BIICs at −80 ° C is effective; however, it is recommended to place BIICs cryovials at lower temperatures (−150 ° C or liquid nitrogen) 24 h after the −80 °C freezing for even greater stability.
7. Take care not to leave cryovials under thawing for too long time because of the presence of DMSO in the freezing medium which is toxic for the cells in the liquid state; refreeze the cryovials at −80 ° C following the same recommendations as for the freezing step of BIIC (*see* **Notes 5** and **6**).

8. The TFF can be used in manual mode through a mounting requiring a peristaltic pump and analog manometers input (feed) and output hollow fiber (retentate). However, we strongly recommend the use of automated systems to accurately control flow rates and especially the transmembrane pressure (which should not exceed 1–1.2 bars).

References

1. Gaudet D, Methot J, Dery S et al (2013) Efficacy and long-term safety of alipogene tiparvovec (AAV1-LPLS447X) gene therapy for lipoprotein lipase deficiency: an open-label trial. Gene Ther 20:361–369
2. Grimm D, Kern A, Rittner K et al (1998) Novel tools for production and purification of recombinant adenoassociated virus vectors. Hum Gene Ther 9:2745–2760
3. Salvetti A, Oreve S, Chadeuf G et al (1998) Factors influencing recombinant adeno-associated virus production. Hum Gene Ther 9:695–706
4. Xiao X, Li J, Samulski RJ (1998) Production of high-titer recombinant adeno-associated virus vectors in the absence of helper adenovirus. J Virol 72:2224–2232
5. Dias Florencio G, Precigout G, Beley C et al (2015) Simple downstream process based on detergent treatment improves yield and in vivo transduction efficacy of adeno-associated virus vectors. Mol Ther Methods Clin Dev 2:15024
6. Chahal PS, Schulze E, Tran R et al (2014) Production of adeno-associated virus (AAV) serotypes by transient transfection of HEK293 cell suspension cultures for gene delivery. J Virol Methods 196:163–173
7. Martin J, Frederick A, Luo Y et al (2013) Generation and characterization of adeno-associated virus producer cell lines for research and preclinical vector production. Hum Gene Ther Methods 24:253–269
8. Ye GJ, Scotti MM, Thomas DL et al (2014) Herpes simplex virus clearance during purification of a recombinant adeno-associated virus serotype 1 vector. Hum Gene Ther Clin Dev 25:212–217
9. Urabe M, Ding C, Kotin RM (2002) Insect cells as a factory to produce adeno-associated virus type 2 vectors. Hum Gene Ther 13:1935–1943
10. Wasilko DJ, Lee SE, Stutzman-Engwall KJ et al (2009) The titerless infected-cells preservation and scale-up (TIPS) method for large-scale production of NO-sensitive human soluble guanylate cyclase (sGC) from insect cells infected with recombinant baculovirus. Protein Expr Purif 65:122–132
11. Cecchini S, Virag T, Kotin RM (2011) Reproducible high yields of recombinant adeno-associated virus produced using invertebrate cells in 0.02- to 200-liter cultures. Hum Gene Ther 22:1021–1030
12. Buclez PO, Dias Florencio G, Relizani K et al (2016) Rapid, scalable, and low-cost purification of recombinant adeno-associated virus produced by baculovirus expression vector system. Mol Ther Methods Clin Dev 3:16035

Chapter 6

Multimodal Production of Adeno-Associated Virus

Ivette M. Sandoval, Nathan M. Kuhn, and Fredric P. Manfredsson

Abstract

Adeno-associated virus (AAV) is an increasingly popular tool in the research laboratory, and use of this viral vector clinically is occurring at an accelerated pace. Nevertheless, despite its popularity, AAV is a relatively cumbersome virus to produce; however, significant efforts have been invested to develop, optimize, and simplify methodology that allows the generation of high-quality AAV with significantly increased production yields. Here we describe multiple modalities for production and purification of AAV particles produced in HEK293 cell cultures using an iodixanol density gradient. We include two methods adapted for harvesting virus from the culture media: tangential flow filtration (TFF) and polyethylene glycol precipitation (PEGylation). Moreover, we also describe the protocol for anion exchange chromatography, which can be used after the iodixanol gradient as an additional purification step. Last, we provide various protocols for determining virus titer.

Key words AAV, AAV production, Iodixanol gradient, Tangential flow filtration, PEGylation, Digital droplet PCR

1 Introduction

Adeno-associated virus (AAV) is an increasingly popular research tool that can be used to modulate expression in a host of experimental setting, ranging from in vitro cultures to nonhuman primates. Not only is AAV a powerful tool used to overexpress a protein in vivo, but it is a readily utilized tool to reduce or modulate expression of endogenous genes [1]. Despite its strong popularity, AAV is a relatively cumbersome virus to produce. However, significant efforts have been invested to develop and optimize methodology that allows the generation of high-quality AAV with increased production yields.

AAVs consist of a single-stranded DNA (ssDNA) genome containing two open reading frames, Rep and Cap, flanked in between two inverted terminal repeats (iTRs). Rep is translated into four

Fredric P. Manfredsson and Matthew J. Benskey (eds.), *Viral Vectors for Gene Therapy: Methods and Protocols*, Methods in Molecular Biology, vol. 1937, https://doi.org/10.1007/978-1-4939-9065-8_6,

proteins: Rep78, Rep68, Rep52, and Rep40, involved in AAV DNA replication and packaging, and Cap is translated into three structural capsid proteins: VP1, VP2, and VP3. Typically, generation of recombinant AAV vectors is done by replacing the viral genes with the gene of interest, while supplying necessary viral proteins for packaging in trans. Therefore, three vector plasmids are usually required: a transfer plasmid containing the desired transgene cloned in between the two iTRs, a second plasmid containing viral genes Rep and Cap, and a third plasmid containing adenovirus genes coding for helper proteins necessary for AAV assembly. Alternatively, the Rep and Cap and the adenovirus helper genes can be combined into a single plasmid. The plasmids are then introduced into cells, and subsequently AAV particles can be harvested and purified from the growth media and/or from cell pellets.

In this chapter we describe multiple modalities for purification of AAV particles produced in HEK293 cell cultures. We provide a protocol for purification of AAVs by density gradient ultracentrifugation using iodixanol, and a protocol for purification by column anion exchange chromatography. We also include two methods adapted for harvesting virus from the culture media: tangential flow filtration (TFF) and polyethylene glycol precipitation (PEGylation).

Last, we provide various protocols for determining virus titer. It is important to note that other methods, such as the use of a baculovirus system in insect cells [2, 3] or the use of cesium chloride density gradient, are also frequently used [4]. The method of choice for purification will be based on the ultimate use of the virus; one or multiple procedures can be utilized in tandem to produce AAV preparations of varying titers and purities (Fig. 1).

The methodology discussed is suitable for large-scale production, as well as small, proof-of-principle batches of AAV.

2 Materials

2.1 Genome and Helper DNA Preparation

1. Bacterial culture reagents: Terrific broth and/or Luria-Bertani (LB) broth and agar plates (*see* **Note 1**). Antibiotic for plasmid selection.
2. Sure 2 competent cells or similar (Agilent Technologies, Santa Clara, CA, USA; *see* **Note 2**).
3. Plasmid purification kit or reagents for cesium chloride (CsCl) density centrifugation [5–7].
4. Shaker incubator.
5. Centrifuge, tubes, and rotors.

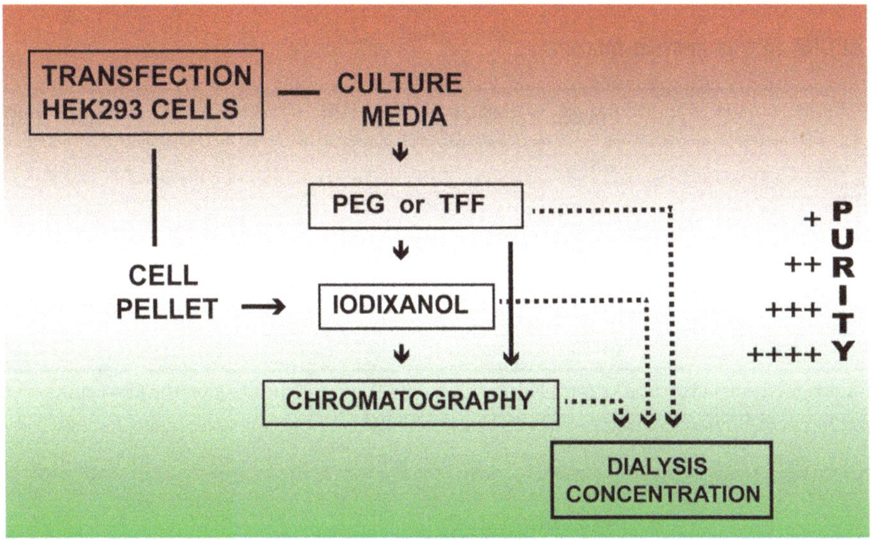

Fig. 1 Schematic of various approaches to purification of AAV particles from HEK293 cells. AAV particles can be purified from cell pellets or from the culture media. Each method can be used individually. However, by combining more than one approach higher purity preparations can be achieved. Likewise, yield can be increased by simultaneously purifying AAV particles from media and cell pellets from a single transfected culture

2.2 Tissue Culture and Transfection

1. Human Embryonic Kidney (HEK293) cells.
2. Cell growth medium: Dulbecco's Modified Eagle's medium (DMEM) supplemented with 10% fetal bovine serum (FBS) and 1% penicillin/streptomycin.
3. Sterile 1× PBS and 1× PBS/EDTA (5 mM).
4. PEI solution. Mix 0.081 g PEI (Polyethylenimine; MW 25,000) with 200 mL preheated ddH_2O. Heat at 80 °C until the PEI is dissolved. Adjust pH to 8.0 with HCl (*see* **Note 3**), adjust final volume to 250 mL and filter sterilize. Place in clean glass container and subject to two freeze-thaw cycles. After the second thaw, aliquot according to transfection scale, and store at −80 °C until use.
5. Genomic DNA and helper DNA (*see* Table 1).
6. Cell culture dishes vessels (e.g., Nunc Triple Flask, Thermo Fisher Scientific, Waltham, MA, USA; *see* **Note 4**).
7. 250 mL graduated cylinder.

2.3 Tangential Flow Filtration

1. Media from transfected cells.
2. Tabletop centrifuge capable of spinning large volumes at 3000 × *g* (*see* **Note 5**).
3. Pall mini profile capsule 0.5 μm filter (Pall Corporation, New York, NY, USA) and 13 mm tubing.

Table 1
Amount of DNA in transfection mixture

	Double transfection	Triple transfection
AAV genome	300 μg	300 μg
Helper	1170 μg (e.g., pXYZ5, pKRAP1)	690 μg (pXX6)
pCap	N/A	300 μg
NaCl (1.5 M)	2.6 mL	2.6 mL
ddH_2O	to 25 mL	to 25 mL

This table is designed for transfection of 6 triple-layer flasks (3000 cm^2). Depending on the packaging system, either a single (left column) or triple (right column) transfection is required

4. Peristaltic Pump.
5. Pall minimate system (Pall Corporation, New York, NY, USA) with one or more Tangential Flow Filtration 100 k Capsules with connectors and 3.2 mm tubing.
6. 2 L ddH_2O.
7. 1 N NaOH.
8. 15% glycerin +0.05% azide.

2.4 PEGylation

1. Media from transfected cells.
2. 2.5 M NaCl (73.05 g in 500 mL ddH_2O).
3. 40% PEG8000 (Polyethylene glycol; MW 8000) solution. Mix 350 mL 2.5 M NaCl with 200 g PEG8000 under low to medium heat, adjust volume to 500 mL with ddH_2O. Filter sterilize with 0.22 μm filter (*see* **Note 6**). Store at 4 °C.
4. Large flasks and centrifuge bottles, table top centrifuge with large capacity (*see* **Note 4**).
5. AAV lysis buffer: 150 mM NaCl, 50 mM Tris–HCl pH 8.5, filter sterilize.
6. Benzonase® Endonuclease (≥250 units/μL; Sigma-Aldrich, St. Louis, MO, USA).

2.5 Cell Harvest and Iodixanol Gradient

2.5.1 Harvest

1. Sterile 1× PBS and 1× PBS/EDTA (5 mM).
2. 250 mL sterile conical tubes (Corning, Corning, NY, USA).
3. Tabletop centrifuge with large capacity bucket rotor, and adapters for conical tubes (e.g., Eppendorf Centrifuge 5920R with S-4x1000 rotor (Eppendorf, Hamburg, Germany)).

2.5.2 Cell Lysis by Freeze/Thaw of Pellet

1. AAV lysis buffer (*see* Subheading 2.4, **item 5**).
2. Protease inhibitor cocktail (cOmplete™, Mini Protease Inhibitor Cocktail) (Roche, Basel, Switzerland).
3. Vortex.
4. Water bath.
5. Dry ice/ethanol bath.

2.5.3 Benzonase® Digest

1. Benzonase® Endonuclease.
2. 2.5 M $MgCl_2$.
3. Water bath incubator.
4. Super speed centrifuge.

2.5.4 Iodixanol Gradient

1. 5× TD buffer: 5× PBS, 5 mM $MgCl_2$, 12.5 mM KCl, ddH_2O. Filter sterilize.
2. 5 M NaCl.
3. 15, 25, 40, and 60% iodixanol OptiPrep™ Density Gradient Medium (Sigma-Aldrich, St. Louis, MO, USA) solutions (*see* Table 2).
4. Ultracentrifuge tubes Quick-Seal®, Polypropylene, 39 mL, 25 × 89 mm and tube rack (Beckman Coulter, Brea, CA, USA).
5. 20 mL syringe with 18 gauge needle to load the virus mix into the ultracentrifuge tubes.
6. 5 mL syringe with 18 gauge needle to collect AAV fraction from gradient.
7. 18 gauge needle to release pressure from sealed ultracentrifuges tube.
8. Stand with clamp.

Table 2
Composition of various iodixanol solutions

	Optiprep	5 M NaCl	5XTD	H20	Total volume
15%	12.5	10 mL	10 mL	17.5 mL	50 mL
25%	20.8	–	10 mL	19.2 mL	50 mL
40%	34	–	10 mL	6 mL	50 mL
60%	50	–	–	–	50 mL

Please note that the iodixanol solution is sold as 60%

Preparation of Iodixanol Gradient by Hand

1. 10 or 20 mL syringe.
2. 18 gauge needles.
3. 18 gauge blunt fill needle (Becton Dickinson, Franklin Lakes, NJ, USA).
4. Autoclaved long Pasteur pipettes.

Preparation of Iodixanol Gradient by Pump

1. Watson-Marlow 205S Pump (Watson-Marlow, Falmouth, Cornwall, UK). Tygon Tubing (Bio-Rad, Hercules, CA, USA). Silicone Tubing (Bio-Rad, Hercules, CA, USA).
2. Drummond calibrated micropipettes 200 μL (Drummond Scientific, Broomall, PA, USA).
3. Kimble micro capillary pipettes 100 μL (DWK Life Sciences, Millville, NJ, USA).
4. ddH_2O.
5. 250 mL conical tube as reservoir and a stand.

Sealing of Tubes and Centrifugation

1. Tube sealer ("Quick Charge Soldering Iron Kit"; Iso-Tip, Altoona, WI, USA).
2. Ultracentrifuge rotor (Beckmann Typ1 70 Ti or equivalent, (Beckman Coulter, Brea, CA, USA).
3. Ultracentrifuge.

2.6 Anion Exchange Sepharose Column Chromatography

1. Peristaltic pump P-1 (GE Healthcare, Boston, MA, USA).
2. GE HiTrap Q High Performance Sepharose column 5 mL (GE Healthcare, Boston, MA, USA; *see* **Note 7**).
3. Tubing: Polyethylene 1.8 mm o.d., and silicone 4.5 mm o.d./ 2.1 mm i.d. (GE Healthcare, Boston, MA, USA).
4. Tubing and pump connectors: Flangeless/M6 Male connector. 2 units, Pump P-1 tube connector, Luerlock Female/M6 Female adaptor (GE Healthcare, Boston, MA, USA).
5. Ring stand and clamps.
6. 60 mL syringe with Luerlock tip.
7. Collection vessels: beaker for waste and 50 mL conical tube.
8. Low salt buffer: 20 mM Tris, 15 mM NaCl, pH 8.5.
9. High salt buffer: 20 mM Tris, 1 M NaCl, pH 8.5.
10. Elution buffer: 20 mM Tris, 350 mM NaCl, pH 8.5.

2.7 Buffer Exchange/ Concentration

1. Apollo centrifugal spin concentrator QMWL 150KDa (Orbital Biosciences, Topsfield, MA, USA) or equivalent.
2. Exchange buffer: 1× PBS (Ca^{2+} Mg^{2+}) or Lactated Ringers buffer.
3. Sigmacoted microcentrifuge tubes and pipette tips (*see* **Note 8**).

2.8 Virus Titering (See Note 9)

2.8.1 Dot-Blot (See Note 10)

1. Purified virus from Subheading 3.7.
2. Genome DNA.
3. Probe against genome DNA.
4. Streptavidin IRDye-800CW (LI-COR Biosciences, Lincoln, NE, USA).
5. Hybond-N membrane (Amersham, Little Chalfont, United Kingdom).
6. Whatman filter paper.
7. Alkaline buffer: 400 mM NaOH, 10 mM EDTA.
8. Hybridization buffer: 7% SDS, 250 mM $NaHPO_4$ pH 7.2, 1 mM EDTA.
9. Wash buffer: 1% SDS, 40 mM NaHPO4 pH 7.2, 1 mM EDTA.
10. Dot-blot apparatus.
11. UV crosslinker.
12. Hybridization oven and tubes.
13. Vacuum.
14. Odyssey Infrared Imaging System (LI-COR Biosciences, Lincoln, NE, USA).

2.8.2 Digital Droplet PCR (ddPCR)

1. Bio-Rad QX 200 Droplet Digital PCR system, droplet generator, plate sealer, and thermal cycler (Bio-Rad, Hercules, CA, USA).
2. ddPCR supermix for probes or EvaGreen (no dUTP) (Bio-Rad, Hercules, CA, USA).
3. ddPCR primers (and probe) for viral genome (*see* **Note 11**).
4. ddPCR droplet oil (Bio-Rad, Hercules, CA, USA).
5. ddPCR cartridge and gasket (Bio-Rad, Hercules, CA, USA).

3 Methods

3.1 Genome and Helper Preparation

3.1.1 Prepare Bacterial Cell Culture for Each One of Your Genome and Helper Plasmids

1. Inoculate 4 mL of culture media containing the appropriate antibiotic with a single colony of transformed bacteria. Incubate the culture at 30 °C (*see* **Notes 1** and **2**), with vigorous shaking for 6–8 h.
2. Inoculate 500 mL (*see* **Note 1**) of terrific broth containing the appropriate antibiotic with 1 mL of the previous culture and incubate overnight (12–16 h) at 30 °C with vigorous shaking.
3. Harvest bacterial cells (*see* **Note 12**) by centrifugation at 4000 × *g* for 15 min at 4 °C. The pellet can be processed immediately or stored frozen at −20 °C (**potential stopping point**).

3.1.2 Purify Plasmid DNA

1. DNA quality is critical to obtain high titer viral preparations. We highly recommend purifying the DNA by CsCl density centrifugation [8] as this method yields highly pure and supercoiled DNA preparations, thereby increasing transfection efficiency.
2. Anion exchange resin column based methods, such as Maxi and Giga prep commercial kits, are not optimal but they are much easier to implement and less time consuming than CsCl gradients. If these will be your methods of choice, make sure the resuspension buffer has been freshly supplemented with RNAse in order to avoid contamination of your plasmid with RNA. Also, after elution from the column we recommend performing a final extraction using a standard phenol:chloroform protocol to ensure no protein contamination remains.

3.1.3 DNA Quality Control

1. Digest your plasmid(s) with a restriction enzyme and run in an agarose gel for quality control. If *any* background signal is observed when visualized under the UV trans-illuminator, the DNA is not pure enough and this will affect the yield and quality of your AAV prep.
2. The plasmids containing the transfer genome must contain both inverted terminal repeats (iTR) for proper packaging into viral capsid. For AAV 2 iTRs the integrity is typically checked with a SmaI restriction enzyme digestion.
3. Store plasmid DNA in -20 °C (**potential stopping point**).

3.2 Tissue Culture and Transfection (See Note 13)

1. Carry out all procedures using aseptic culture technique in a biosafety cabinet.

3.2.1 Passage of HEK293 Cells

1. Human Embryonic Kidney (HEK293) cells are maintained in Dulbecco's Modified Eagle's medium (DMEM) supplemented with 10% fetal bovine serum (FBS) and 1% penicillin/streptomycin, at 37 °C in a 5% CO_2 incubator.
2. For maintenance grow cells to confluency and passage cells 1/10 twice a week (*see* **Note 13**).
3. For transfection grow sufficient cells to seed the desired amount of flasks with 1/3 of the number of cells at confluency. For example: to seed 6 triple flasks for transfection, grow 2 triple flasks to confluency and split cells evenly (equivalent to a 1:3 ratio). Culture should reach ~70% confluency in approximately 36 h.

3.2.2 PEI Transfection (See Note 14)

1. Confluency of HEK293 cells is paramount to achieving a good titer. Transfect when the confluency is ~70–80% for optimal results (*see* **Note 15**).

2. Thaw PEI solution and DNA (*see* Table 1 for amounts, *see* **Note 16**) and heat PEI to 65 °C to solubilize precipitates. Mix thoroughly by vortexing. Let cool to room temperature before use.
3. Prepare mixture as described in Table 1. Mix the PEI and DNA fractions together by adding PEI mixture to DNA mixture in a dropwise manner. Vortex vigorously for >30 s (*see* **Note 17**).
4. Incubate PEI:DNA mixture for 20 min at room temperature. Longer incubations will not enhance transfection efficacy (*see* **Note 18**).
5. Add 8.8 mL of the PEI:DNA mixture to graduated cylinder and decant media from tissue culture flask into cylinder. Return the media/transfection mix to the flask, paying attention to not disturb the cell layer.
6. Place in incubator until harvest or media change.

3.3 Tangential Flow Filtration

3.3.1 Media Preparation

1. Collect media at desired times. Typically, at the end of the cellular harvest, or at multiple time points [9].
2. Centrifuge media at 3000 × *g* to pellet cells. Transfer media to clean bottles and store at 4 °C (*see* **Note 19**).
3. To prevent clogging of the TFF cassette the media has to be pre-filtered to remove large debris. Assemble a Pall mini profile capsule filter. To do so, connect a tube running from the media containing bottle, through a peristaltic pump and to the input port of the filter. Connect another tube from the exit port of the filter to a collection bottle (*see* **Note 20**). Rinse and flush filter unit with 1 L of 0.2 μm filtered ddH_2O before filtering the virus containing media. Remove all water from tubing before filtering media. Pump all media through the filter into clean bottles. When all the media is filtered from the initial bottle and through the pump, add 200 mL of dH_2O to the initial bottle to help push the remaining media (in tubing) into the collection bottle.

3.3.2 TFF Assembly and Preparation (Fig. 2)

1. Use connectors supplied with the TFF cassette to assemble multiple cassettes in parallel (*see* **Note 21**; *see* Fig. 2).
2. Use the shortest length of tubing possible running from the retentate output back to the virus containing graduated chamber. Attach clamps to enable flow adjustments.
3. Run tubing from the output (bottom) of virus containing graduated chamber, through a peristaltic pump, to the "feed" input of the TFF cassette.
4. To flush the assembly, connect tubing from both the retentate and filtrate ports and run tubing into a waste beaker (make sure to disconnect any tubing going from retentate port back to

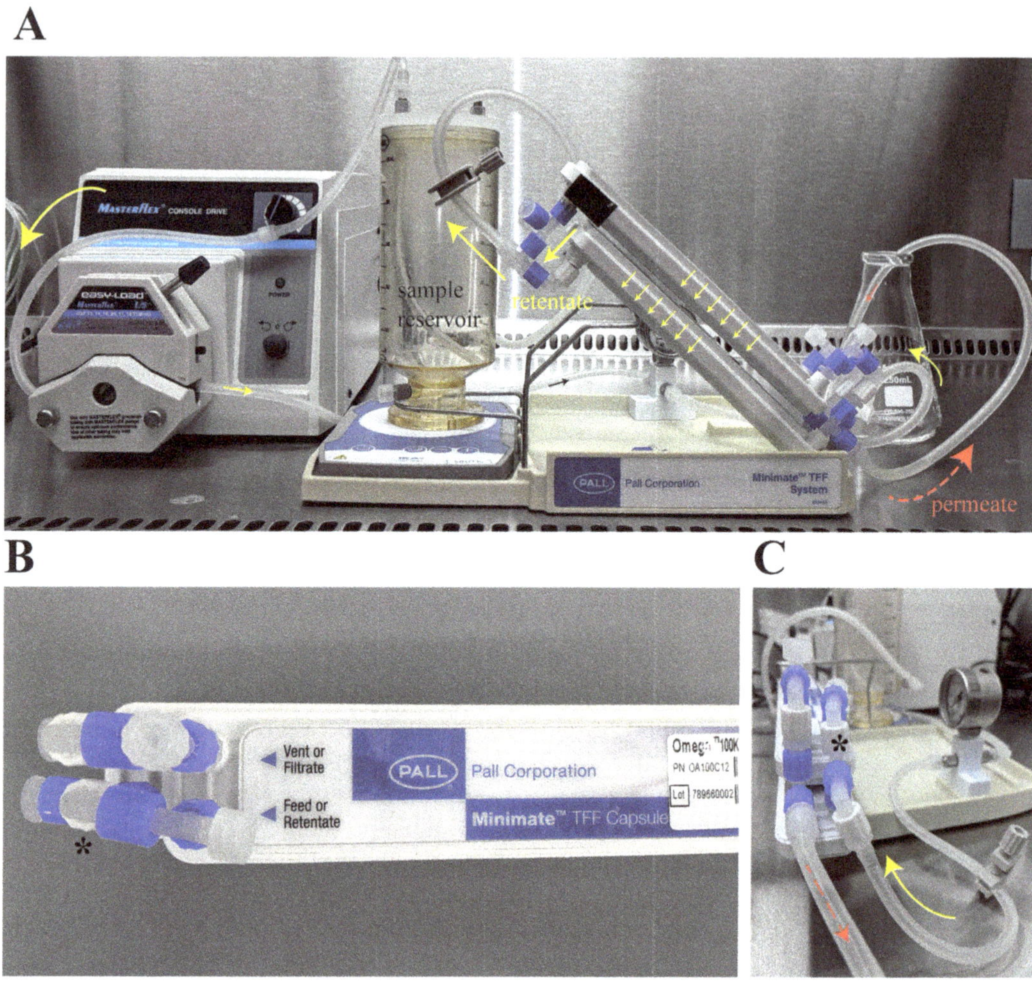

Fig. 2 TFF using the Minimate™ capsule equipment setup. (**a**) One piece of tubing connects the sample reservoir to the feed port of the Minimate™ capsule passing through the pump. The product flows parallel (tangential to) the membrane. The pressure generated by the flow stream causes liquid flow through the membrane (permeate). Another piece of tubing circulates the retentate back to the sample reservoir. The filtrate (permeate) is discarded. Flow direction of the sample is indicated by yellow arrows. Top (**b**) and side (**c**) view of the capsule showing feed port (marked with an asterisk) and exit port for permeate (red arrow)

graduated chamber). Fill chamber with 500 mL of 0.2 μm filtered ddH_2O. Turn on pump to a flow rate of about 40 mL/min. Slowly tighten the clamp on the tubing coming from the retentate port. This will increase back pressure and accordingly increase flow through filtrate lines. Tighten clamp until the flow from the retentate port and the filtrate port are approximately equal (do not exceed feed pressure of 2 bar; 30 psi). In this manner, pump at least 250 mL through both retentate and filtrate lines simultaneously with equal pressure into a waste beaker. Next adjust clamps to pump water through just retentate line. Pump at least 250 mL. Adjust clamps to pump water through just filtrate line. Pump 250 mL of fluid through filtrate line. All wash should be collected in waste beaker (use a 2 L beaker). When flushing is almost complete open the vent ports on each filter unit and release about 10 mL of filtrate to flood the filter chamber. Close vent port when finished.

3.3.3 TFF

1. Adjust the setup so that the retentate port tubing is connected back to the graduated chamber and the filtrate port is flowing into the waste beaker (firmly secure the end of the hose into the waste beaker so that it will not come loose, it is best to have it pointed into the waste beaker instead of straight on the side so that you can see flow rate).
2. Make sure that all connections are tight.
3. Turn on flow to a rate of approximately 50 mL/min.
4. Control flow rate with the pump speed and the clamp on the hose exiting the retentate port, and maintain sufficient volume in the virus containing reservoir (*see* **Note 22**).
5. Concentrate media down to below the 10 mL graduation (*see* **Note 23**).
6. Once the media is concentrated to desired volume close the filtrate port and circulate the concentrated media through the filter unit for 10–15 min in order to dislodge any virus that has coated the filter unit.
7. To collect the concentrated virus media, turn off the pump, remove the retentate tube from the connection to the graduated chamber and direct flow to a 50 mL conical, start pump and pump the retentate into the conical until all media has flowed out (i.e., air coming out).

3.3.4 Cleaning the TFF Cassette(s)

1. First flush the system with 500 mL of the collected filtrate. Remove the short retentate hose and replace with the longer hose and direct to the waste beaker along with the filtrate hosing. Adjust the clamp on the retentate hosing until the flow from the retentate and the filtrate are equal and pump all of filtrate media through the filter units.
2. Next add 500 mL of dH_2O to the chamber. Close the clamp on the filtrate hose and pump at least 250 mL through the retentate hosing. Next open the filtrate clamp and close the retentate clamp and pump at least 250 mL of H_2O through the filtrate hosing. Pump all H_2O from chamber.
3. Heat 1.0 N NaOH to 45 °C. Add approximately 50 mL to the chamber. Connect the filtrate tubing to the back, lower connection port of the chamber. Remove long retentate tubing

and reconnect the short tubing, and connect this tubing to the chamber. Pump the NaOH through the filter units and tubing. Initially adjust the retentate clamp to provide some back pressure and pump the NaOH for about 5 min. Next, completely close the filtrate clamp and pump the NaOH through the assembly for 1 h.

4. Seal all ports with the caps and store at 4 °C (*see* **Note 24**).

3.4 PEGylation (See Note 25)

1. Collect media at desired times. Typically, at the end of the cellular harvest or at multiple time points [9].
2. Mix 200 mL of 40% PEG/NaCl solution with 800 mL of media. Incubate on ice for 2–72 h (*see* **Note 26**).
3. Centrifuge at 3000 × *g* for 30 min at 4 °C.
4. Decant media and add 13 mL of lysis buffer to pellet (*see* **Note 27**).
5. Resuspend pellet by shaking at 4 °C overnight.
6. Proceed to benzonase digest (Subheading 3.5.3).

3.5 Cell Harvest and Iodixanol Gradient

3.5.1 Harvest

1. 72 h post-transfection the cells are ready for harvesting. Decant media (save for virus collection or dispose as biohazard).
2. Wash cells with 40 mL of PBS, discard as biohazard waste.
3. Add 40 mL of PBS-EDTA and ensure that all surfaces are covered. Incubate cells until they begin to detach (~5 min).
4. Collect cells (*see* **Note 28**) and transfer to 250 mL conical tube (*see* **Note 29**).
5. Add an additional 40 mL PBS-EDTA to flask to wash all the cells off, and decant into the same 250 mL conical tube.
6. Centrifuge tubes at 3000 × *g* for 5 min.
7. Wash pellet in PBS to remove EDTA.
8. Centrifuge tubes at 3000 × *g* for 5 min and decant the liquid. At this point the pellet can be stored at −80 °C, or you can proceed immediately to the freeze/thaw step (*see* **Note 30**).

3.5.2 Cell Lysis by Freeze/Thaw of Pellet

1. Here you can combine the pellet from two 250 mL conical tubes. Thaw pellets and add 13 mL lysis buffer and protease inhibitors and dislodge pellet (*see* **Note 31**). After pellet is dislodged, transfer liquid to the second 250 mL conical and repeat.
2. Transfer the material to a 50 mL conical tube. Vortex to ensure that pellets are completely resuspended.
3. Place the tube in dry ice/ethanol bath and leave for 15 min (or until pellet is completely frozen) (*see* **Note 32**).

4. Place tube in a 37 °C water bath for 15 min, or until the tube is completely thawed.
5. Repeat for a total of 3 freezes. Once the material is frozen the third time, it can be placed in -80 °C until further use.

3.5.3 Benzonase Digest and Iodixanol Gradient

1. Thaw the material, add 1 μL benzonase (375 units total) and 6 μL of 2.5 M $MgCl_2$. Incubate for 30 min at 37 °C.
2. Centrifuge the 50 mL conical at 8000 × *g* for 30 min to pellet debris. Immediately transfer the supernatant to your gradient or into a clean 50 mL conical.
3. While you are waiting for the benzonase digest and centrifugation to complete, prepare everything you need for the iodixanol gradient (*see* **Notes 33** and **34,** Fig. 3).

3.5.4 Preparation of Iodixanol Gradient by Hand

1. Prepare iodixanol gradient using the volumes and concentrations of iodixanol shown in Table 2.
2. There are numerous ways to prepare the gradient (*see* **Notes 33** and **34**) depending on the amount of vectors that you produce. Here we describe setting up gradients by hand (<4 samples) or by pump (>4 samples). In either case the gradient gets filled from the bottom.
3. Place a 10 or 20 mL syringe with a blunt fill needle into a Pasteur pipette which is inserted into the ultracentrifuge tube; make sure the tip of the Pasteur pipette is close to the bottom of the tube.
4. Place the viral supernatant from Subheading 3.5.3, **step 2**. into the syringe and wait until all liquid has entered the tube.
5. Add 9 mL of the 15% iodixanol solution to the syringe and wait until all liquid has entered tube (*see* **Notes 35** and **36**).
6. Add 6 mL of the 25% iodixanol solution to the syringe and wait until all liquid has entered tube.
7. Add 5 mL of the 40% iodixanol solution to the syringe and wait until all liquid has entered tube.
8. Add 9 mL of the 60% iodixanol solution (*see* **Note 37**) to the syringe and wait until all liquid has entered tube.

3.5.5 Preparation of Iodixanol Gradient by Pump

1. We use a pump to transfer our gradient from a 250 mL conical on the right to the ultracentrifuges tubes on the left (*see* Fig. 2A).
2. Prepare pump by attaching capillaries to tubing (also make sure there is adapter tubing). It is probably better if all tubing is even in length to make sure they reach the bottom level of the 250 mL conical. Set pump at 37 RPM and in counter clockwise direction.

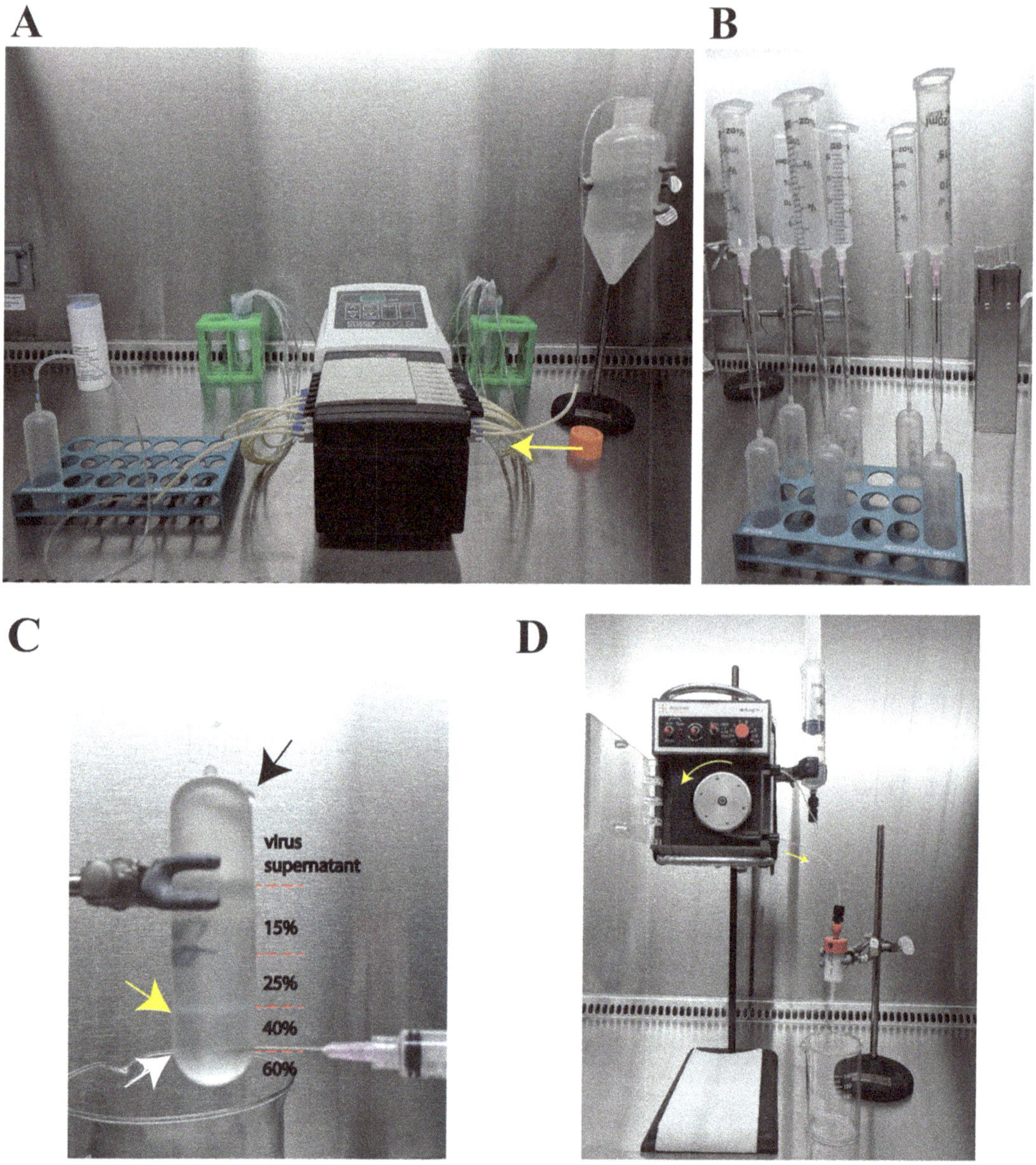

Fig. 3 Iodixanol gradient setup. (**a**) Setup for preparing gradient using pump. Iodixanol flows right to left from a 250 mL conical tube to the ultracentrifuge tube at a set flow rate. Tubing is connected in both ends to capillary glass tubes that reaches the bottom of either tube. Figure shows setup for a single sample, but up to eight samples can be prepared simultaneously. (**b**) Setup for preparing gradient by hand. A syringe with needle is placed inside a Pasteur pipette as shown; the Pasteur pipette carries the crude virus solution or the iodixanol solutions to the bottom of the tube. (**c**) Sealed tube with gradient after centrifugation showing iodixanol layers. Syringe must be inserted at the 40:60 interphase (white arrow) to collect AAV fraction. 25:40 interphase contains protein debris (yellow arrow) and must be avoided. Tube must be punctured (black arrow) to release vacuum pressure before extracting the AAV fraction. (**d**) Setup for column chromatography. The reservoir (60 mL syringe) is attached to the column through a single piece of tubing that passes through the pump. Flow direction is indicated by yellow arrows

3. On the right side (input side) use Drummond glass capillaries. On left side (gradient) use Kimble glass capillaries (be careful, these are more fragile).
4. Run ~100 mL of water through all tubing to make sure that everything is working properly and that there are no leaks, and run the lines dry.
5. Add the virus to the bottom of the ultracentrifuge tube as described in Subheading 3.5.4, **step 3**.
6. Place the right capillaries in the 250 mL conical. Make sure they are all at the bottom.
7. Put the exact amount of 15% solution (for as many gradients that you are running) into 250 mL conical (right side of pump) and turn on. Watch as solution travels to the left side and turn off the pump as it gets close to the capillary. At this point we are not worried about a few air bubbles.
8. Carefully insert the left capillaries on to the bottom of each tube. The tubing should not touch the inside of the ultracentrifuge tube.
9. Turn on the pump and keep an eye on the gradients. They should rise at the same rate.
10. When the iodixanol reaches the bottom of the 250 mL conical (~0.5 cm above bottom) turn off the pump (*see* **Note 38**).
11. Repeat with exact amount of 25, 40, 60% iodixanol as indicated in Table 2.
12. For the 60% layer, run the fluid until it gets to the capillary (i.e., completely empty the 250 mL conical and empty the tubing). Stop the pump just as air reaches the capillaries.
13. Carefully remove all capillaries from the gradients and make sure there is no cross contamination. Remove the left side capillary from the tubing and discard.

3.5.6 Sealing of Tubes and Centrifugation

1. Overlay the AAV layer (now at the top of the ultracentrifuge tube) with lysis buffer taking care not to disturb the gradient. Try to remove any bubbles by squeezing the tube (to push the bubbles out) and aspirating the air with a pipettor and a Pasteur pipette.
2. Heat seal the ultracentrifuge tube and do a squeeze test (press the tube hard with fingers to ensure that there are no leaks).
3. Place the ultracentrifuge tube, place into a Ti70 ultracentrifuge rotor.
4. Ensure that each tube has an ultracentrifuge adapter.
5. Centrifuge the gradients for 60000 RPM (265000 × *g*) for 2 h at 15 °C. Use max acceleration and minimum deceleration.

3.5.7 Collection of Gradient

1. Following centrifugation, a majority of the virus will be at the 60/40 interface (*see* Fig. 3c, *see* **Note 39**).
2. Using a syringe and a 18 ga. needle carefully pierce the tube just below the seam at the bottom (*see* **Notes 39** and **40**; just at the level where the tube rounds and just below the 60/40 interface). There is no direct visual indication of where the virus is.
3. Puncture the top of the tube with a 18 ga needle. Make sure to cover the needle with a kimwipe as there is built up pressure. This will prevent a vacuum to form.
4. With the bevel facing upward carefully aspirate approximately 5 mL of the 60/40 interface (below and above). As the interface starts traveling downward tilt the needle accordingly to keep it at the same relative point. Always keep an eye on the white protein layer as this should be avoided.

3.6 Q-Sepharose Anion Exchange Chromatography Purification (See Note 7)

3.6.1 System Set Up According to Manufacturer's Recommendations (Fig. 3d)

1. Insert the barbed ends of the pump connectors to both sides of a 140 mm long piece of silicone tubing. Connect polyethylene tubing to both sides of the silicone tubing into the compression end of the connector fitting (insert the tubing ends through the connector nut and the sealing plug and finger-tighten the connector nut).
2. Attach a compression connector Flangeless/M6 Male to both open ends of the polyethylene tubing.
3. Attach a Luerlock Female/M6 Female adaptor to one of the Flangeless/M6 Male connectors and connect to a 60 cc syringe (after removing the plunger).
4. Clamp the syringe to a ring stand approximately at eye level and insert the silicone tubing into the pump. Make sure the bottom of the syringe is higher than the top end of the pump tubing.

3.6.2 System Priming and Column Equilibration

1. Set the pump speed at 10× and #10 (approximately 5 mL/min). Add 12 mL of Low Salt buffer (Buffer A) to the syringe and run the pump until the buffer begins to drip from the free end of the column connector. This will remove all air in the tubing.
2. From this point on be mindful to avoid introducing air into the column during all steps of the process or to allow the column to run dry.
3. Remove the stopper from the top of the column. Run the pump momentarily until a couple of drops of Buffer A drip atop of the column from the open end of the tubing. Screw the column connector into the top of the column and to the free end of the column connector.

4. Remove the snap-off tab at the end of the column outlet and wash the column with the remaining volume of Buffer A.
5. Wash the column with 25 mL (5 column volumes) of High Salt buffer.
6. Equilibrate the column with 50 mL (10 column volumes) of Buffer A. Proceed to load sample.

3.6.3 Sample Preparation and Purification

1. Remove sample fraction (approximately 5 mL) from the iodixanol gradient as described in Subheading 3.5.4, **step 4** and place into a 50 mL conical tube. Pool the desired number of fractions and dilute by mixing with equal volume of Buffer A (low salt). A maximum of fractions equivalent to 5 cell factories may be loaded per 5 mL column.
2. Add the diluted sample to the syringe and apply to the column at the same flow rate (5 mL/min).
3. Wash the column with 50 mL of Buffer A (10 column volumes) to remove residual iodixanol.
4. Prepare the Apollo concentrator by washing both the top and bottom surfaces with approximately 15 mL of 1× PBS (Ca^{2+} Mg^{2+}) or Lactate Ringers, discard buffer. Add 10 mL of the same buffer to the bottom of the concentrator.
5. Add 20 mL of elution buffer to the syringe and apply to the column at the same flow rate.
6. Collect rAAV sample by eluting fraction directly into the Apollo concentrator. Collect only the first 15 mL of the eluate. Proceed to sample concentration and buffer exchange.

3.7 Buffer Exchange/Concentration

1. Wet/wash the concentrator column by filling the top with PBS. Let sit until needed and then discard the fluid.
2. Add ~5 mL PBS ($Ca^{2+}Mg^{2+}$) to bottom.
3. Add virus to top of concentrator and top off with modified PBS (PBS with Ca^{2+} and Mg^{2+}).
4. Centrifuge at 3500 x *g* for 15 min at room temperature (*see* **Note 41**).
5. Empty liquid below the concentrator and top off the liquid above the filter and centrifuge again.
6. Repeat this **step 4** times or more if needed.
7. During the last step, you will have to adjust the bottom volume to achieve your targeted volume of virus. 4.7 mL of PBS at bottom of tube results in ~500 μL of virus remaining above filter (*see* **Note 42**).
8. Carefully remove the virus and place in sigmacoted tube.

3.8 Virus Titering

3.8.1 Dot-Blot

1. Prepare the probe in advance by standard PCR using biotinylated dCTPs (*see* **Note 43**).
2. Prepare a standard curve by performing a serial twofold dilution of a plasmid containing the same (or equivalent) viral genome to be titered. Start with 500 μL of plasmid DNA diluted in alkaline buffer, at a concentration of 6 μg/μL. Dilute 1:2 by mixing 250 μL of DNA solution with 250 μL of alkaline buffer, mix thoroughly by vortexing. Repeat sequentially until at least 8 different standard dilutions have been made.
3. Prepare virus dilution curve: Start with a 25-fold dilution by mixing 10 μL of virus preparation into 240 μL of alkaline buffer. Dilute twofold by mixing 120 μL of virus solution with 120 μL of alkaline buffer, mix thoroughly each time by vortexing. Repeat sequentially until at least 6 viral dilutions have been made.
4. Set the hybridization oven to 65 °C and pre-warm 10 mL of hybridization buffer.
5. Wet Amersham Hybond-N membrane and whatman filter paper with deionized water and assemble the dot-blot apparatus.
6. Load 100 μL of each dilution, samples and standards, into the wells. The standard curve will range from 600 ng to at least 4.7 ng. It is recommended to load standard in duplicates. Apply vacuum until all the wells have been vacuumed dry, approximately 1 min. Open the clamps to remove membrane and cut a notch on the left top corner of the membrane for orientation.
7. Cover membrane (still on the filter paper) with clingwrap and crosslink DNA to membrane using UV crosslinker (we use the autocross link function on machine which delivers 120,000 microjoules of UV energy).
8. Place membrane (with DNA side facing up) inside the hybridization tube containing 10 mL of warm buffer and set inside the hybridization oven. Incubate for 2 h.
9. Denature probe by incubating at 95 °C for 5 min, then place immediately on ice for 5 min. Add 300–500 ng of probe directly to the hybridization tube containing buffer and membrane. Incubate at 65 °C overnight (16-18 h).
10. Next day, wash membrane in hybridization oven 3 x 15 min in wash buffer.
11. Transfer membrane to a western blot tray. Incubate for 2 h at room temperature in 5–10 mL of wash buffer containing Streptavidin IRDye-800CW conjugate (1:10,000 dilution).
12. Wash membrane 3 x 5 min with PBS-tween.

13. Visualize blot and measure signal intensity using Odyssey Infrared Imaging System.
14. Build a standard curve by plotting signal intensity vs. number of genomes. Calculate number of genomes from mass (ng) of DNA loaded into the membrane using Avogadro's number. For calculations, keep in mind that plasmid DNA—used as a standard—is double stranded, while AAV genome DNA is single stranded.

3.8.2 Digital Droplet PCR (See Note 9)

1. Serially dilute the virus 1:500000 (1:50 + 1:100 + 1:100) (*see* **Note 44**).
2. Prepare ddPCR mastermix according to manufacturer's instructions, for a total reaction volume of 25 μL. For each sample, run several concentrations. Also include at least one no template control for each run.
3. Generate droplets and seal the plate.
4. Set up thermocycling parameters as determined for the specific primer set. For the CAG probe listed in *see* **Note 11** we utilize an annealing temperature of 62 °C (*see* **Note 45**).
5. Analyze droplets. The Biorad droplet reader will provide data in terms of genomes/μL of reaction mix. You will have to use this number to determine the titer of your vector batch: Multiply this value by 25 (total reaction volume)—this is the absolute number of genomes that was included in reaction mixture. Divide this number by the volume of virus that was used in the reaction mixture, and multiply this with your dilution (500,000), and adjust to the normally used vector genomes/mL by multiplying by 1000. This will yield the concentration of the vector batch.

3.9 Storage and Handling of Virus

1. A detailed discussion of the handling of virus can be found here [10]. However, if we plan to keep virus for ~<2 years we do not freeze the virus as this will affect efficacy. Rather we store the tube (sealed with parafilm) at 4 °C for extended periods of time without any detriment to biological activity. Moreover, any surfaces that come into contact with the virus must be sigmacoted.
2. Sigmacote tips by aspirating the full volume of the tip with Sigmacote, expunge, and let dry in safety hood.
3. Sigmacote tubes by filling the tubes with sigmacote, let sit for a few seconds, and then remove all liquid. Let dry in safety hood.

4 Notes

1. We highly recommend using Terrific Broth (TB) for growing the large cultures. TB is highly enriched providing large quantities of nutrients, growth factors, and nucleic acids precursors, allowing cells to have an extended growth phase. TB also contains buffering capacity preventing cell death due to a drop in pH during growth. LB is suitable for growing small-scale cultures. Scale the culture based on the amount of plasmid needed for your AAV preparation (*see* Table 1). DNA yield will highly depend on plasmid copy number, the growth phase of the bacterial culture, and the purification method. As a reference, we can typically obtain 2–4 mg of pXYZ5 and 4–6 mg of pXX6 per 1 L of culture.
2. Because of the palindromic nature of the terminal repeats, they are prone to recombination. To prevent this, genomic DNA must be grown in recombination deficient bacterial cells. Moreover, in order to reduce the rate of growth the cultures are grown at 30 °C.
3. The pH of this solution will change pretty rapidly upon addition of acid or base. Add only a few drops at a time. Allow the solution to mix and recheck the pH to prevent over or undershooting the desired pH.
4. Scale of tissue culture and transfection depends on the needed yield of viral vector. Our standard scale of production consists of 6 triple flasks (3000 cm^2) which yields ~ 200 μL + of virus with a titer of E^{12} to E^{13} vg/mL. Smaller or larger formats simply requires the resultant adjustment to reagents based on the transfected surface area.
5. In our workflow we spin up to 1 L centrifuge flasks.
6. Do not overheat. Be patient: both solubilization and filtration takes a significant amount of time.
7. We have utilized Q-sepharose columns to purify AAV5. However, other resins and buffer conditions can be used for other serotypes with the same general setup [11–13].
8. Following concentration we do not expose the AAV to any surface without siliconization of that surface (this includes tips and tubes, *see* [10]for a detailed protocol). Inclusion of Pluronic F68 in the final formulation is also used to minimize sticking to surfaces [14].
9. There are numerous methods to determine the titer of AAV. This includes physical titer determinations via measuring viral genomes (e.g., dot-blot, qPCR, and ddPCR), infectious titer measurements determined by treating a cell-line that is permissible to AAV infection [15], or visualizing and quantifying AAV

particles using electron microscopy [9]. It is important to note that AAV titers can vary widely depending on the methodology used as well as between different laboratories. Absolute quantitation of AAV using ddPCR holds promise as this methodology is extremely precise and does not need a standard curve [16, 17]. Moreover, AAV reference standard material has been generated: AAV2 and AAV8 standards are available from American Type Culture Collection (ATCC; Manassas, VA, USA) [18, 19], and these can be utilized to normalize any in house methodology.

10. Our dot-blot protocol uses detection of probes labeled with near-infrared dyes; however, additional means of detection such as radiolabeled probes can be used [15].
11. Primers and probes are designed using the same methodology as those used for qPCR; however, cycling conditions will change depending on which technology used. For the standard CAG (aka CMV/CBA hybrid) promoter [20] we utilized the following oligonucleotides: Forward primer: 5′ GCCTTTTAT GGTAATCGTGCG 3′, reverse primer: 5′ AGGCTGGAG AGGGAGAAG 3′, Probe: 5′ TCCTTTGTCCCAAATCTG TGCGGAG 3′.
12. We recommend removing an aliquot (2 mL) and purifying the plasmid DNA using a miniprep kit and analyze by restriction digest with appropriate restriction enzymes before proceeding to large-scale purification.
13. Closely monitor your cells and keep track of the number of passages. Healthy cells should reach confluency after ~3–4 days. We have found that cell health is a major determinant for success while preparing AAV.
14. If you intend to do TFF, replace media with serum-free media 24 h post-transfection.
15. Confluency can be very subjective, and can vary between researchers. To overcome this, we are utilizing a macro on our microscope software that calculates percent of surface coverage as a measure of cell confluency.
16. Be careful when thawing the PEI tube as it may crack if heated too quickly. Thaw for some time at room temperature before placing in 65 °C.
17. We have noted that poor DNA quality will result in the formation of very large DNA precipitates in this step. If this happens it is likely that transfection will be relatively inefficient.
18. Incubating the transfection mixture for too long (several hours) may result in decreased efficacy.
19. We have successfully stored media for >1 week without any significant effect on biological activity of virus [9].

20. Pay very close attention to the interface of the tubing and the ports on the filter as the tubing will easily pop off with too much pressure. Best practice is to clamp tubing in place with a screw clamp. Moreover, watch for leaks where the tubing runs through the peristaltic pump, as the tubing is thick and can shear with prolonged use.
21. Using multiple TFF cassettes will drastically enhance overall flowrate (*see* https://laboratory.pall.com/en/tangential-flow-filtration.html).
22. It is our experience that filtrate flow will decrease over time (presumably via coating of the membrane). To increase filtrate flow, temporarily increase the pressure (pump speed) to "unclog" the cassette.
23. In order to fit the retentate in a single iodixanol gradient, the total final volume needs to be less than 15 mL. To help with this we placed a mark on the reservoir.
24. For up to 3 days, the cassette can be stored in ddH_2O. For up to 6 months store in 1.0 N NaOH. For storage longer than 6 months, maintain in 15% glycerin with 0.05% azide.
25. Alternative the material can be purified using polyethylene glycol (PEG) / aqueous two-phase partitioning followed by dialysis [21] although this results in a preparation of lower purity.
26. We have tested several different incubation periods without any apparent effect on the final titer.
27. The volume of lysis buffer will depend on the scale of production and how many vessels you use for pelleting. We typically centrifuge all PEGylated media in a 1 L centrifuge bottle, and this pellet will be used in a single iodixanol gradient.
28. You may have to vigorously shake flasks to dislodge cells from all layers.
29. Each 250 mL conical will hold cells from 3 triple flasks.
30. It is our experience that the cell pellet is difficult to dislodge from 250 mL conical after freezing without disturbing the pellet. It is therefore advisable to transfer the pellets to the 50 mL conical before storing the pellet.
31. Avoid disturbing the pellet. Ideally, the pellet should be intact while transferring it.
32. Make sure to use an alcohol resistant marker to label tubes.
33. There are numerous ways to prepare the gradient. We have devised various workflows depending on how many vectors are produced at any given time (Fig. 3).

34. Each gradient can hold material from 3000 cm^2 of cells (e.g., 6 triple flasks).
35. Shake all iodixanol solutions before use.
36. Different protocols call for different volumes. We have also used 7.5 mL 15%, 5 mL 25%, 7.5 mL 40%, and 5 mL 60% [22].
37. Iodixanol is purchased at 60%.
38. You do not want to get air into the tubing as this may disturb the gradient. If this happens, you can reverse the flow of the pump for a few seconds.
39. The interfaces are very difficult to see following centrifugation. However, the plastic seam on the Beckman tube lines up very well with the 60/40 interface. We thus use this as a guide for inserting the needle (Fig. 3c).
40. We recommend practicing on a tube prior to the first collection. Once inserted, the needle cannot be removed as this will call the gradient to collapse. If needle gets clogged, leave in place, and use a new syringe and needle.
41. 15 min may not be sufficient time to equilibrate the column, if that is the case, additional time can be used. The inclusion of 4.7 mL buffer below the filter ensures that the column will not run dry.
42. We measure the volume of virus using a pipettor and a siliconized tip (that we reuse for the same virus until we have achieved the desired volume).
43. Template and primers used to generate the probe will depend on vector to be titered. Make the probe ~500–700 bp long. For PCR it is preferred to use a low fidelity polymerase, as they incorporate biotinylated nucleotides with more efficiency. Biotin labeled dCTP should be used at a 1:1 ratio of total dCTP. After PCR use a clean-up kit to remove unincorporated nucleotides. Probes can be store at −20 °C for long periods of time until ready to use.
44. Bio-Rad recommends 1–120,000 copies/20 μL reaction. Due to the sensitivity of ddPCR, we recommend that you use dedicated pipettors and filtered tips. When pipetting oil, it is recommended that you use the smallest pipettor size possible.
45. Each primer/probe set should be validated on a known viral preparation before general use.

Acknowledgments

The methodology outlined herein represents the accumulated knowledge accrued over years of trial and error, and could not be possible without the help of countless people. We would especially like to recognize: Mark Potter, Vince Chiodo, Shannon Boye, and Sanford Boye at the University of Florida. Kevin Nash at the University of South Florida.

References

1. Manfredsson FP (2016) Introduction to viral vectors and other delivery methods for gene therapy of the nervous system. Methods Mol Biol 1382:3–18
2. Kondratov O, Marsic D, Crosson SM et al (2017) Direct head-to-head evaluation of recombinant adeno-associated viral vectors manufactured in human versus insect cells. Mol Ther 25:2661–2675
3. Urabe M, Ding C, Kotin RM (2002) Insect cells as a factory to produce adeno-associated virus type 2 vectors. Hum Gene Ther 13:1935–1943
4. Strobel B, Miller FD, Rist W et al (2015) Comparative analysis of cesium chloride- and iodixanol-based purification of recombinant adeno-associated viral vectors for preclinical applications. Hum Gene Ther Methods 26:147–157
5. Modak SP, Imaizumi MT, Chappuis M et al (1978) Preparative density gradient centrifugation of RNA and DNA in cesium sulfate-urea mixture. Mol Biol Rep 4:55–60
6. Nilsen TW (2013) RNA-friendly plasmid preparation. Cold Spring Harb Protoc 2013:174–179
7. Heilig JS, Elbing KL, Brent R (2001) Large-scale preparation of plasmid DNA. Curr Protoc Mol Biol Chapter 1:Unit1 7
8. Burger C, Nash KR (2016) Small-scale recombinant adeno-associated virus purification. Methods Mol Biol 1382:95–106
9. Benskey MJ, Sandoval IM, Manfredsson FP (2016) Continuous collection of adeno-associated virus from producer cell medium significantly increases total viral yield. Hum Gene Ther Methods 27:32–45
10. Benskey MJ, Manfredsson FP (2016) Intraparenchymal stereotaxic delivery of rAAV and special considerations in vector handling. Methods Mol Biol 1382:199–215
11. Nass SA, Mattingly MA, Woodcock DA et al (2018) Universal method for the purification of recombinant AAV vectors of differing serotypes. Mol Ther Methods Clin Dev 9:33–46
12. Potter M, Lins B, Mietzsch M et al (2014) A simplified purification protocol for recombinant adeno-associated virus vectors. Mol Ther Methods Clin Dev 1:14034
13. Qu W, Wang M, Wu Y et al (2015) Scalable downstream strategies for purification of recombinant adeno-associated virus vectors in light of the properties. Curr Pharm Biotechnol 16:684–695
14. Fischer MD, Hickey DG, Singh MS et al (2016) Evaluation of an optimized injection system for retinal gene therapy in human patients. Hum Gene Ther Methods 27:150–158
15. Zolotukhin S, Potter M, Zolotukhin I et al (2002) Production and purification of serotype 1, 2, and 5 recombinant adeno-associated viral vectors. Methods 28:158–167
16. Sanders R, Huggett JF, Bushell CA et al (2011) Evaluation of digital PCR for absolute DNA quantification. Anal Chem 83:6474–6484
17. Lock M, Alvira MR, Chen SJ et al (2014) Absolute determination of single-stranded and self-complementary adeno-associated viral vector genome titers by droplet digital PCR. Hum Gene Ther Methods 25:115–125
18. Ayuso E, Blouin V, Lock M et al (2014) Manufacturing and characterization of a recombinant adeno-associated virus type 8 reference standard material. Hum Gene Ther 25:977–987
19. Lock M, McGorray S, Auricchio A et al (2010) Characterization of a recombinant adeno-associated virus type 2 Reference Standard Material. Hum Gene Ther 21:1273–1285
20. Benskey MJ, Kuhn NC, Galligan JJ et al (2015) Targeted gene delivery to the enteric nervous system using AAV: a comparison across serotypes and capsid mutants. Mol Ther 23:488–500
21. Guo P, El-Gohary Y, Prasadan K et al (2012) Rapid and simplified purification of recombinant adeno-associated virus. J Virol Methods 183:139–146
22. Boye SL, Bennett A, Scalabrino ML et al (2016) Impact of Heparan Sulfate Binding on Transduction of Retina by Recombinant Adeno-Associated Virus Vectors. J Virol 90:4215–4231

Chapter 7

Generation of High-Titer Pseudotyped Lentiviral Vectors

Shuang Hu, Mingjie Li, and Ramesh Akkina

Abstract

Lentiviral vectors (LVs) are widely used in gene transfer protocols due to many advantages that include stable gene expression, higher transgene payloads, and, importantly, the ability to pseudotype the vectors with a diverse number of heterologous viral envelopes with broad or restricted cell tropism depending on the need. The pseudotyping process also allows for incorporation of specific antibodies/ligands to engineer LVs. These features greatly facilitate customization of lentiviral vectors for cell/tissue specific gene delivery. The VSV-G protein containing envelope remains the most widely used among the viral glycoproteins used for LV pseudotyping due to its versatile host range and stability. However, many other viral envelopes are being identified for special applications of LVs. Here we describe the methodology to generate pseudotyped LVs using a four-plasmid transient transfection system focusing on aspects to generate high-titer vector stocks.

Key words Lentiviral vector, Viral vector pseudotyping, Vector pseudotyping with VSV-G, Vector production and concentration, Vector titration

1 Introduction

Of the many viral-derived gene transfer vectors used today, lentiviral (LV) vectors occupy a special position. Paradoxically, although derived from a deadly virus human immunodeficiency virus-1 (HIV-1) they are endowed with many unique and desirable characteristics that make them vectors of choice for many basic research and clinical gene therapy applications [1–4]. Advantages include gene integration into the target cell leading to stable expression, capacity for larger payloads and ability to enter nondividing cells, minimal immunogenicity and relative safety. Another most desirable feature of these vectors is the ability to be packaged into heterologous viral envelopes in a process called pseudotyping [2, 5]. The parent virus HIV-1 has restricted tropism and primarily infects CD4+ cells such as helper T cells and macrophages. The

Authors Shuang Hu and Mingjie Li contributed equally to this work.

Fredric P. Manfredsson and Matthew J. Benskey (eds.), *Viral Vectors for Gene Therapy: Methods and Protocols*, Methods in Molecular Biology, vol. 1937, https://doi.org/10.1007/978-1-4939-9065-8_7,

ability to pseudotype the LVs with almost any envelope of choice provides them with a capacity to essentially target any cell in the body for gene transfer.

Early studies on pseudotyping retroviral vectors involved murine leukemia virus (MLV)-based vectors with vesicular stomatitis virus-glycoprotein (VSV-G), an envelope derived from a livestock pathogen, the vesicular stomatitis virus [6]. The VSV envelope has a broad cell tropism and can bind to a wide array of cell types. The receptor for VSV-G is identified as the low-density lipoprotein receptor (LDLR) which is widely distributed among many cell types [7]. That HIV-1-based vectors could be pseudotyped with VSV-G has revolutionized the use of these vectors as a main stay of molecular biology and gene therapy studies tool box [2, 4, 8–10]. In addition to broad host tropism, VSV-G also confers high stability to the vector, thus facilitating concentration to very high titers. A number of cell types were shown to be efficiently transduced with the VSV-G LVs that include human hematopoietic stem cells (HSC) and neurons [10–12]. Many clinical trials have used the LVs for gene transfer [13]. However, some disadvantages and limitations exist with VSV-G envelope. Transduction levels are relatively low in cells such as resting lymphocytes due to low LDLR expression. There is cell toxicity associated with highly concentrated vectors. Due to broad host tropism, gene transfer in a heterogeneous cell population is not just limited to the desired target cells [2].

Efficient in vivo systemic use of these vectors has also been hampered by the sensitivity of VSV-G to serum inactivation. To overcome some of these limitations, envelope proteins from other related vesiculoviruses such as Chandipura (CNV-G), Piry (PRV-G), and Cocal viruses have been tested [14, 15]. These have shown higher resistance to serum inactivation and higher transduction levels with certain cell types, for example, the Cocal envelope shows higher transduction on human HSC and CD4+ T cells whereas CNV-G and PRV-G show higher transduction with neuronal cell types. In addition to the envelopes derived from vesiculoviruses, the LVs have been shown to be readily pseudotypable by a large variety of other viral envelopes [2, 5]. Envelopes from viruses belonging to retrovirus, paramyxovirus, orthomyxovirus, rhabdovirus, arenavirus, and alphavirus families/genera have been successfully used for LV pseudotyping. Some of these pseudotypes are highly selective to certain cell types, thus customizing the LV for tailored application to specific tissue targets. Clever applications have also resulted in unique gene delivery to specific organ system. For example, gene delivery to different brain regions was shown to be possible by retrograde axonal transport when LVs pseudotyped with chimeric glycoproteins derived from rabies and VSV-G protein were used [12]. Other innovative pseudotyping

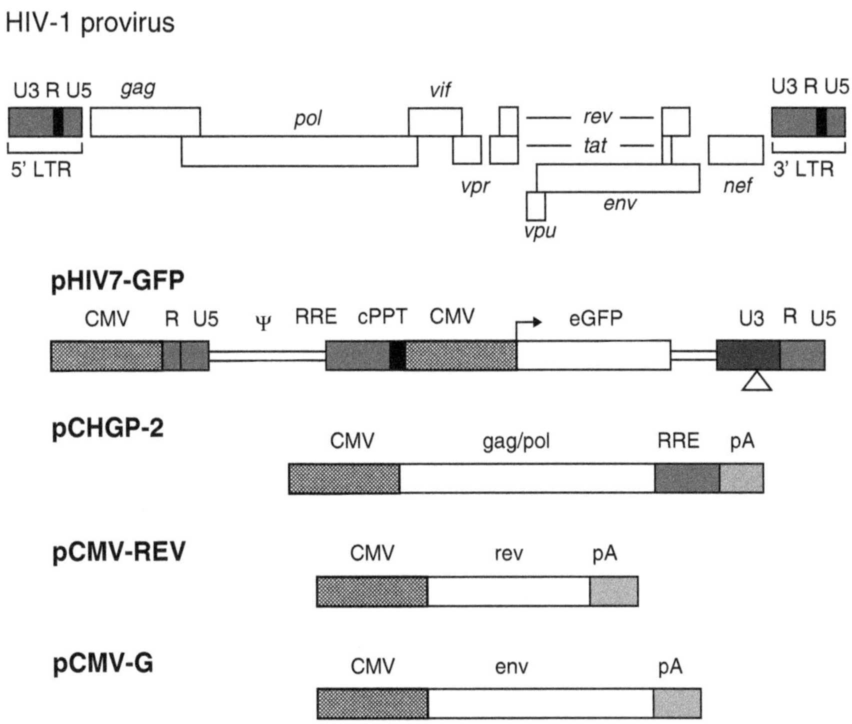

Fig. 1 Schematic of the HIV-1 genome, lentiviral transfer vector, and packaging plasmid constructs. The parent HIV-1 provirus with all the viral genetic elements is shown on top. The transfer vector plasmid pHIV7-GFP depicts a typical gene transfer construct with CMV promoter and GFP reporter. The pCHGP-2 encodes gag and pol proteins in addition to RRE. The pCMV-REV encodes rev protein whereas the pCMV-G encodes VSV-G envelope protein. Other vesiculovirus envelopes such as from Chandipura virus (CNV-G) and Piry virus (PRV-G) can be substituted for VSV-G. Other envelopes with specific tropisms (Rabies gp, for example) can be incorporated as desired in the vector generation protocol

approaches have used specific antibodies to target specific cell subpopulations [16].

To generate a safe replication defective and self-inactivating (SIN) lentiviral vector, the HIV-1 genome has been essentially gutted to remove all irrelevant elements for gene transfer and the essential sequences are supplied in four separate plasmids to prevent any possibility of generating a replication competent virus [4, 9]. The current lentiviral vector generating system uses a transient plasmid transfection system employing co-transfection of multiple plasmids into producer cells. These consist of a transfer vector (gene payload) plasmid and the plasmids encoding the helper proteins required for vector packaging (Fig. 1) [3, 4, 17]. Stable packaging cell lines have been developed, but the vector titers are generally lower than those obtained by transient transfection and less flexible for modifying packaging elements [1]. Commercially available transfection reagents help achieve high efficient transfection efficiency, but are not cost-effective for producing large-scale vector preparations [1, 18]. In contrast, the less expensive calcium phosphate precipitation based methods are highly efficient in HEK293T cell transfection, thus more cost-effective. For pseudotyping, the envelope of choice is supplied as an independently expressed protein for packaging and virus release. Here we describe the use of the envelope proteins VSV-G, chandipura virus-glycoprotein (CNV-G) or piry virus-glycoprotein (PRV-G) for pseudotyping LVs. This protocol can be used to substitute any other envelope encoding plasmids.

2 Materials

2.1 General Laboratory Materials and Equipment

1. BSL-2 laboratory with biosafety cabinet.
2. Personal protective equipment (PPE)- lab coat, gloves, face mask, protective glasses.
3. Refrigerator, −20 °C and −80 °C freezers.
4. Incubator, 37 °C with 5% CO_2.
5. Cell culture treated dishes (100 × 20 mm).
6. 1.5 and 2 mL microcentrifuge tubes.
7. 15 and 50 mL conical vials.
8. 5, 10, and 25 mL serological pipettes.
9. 1× phosphate buffered saline (PBS).
10. Heat-inactivated fetal bovine serum (FBS).
11. Shaker, 37 °C.
12. Nanodrop.
13. Benchtop centrifuge.
14. Ultracentrifuge.
15. Flow cytometry.
16. Single channel manual pipettes and tips: p20, p200, and p1000.

2.2 Generation of Packaging Plasmids

1. *E. coli* culture—DH5α (Thermo Fisher Scientific, Cat. #18263012).
2. 0.2 gene MicroPulser cuvette (BIO-RAD, Cat. #1652081).
3. Electroporator (BIO-RAD).
4. LB medium.
5. Miniprep (Qiagen, Cat. #27104) and Maxiprep kits (Qiagen, Cat. #12191).
6. SOC (Thermo Fisher Scientific, Cat. #15544-034).

2.3 Generation of Pseudotyped Lentiviral Vector

1. HEK293T cells (human embryonic kidney cells containing SV40 large T antigen).
2. Dulbecco's Modified Eagle's Medium (DMEM; ATCC Cat. #30-2002) supplemented with 10% fetal bovine serum (FBS).
3. Trypsin-EDTA (Sigma, Cat. #T3924).
4. Plasmid DNA.
5. TE 79/10 (1 mM Tris–HCl/0.1 mM EDTA, pH 7.9).
6. 2× HBS (HEPES buffered saline): 0.05 M HEPES, 0.28 M NaCl, 1.5 mM Na2HPO4, adjust to pH 7.05 and sterilize through 0.2 μm filter.
7. 2 M $CaCl_2$.
8. 0.6 M butyric acid (Sodium butyrate) (Sigma, Cat #B5887).

2.4 Concentration of Vector Preparation

1. 60 mL syringe.
2. 0.2 μm syringe filter.
3. 38.5 mL Thinwall polypropylene tube (Beckman Coulter, Cat# 326823).
4. SW28 rotor (Beckman Coulter).

2.5 Vector Quantification and Titration

1. HT1080 (a human fibrosarcoma cell line).
2. Polybrene (Hexadimethrine bromide) (Sigma, Cat #H9268).
3. 3.7% formaldehyde in PBS.
4. Round-bottom polystyrene tubes for FACS.
5. Flow cytometry machine.
6. QIAamp DNA Mini Kit (Qiagen, Cat # 51304).
7. PCR primers and TaqMan probes.
8. Quantitative real-time PCR machine.
9. 96-well plate.
10. Lenti-X p24 Rapid Titer Kit.
11. ELISA plate reader at OD 450 nm and OD 490 nm.
12. Viral RNA extraction kit.
13. iTaq Universal Probes One-step Kit.

3 Methods

3.1 Preparation of Plasmids for Vector Production

Four plasmids are used for the production of lentiviral vector preparations, including a lentiviral transfer plasmid (e.g., pHIV-7-GFP), pCMV-Rev, pCHGP-2, and an envelope plasmid from pCMV-VSV-G, pCMV-CNV-G (glycoprotein from Chandipura virus), or pCMV-PRV-G (glycoprotein from Piry virus) for envelope synthesis. All plasmids are transformed to electrocompetent E. coli culture (DH5α), followed by expansion and purification through the Maxiprep kit.

1. Mix 10 ng of plasmid in a volume of 2 μL with 20 μL of DH5α *Escherichia coli* in a 0.2 gene MicroPulser cuvette kept on ice.
2. Slowly place the cuvette onto the electroporator. Electroporate at 2 kV, 25 μF, and 200 Ω pulse controller setting.
3. Recover the bacteria containing target plasmid in 1 mL SOC medium and culture in a tube shaker (300 RPM, 37 °C).
4. After 45 min plate the DH5α *E. coli* onto an LB agar plate containing a specific antibiotic and culture at 37 °C overnight.
5. The next day, pick several isolated colonies by pipette tips. Culture each individual colony in 5 mL of LB broth for 8 h.
6. Sediment the bacterial cultures by centrifugation (2000 RPM, 5 min). Isolate plasmid DNA from the cell pellet using Miniprep kit according to manufacturer's protocol.
7. Check the purity of the plasmid DNA by agarose gel electrophoresis.
8. Expand and culture the recombinant DH5α *E. coli* transformed with pure plasmid DNA in a large volume of LB medium (1 L in total) for 16–20 h.
9. Take 800 μL of each recombinant *E. coli* culture, mix it with 200 μL glycerol, and then store at −80 °C as stock for long-term usage. Pellet the rest of the LB bacterial culture by centrifugation (2000 RPM, 5 min). Use this pellet to do plasmid Maxiprep according to manufacturer's protocol.
10. Recover the purified plasmid in water at a concentration between 2 and 4 μg/μL and store at −20 °C.

3.2 Generation of Pseudotyped Lentiviral Vector

3.2.1 Packaging of Vectors

Vector packaging is performed by co-transfection of the packaging and transfer plasmids into HEK293T cells. The packaging cell line, HEK293T is maintained in DMEM supplemented with 10% FBS in 37 °C incubator with 5% CO_2.

1. Plate HEK293T cells at 30–40% confluence to 100-mm tissue culture dishes in culture medium (*see* **Note 1**).
2. After 20–24 h culture, check the cell density. The cells should be about 80% confluent at the time of transfection.
3. Prepare a 15 mL tube. Add TE 79/10 to the tube. The volume of TE 79/10 = 440 μL—the volume of DNA.
4. Add 15 μg of the packaging construct pCHGP-2, 15 μg of transfer vector (e.g., pHIV-7-GFP), 5 μg of the respective pseudotyping envelope (pCMV-VSV-G, pCMV-CNV-G, or pCMV-PRV-G), and 5 μg of pCMV-Rev [14, 18]. Gently mix.

5. Add 60 μL 2 M $CaCl_2$ solution. Gently mix.
6. Prepare another 15 mL tube. Add 0.5 mL 2× HBS.
7. Transfer the DNA-$CaCl_2$ mixture to the tube containing 2× HBS with constant agitation.
8. Keep the reaction at room temperature until precipitation is observed (about 30 min).
9. Mix the precipitation well by vortexing. Carefully add the 1 mL of suspension onto cells for transfection.
10. After 4 h incubation at 37 °C with 5% CO_2, replace the medium with 6 mL complete medium. Add 60 μL 0.6 M butyric acid (transcription activator). Return to culture.
11. At 24 h post transfection, collect the supernatant and store at −80 °C. Add fresh DMEM with butyric acid back to the plate. Similarly, two more collections are performed at 48 and 72 h post transfection, with one media change in between.

3.2.2 Concentration of Vector Preparation

1. Put the frozen vector supernatant in 37 °C water bath until 90% thawed and then place on ice to thaw completely.
2. Centrifuge the supernatant at 2000 rpm (900 × *g*) for 10 min to remove any cell debris in vector collections.
3. Filter the supernatant with 0.2 μm syringe filter (*see* **Note 2**).
4. Transfer the supernatant to polypropylene tubes (*see* **Note 3**). Concentrate the supernatant by ultracentrifugation at 24,000 RPM (108,000 × *g*) and 4 °C for 2 h with Beckman SW28 swing rotor.
5. Carefully pour out the supernatant so that the pellet is left intact. More filtered vector preparation can be added back to the same tubes for additional 3 to 4 rounds of centrifugation, depending on the volume of vector preparation.
6. After the final centrifugation, resuspend the pellet that contains the concentrated vector in DMEM, aliquot into small volume, and store at −80 °C for future use.

3.2.3 Vector Quantification and Titration

Functional vector titers can be obtained by using serial dilutions from vector stocks to transduce HEK293T cells (or other cell lines that are generally permissive to transduction, such as HT1080) and detect reporter gene expression by FACS analysis. Integration titer can be determined by detecting vector genome sequence in the transduced cells by real-time PCR. Other quantification methods are also used in different scenarios, such as detecting p24 content of vector preparation using the Lenti-X p24 Rapid Titer Kit (Clontech) or detecting vector RNA copy number by qRT-PCR (quantitative reverse transcription PCR) with primers and probe on HIV-1

LTR region as described previously [11, 18, 19]. The following shows how to determine the functional or integration titer.

1. Seed 5×10^4/well of HT1080 cells in 12-well plate in 1 mL DMEM medium supplemented with 10% FBS.
2. After overnight culture, trypsinize cells from one well and count the number of the cells.
3. Add serially diluted vector stock to the cultured cells.
4. Add 1 uL 4 mg/mL polybrene in each well containing vector. Continue culture for 48 h.
5. Trypsinize the cells. Centrifuge the cells at $900 \times g$ for 6 min. For vectors with a fluorescent reporter gene (e.g., eGFP), go to next step for FACS analysis. For vectors without a reporter, go to **step 8** for real-time qPCR.
6. Following centrifuge, remove the supernatant and resuspend the pellet with 300 μL of 3.7% formaldehyde in PBS.
7. Determine the percentage of eGFP positive cells by FACS analysis. The titer will be represented as transduction unit per milliliter concentrated vector (TU/mL).

$$\text{Titer} = \frac{\text{cell number} \times \%\text{ of GFP}^+ \text{ cells} \times \text{dilution}}{\text{vector volume (mL)} \times 100}$$

   in this formula, the cell number stands for the cell count when the vector was added.

8. For vectors without a fluorescent reporter gene, extract genomic DNA from HT1080 cells using QIAamp DNA Mini Kit (Qiagen) according to the manufacturer's protocol. Amplify vector sequence in genomic DNA using QuantStudio 3 System with primers (in HIV-1 PBS/psi region) [20] 5′-CCGTTGT CAGGCAACGTG-3′ and 5′-AGCTGACAGGTGGTGGCAA T-3′, and TaqMan probe 5′-FAM-AGCTCTCTCGACGCA GGACTCGGC-TAMRA-3′. The albumin gene that is a single copy gene in the genome is also amplified with primers 5′-TGAAACATACGTTCCCAAAGAGTTT-3′ and 5′-CTCT CCTTCTCAGAAAGTGTGCATAT-3′, and probe 5′-FAM-T GCTGAAACATTCACCTTCCATGCAGA-TAMRA-3′ as an internal control. Determine the copy numbers of vector and albumin by PCR in 96-well plate according to the manufacture's instruction with the following program: 50 °C, 2 min, 95 °C, 10 min, 95 °C, 15 s; 60 °C, 2 min; 35 cycles. The titer will be represented as integration units per milliliter concentrated vector (IU/mL).

$$\text{Titer} = \frac{\text{cell number} \times \text{copy number of vector} \times 2 \times \text{dilution}}{\text{copy number of albumin} \times \text{vector volume (mL)}}$$

4 Notes

1. We usually plate at least 5 dishes per vector for preparing concentrated high-titer vector.
2. It may also be filtered with 0.2 μm bottle-top filter for preparing large quantity and reducing the loss of the vector.
3. The polypropylene tubes are autoclavable. Using autoclaved centrifuge tubes and 0.2 μm filter can minimize contamination during handling the vector supernatant. This is especially important if the transduced cells are used for long-term culture or engrafting animals.

Acknowledgments

Work reported here was supported by NIH grants to RA. We would like to thank Lauren Kinner-Bibeau for assistance in manuscript preparation.

References

1. McCarron A, Donnelley M, McIntyre C et al (2016) Challenges of up-scaling lentivirus production and processing. J Biotechnol 240:23–30
2. Joglekar AV, Sandoval S (2017) Pseudotyped lentiviral vectors: one vector, many guises. Hum Gene Ther Methods 28:291–301
3. Merten OW, Hebben M, Bovolenta C (2016) Production of lentiviral vectors. Mol Ther Methods Clin Dev 3:16017
4. Segura MM, Mangion M, Gaillet B et al (2013) New developments in lentiviral vector design, production and purification. Expert Opin Biol Ther 13:987–1011
5. Cronin J, Zhang XY, Reiser J (2005) Altering the tropism of lentiviral vectors through pseudotyping. Curr Gene Ther 5:387–398
6. Burns JC, Friedmann T, Driever W et al (1993) Vesicular stomatitis virus G glycoprotein pseudotyped retroviral vectors: concentration to very high titer and efficient gene transfer into mammalian and nonmammalian cells. Proc Natl Acad Sci U S A 90:8033–8037
7. Finkelshtein D, Werman A, Novick D et al (2013) LDL receptor and its family members serve as the cellular receptors for vesicular stomatitis virus. Proc Natl Acad Sci U S A 110:7306–7311
8. Naldini L, Blomer U, Gallay P et al (1996) In vivo gene delivery and stable transduction of nondividing cells by a lentiviral vector. Science 272:263–267
9. Zufferey R, Dull T, Mandel RJ et al (1998) Self-inactivating lentivirus vector for safe and efficient in vivo gene delivery. J Virol 72:9873–9880
10. Akkina RK, Walton RM, Chen ML et al (1996) High-efficiency gene transfer into CD34+ cells with a human immunodeficiency virus type 1-based retroviral vector pseudotyped with vesicular stomatitis virus envelope glycoprotein G. J Virol 70:2581–2585
11. Li M, Husic N, Lin Y et al (2010) Optimal promoter usage for lentiviral vector-mediated transduction of cultured central nervous system cells. J Neurosci Methods 189:56–64
12. Kobayashi K, Inoue KI, Tanabe S et al (2017) Pseudotyped lentiviral vectors for retrograde gene delivery into target brain regions. Front Neuroanat 11:65
13. Morgan RA, Gray D, Lomova A et al (2017) Hematopoietic stem cell gene therapy: progress and lessons learned. Cell Stem Cell 21:574–590
14. Hu S, Mohan Kumar D, Sax C et al (2016) Pseudotyping of lentiviral vector with novel vesiculovirus envelope glycoproteins derived from Chandipura and Piry viruses. Virology 488:162–168

15. Trobridge GD, Wu RA, Hansen M et al (2010) Cocal-pseudotyped lentiviral vectors resist inactivation by human serum and efficiently transduce primate hematopoietic repopulating cells. Mol Ther 18:725–733
16. Zhou Q, Schneider IC, Edes I et al (2012) T-cell receptor gene transfer exclusively to human CD8(+) cells enhances tumor cell killing. Blood 120:4334–4342
17. Segura MM, Garnier A, Durocher Y et al (2010) New protocol for lentiviral vector mass production. Methods Mol Biol 614:39–52
18. Li M, Husic N, Lin Y et al (2012) Production of lentiviral vectors for transducing cells from the central nervous system. J Vis Exp 24:e4031
19. Lizee G, Aerts JL, Gonzales MI et al (2003) Real-time quantitative reverse transcriptase-polymerase chain reaction as a method for determining lentiviral vector titers and measuring transgene expression. Hum Gene Ther 14:497–507
20. Sastry L, Johnson T, Hobson MJ et al (2002) Titering lentiviral vectors: comparison of DNA, RNA and marker expression methods. Gene Ther 9:1155–1162

Chapter 8

A Scalable Lentiviral Vector Production and Purification Method Using Mustang Q Chromatography and Tangential Flow Filtration

Stuart Tinch, Kathy Szczur, William Swaney, Lilith Reeves, and Scott R. Witting

Abstract

Lentiviral vectors have rapidly become a favorite tool for research and clinical gene transfer applications which seek to permanently introduce alterations in the genome. This status can be attributed primarily to their ability to transduce dividing as well as quiescent cells. When coupled with internal promotor selection to drive expression in one cell type but not another, the ease with which the vectors can be pseudotyped to either restrict or expand tropism offers unique opportunities previously unavailable to the researcher to manipulate the genome. Although LV can be produced from stable packaging cell lines and/or in suspension culture, by and far, most LV vectors are produced using adherent 293 T cells grown in plasticware and production plasmids transiently transfected with either PEI or Calcium Phosphate. The media is usually changed and un-concentrated vector supernatant collected between 24 and 48 h post-transfection. The supernatant may then be purified by Mustang Q chromatography, concentrated by Tangential Flow Filtration, and finally diafiltered into the final formulation buffer of choice. Here we describe a pilot scale method for the manufacture of a Lentiviral vector that purifies and concentrates approximately 6 L of un-concentrated LV supernatant to approximately 150 mL. Typical titers for most vector constructs range between 1×10^8 and 1×10^9 infectious particles per mL. This method may be performed reiteratively to increase total volume or can be further scaled up to increase yield.

Key words Lentiviral vector production, Purification, Mustang Q chromatography, Tangential flow filtration

1 Introduction

The vast majority of Lentiviral vectors used in research today, including those under development for clinical use, have mostly been derived from the type 1 Human Immunodeficiency Virus (HIV) [1]. Lentiviruses belong to the *Retroviridae* family. Like all members of this family, these viruses and the vectors derived from them encode their genetic information on RNA rather than DNA

Fredric P. Manfredsson and Matthew J. Benskey (eds.), *Viral Vectors for Gene Therapy: Methods and Protocols*, Methods in Molecular Biology, vol. 1937, https://doi.org/10.1007/978-1-4939-9065-8_8,

[2]. These enveloped virions are diploid, consisting of two identical copies of positive polarity viral genomic RNA [2]. Thus, the genomic RNA of these viruses is very similar to messenger RNA found in most cells. During the retrovirus' natural life cycle, the RNA is converted to DNA by the enzyme reverse transcriptase and the DNA is then stably integrated into the genome of the infected cell [2].

There are several possible ways to manipulate or alter a cells genome: direct inoculation of DNA into the cells via microinjection, electroporation, chemical-based transient transfection (e.g., calcium phosphate or polyethylenimine (PEI)), or viral transduction. Microinjection is time consuming, requires an expertise that most labs do not possess, and is not amenable to large-scale approaches. Electroporation and chemical-based transient transfection do not have the expertise limitations associated with microinjection, can be used at scale, and may be used in a clinical setting; however, they typically do not result in permanent alterations of the genome. Since retroviruses can stably integrate into the host cell's genome, they are ideal candidates for gene therapy approaches where a permanent alteration of the genome is desired.

In 1989, the fledgling field of Gene Therapy was born when Dr. Steven Rosenberg successfully introduced neomycin resistance into human tumor-infiltrating lymphocytes using a gamma retrovirus vector [3]. These cells were infused into five patients with advanced melanoma and, as such, became one of the first human gene therapy trials. A subsequent hallmark gene therapy trial used gamma retrovirus vectors to correct the ADA gene in T-cells from two different ADA-SCID patients [4]. Future trials built upon these early successes by using gene corrected umbilical cord blood CD34+ cells in conjunction with bone marrow transplant to obtain a persistent effect [5–7]. However, problems with the clinical use of gamma retroviral vectors emerged. It was reported that two subjects developed leukemia due to the retroviral vector integrating near a proto-oncogene [8]. Further investigation would establish that gamma retroviral vectors do not integrate in a truly random fashion, but have a preference to integrate near the 5′ ends of transcription units and associated CpG islands [9]. Therefore, alternate vector systems and/or further modifications were necessary for the field to move forward.

In contrast to gamma retroviral vectors, lentiviral vectors show a trend toward integration within active transcription units rather that upstream of these transcriptional start sites [10, 11]. This finding, paired with the fact that lentiviral vectors, based on the human immunodeficiency virus (HIV), are able to transduce non-dividing cells, it was inevitable that a switch toward lentiviral vector use in the clinic would occur. However, several safety concerns would remain to be addressed. One improvement was the use of a self-inactivating (SIN) lentiviral vector design as a potential solution to unfavorable integrational mutagenesis [12]. The SIN design produces integrated vectors which lack enhancer–promoter sequences in the U3 region of the long terminal repeats (LRTs) and were shown to have a reduced transforming capacity when compared to vectors with intact LTRs [13]. An additional benefit of the SIN design is that it allows the incorporation of weaker internal promoter to drive transgene expression which further reduces the likelihood of trans-activating neighboring genes. Other significant advancements in lentiviral vector safety involved improving the production system. It was shown that only gag, pol, and rev were the only HIV accessory proteins required for high titer production [14]. Additionally, these proteins may be expressed on separate production plasmids. The separation of the accessory proteins and presence of a SIN design greatly limit the chances of the creation of a replication competent lentivirus vector (RCL). To date, an RCL sourced from third generation lentiviral vector systems has not been reported.

The production of lentiviral vectors has been optimized over the past decade and a half. Most lentiviral vectors are produced using HEK293T or HEK293 cells. The primary difference between these cell lines is the presence of the SV40 T-antigen stably integrated into the HEK293T cell line. The SV40 T-antigen has been shown to yield higher titers; therefore, a majority of protocols utilize HEK293T [15]. Since the HEK293T cells are normally adherent, most production protocols involve growing the cells in tissue culture flasks followed by transient transfection. Large-scale vector manufacturing generally involves a direct scale up from small-scale systems (such as T-flasks) to multilayer Cell Stacks or HyperSTACKS. Transient transfections are typically performed using either PEI or CaPO4 due to the ease and high titers obtained. Electroporation has also been reported. As described above, the current third generation system is most often used. A common variant of this system requires transient transfection of four plasmids expressing the following: a lentiviral vector backbone; gag-pol; rev; and an envelope glycoprotein, typically from vesicular stomatitis virus (VSVg). After several hours, a media change is typically done to remove the transfection mixture. The supernatant is then collected between 48 and 72 h post-transfection in one or more harvests. Supernatant is minimally filtered or clarified to remove cells. Small-scale or research preparations may be concentrated by ultracentrifugation with or without a sucrose gradient and then resuspended in the desired final formulation buffer in a reduced volume. This method, however, is not readily scalable for large-scale manufacturing. For this reason, many large-scale protocols call for the capture of the vector from the supernatant by anion exchange chromatography [16]. During capture, major impurities in the preparation are removed. The eluted material is then further concentrated using tangential flow filtration (TFF). The vector may then be diafiltered into its final formation buffer using TFF [17].

Here we describe a pilot scale method for the manufacture of a Lentiviral vector that purifies and concentrates approximately 6 L of un-concentrated supernatant to approximately 150 mL. The process involves cell expansion, transient transfection, incorporation of Benzonase treatment for plasmid DNA reduction, filtration to remove cells, clarification of the supernatant in preparation for Mustang Q chromatography, and TFF and diafiltration into the product's final formulation buffer. Typical titers for most vector constructs range between 1×10^8 and 1×10^9 infectious particles per mL. This method may be performed reiteratively to increase total volume or can be further scaled up to increase yield.

2 Materials

2.1 Cell Culture Materials

1. HEK293T cells.
2. Baxa Pump (or alternate veristaltic pump).
3. Three Baxa #21 fluid transfer sets (Baxter, cat# H93821).
4. Ashton Pumpmatic 10 mL pipets.
5. Sterile 1, 2, 5, 10, 25, 50, and 100 mL serological pipettes.
6. T225 flask.
7. One 5-Layer Corning CellBind® Cell Stack.
8. Six 10-Layer Corning CellBind® Cell Stacks.
9. Seven filling caps (female Luer end) for Corning Cell Stacks.
10. Three 850 cm^2 roller bottles.
11. Dulbecco's Phosphate-Buffered Saline (DPBS).
12. TrypLE SELECT™ or Trypsin.
13. Single bottles of complete cell culture media (CCM): Add 56 mL of fetal bovine serum (FBS) and 5.6 mL of sodium pyruvate to each 500 mL bottle of DMEM (4.5 g/L glucose, with L-glutamine or GlutaMAX-I). The final concentration is 10% fetal bovine serum (FBS) and 1% sodium pyruvate. Store at 4 °C. *See* **Note 1**.
14. 10 L vector production CCM: Add 60 mL of FBS and 56 mL of sodium pyruvate to each of two 500 mL bottles of FBS. Connect the female Luer end of the 10 L bag of DMEM (4.5 g/L glucose, with L-glutamine or GlutaMAX-I) to the male Luer fitting of a Baxa #21 fluid transfer set. Connect the remaining end of the Baxa #21 fluid transfer set to the female Luer fitting of the Ashton Pumpmatic pipet. Using the Baxa pump, transfer the entire contents of each bottle of FBS into the 10 L bag of DMEM. Store at 4 °C. *See* **Note 1**.

2.2 Transfection Materials

1. 2× HEPES buffered saline, pH 7.05.
2. 2.5 M $CaCl_2$ (*see* **Note 2**).
3. Sterile water for injection.
4. Third generation lentivirus packaging plasmids (e.g., pMDL, pRev, pVSVg, and transgene plasmid). *See* **Note 3**.
5. Benzonase (50 U/mL of final harvest volume).
6. DMEM (4.5 g/L glucose, with L-glutamine or GlutaMAX-I), unsupplemented.
7. 0.22 μm PVDF syringe filter with 10 mL syringe.
8. 1 M $MgCl_2$.

2.3 Supernatant Harvest Materials

1. Sterile tube welder (e.g., Terumo TSCDII).
2. Baxa Pump (or alternate veristaltic pump).
3. Two Baxa #21 fluid transfer sets (Baxter, cat# H93821).
4. Six 1 L bioprocess bags.
5. One 5 L bioprocess bag.
6. One 10 L bioprocess bag.
7. Three modified Leukocyte Reduction Filters (Haemonetics, cat # RCQT): with a sterile tube welder, replace input end with female Luer connection and output end with male Luer connection. These Luer connections can be obtained by using one Fluid Transfer Extension Set (Baxter, H93886) per LRF.
8. Millipak 60 Gamma Gold 0.45 μm filter. A sterile 3 inch piece of #17 silicone tubing with a male Luer fitting is fitted to the inlet end and a second sterile 3 inch piece of #17 silicone tubing with a female Luer is fitted to the outlet end. Zip tie all tubing connections to secure.

2.4 Mustang Q Materials (See Note 4)

1. Sanitization Buffer: 300 mL of 1 M NaOH.
2. Preconditioning Buffer: 300 mL of 1 M NaCl.
3. Equilibration and Wash Buffer: 4800 mL of 25 mM Tris–HCl (pH 8), 150 mM NaCl. Add 120 mL of 1 M Tris–HCL (pH 8) and 120 mL of 5 M NaCl to 4560 mL of sterile water. Store at 4 °C.
4. Elution Buffer: 25 mM Tris–HCl (pH 8), 1.2 M NaCl. Add 30 mL of 1 M Tris–HCL (pH 8) and 288 mL of 5 M NaCl to 882 mL of sterile water. Store at 4 °C.
5. Dilution Buffer: 25 mM Tris–HCl (pH 8). Add 30 mL of 1 M Tris–HCL (pH 8) to 1170 mL of sterile water. Make in a 5 L bioprocess bag to allow for space the volume of Elution buffer coming off the column. Store at 4 °C.

6. Two sanitary silicone gaskets for 1.5 inch Maxi Flanges (Part # GMX455-PX).
7. Pressure transducer (Spectrum ACPM-499-03 N or similar).
8. MQ Tubing Set A: approximately 6 foot length of #73 silicone tubing containing a female Quick-Connect 3/8 inch hose barb on each end.
9. MQ Tubing Set B: approximately 4 foot length of #73 silicone tubing containing a male Quick-Connect 3/8 inch hose barb on each end.
10. MQ Filter Set 1A: Mustang Q (Pall) 60 mL membrane volume capsule with a 4 inch #17 silicone tubing piece containing a female Luer fitting on each of the two hose barbs.
11. MQ Filter Set 1B: 3 inch length of #73 silicone tubing containing a male Quick-Connect 3/8 inch hose barb on one end and a ProConnex 1.5 inch flange with 3/8 inch hose barbs on the other end.
12. MQ Filter Set 1C: 3 inch length of #73 silicone tubing containing a female Quick-Connect 3/8 inch hose barb on one end and a Maxi Flange 1.5 inch flange with 3/8 inch hose barb on the other end.
13. 1 L labtainer or bioprocess bag for capsule drain-off.
14. One 60 mL syringe as a vent to the column.
15. Six male Luer caps.
16. KrosFlo miniKros Pilot peristaltic pump (with integrated pressure monitor) fitted with a #73 tubing pump head (*see* **Note 5**).
17. Assorted sizes of labtainers or bioprocess bags for waste collection of materials run through the capsule.
18. Four hemostats.
19. Sterile forceps.
20. Weighted ring stand with clamps sufficient to support the Mustang Q capsule and tubing in an upright position.

2.5 TFF Materials (See Note 4)

1. Sterile water, 350 mL.
2. Sterile 20% Ethanol, 350 mL.
3. Equilibration Buffer, 350 mL (*see* **Note 6**).
4. Diafiltration Buffer, 150 mL (*see* **Note 6**).
5. Spectrum 190 cm^2 TFF column; 500 kD cutoff.
6. Processing bag: 500 mL CryoMACS® with a Universal Spike #29 (with female end) inserted into each spike port.
7. One 5 L bioprocessing bag for permeate collection.
8. TFF Tubing Set A: 16 inch length of #14 silicone tubing containing a male Luer 1/16 inch hose barb on each end.

9. TFF Tubing Set B: 24 inch length of #16 silicone tubing containing a male Luer 1/8 inch hose barb on each end.
10. TFF Tubing Set C: 9 inch length of #16 silicone tubing containing a male Luer 1/8 inch hose barb on each end.
11. TFF Tubing Set D: 3 inch length of #16 silicone tubing containing a male Luer 1/8 inch hose barb on one end and a female 1/8 inch hose barb on the other end.
12. One 30 mL syringe.
13. Baxa Pump (or alternate veristaltic pump).
14. Baxa #21 fluid transfer set (Baxter, cat# H93821).
15. KrosFlo miniKros Pilot peristaltic pump (with integrated pressure monitor) fitted with a #16 tubing pump head (*see* **Note 5**).
16. Three pressure transducers (Spectrum ACPM-499-03N or similar).
17. KrosFlo miniKros Pilot Filter Stand.
18. Large capacity scale with RS232 data port connection (e.g., Ohaus Defender 3000).

3 Methods

3.1 Initiating Cell Culture (See Note 7)

1. Thaw a vial of 1×10^7 HEK293T cells and seed a T225 flask containing 50 mL of CCM. This equates to a seeding density of approximately 4×10^4 cells per cm^2.
2. Incubate the cells at 37 °C with 5% CO_2 for four days.
3. Remove the spent CCM from the T225 flask and discard.
4. Pipette 30 mL of DPBS to the flask, covering the entire surface area.
5. Remove the DPBS from the T225 flask and discard.
6. Pipette 3 mL of TrypLE SELECT™ or Trypsin to the flask and incubate at 37 °C until the majority of the cells become dislodged or for up to 10 min.
7. Pipette 7 mL of CCM to the flask, pipet up and down several times to break apart any large cell clumps and collect the cells.
8. Count the cells and determine their viability and concentration.
9. Add the volume of cell suspension necessary to seed 6.37×10^7 total viable cells into the Corning 5-Layer Cell Stack.
10. Replace the cap on the Cell Stack with a Corning filling cap with female Luer and connect the cap to the male Luer fitting of a Baxa #21 Fluid Transfer Pump set.

11. Connect the remaining end of the Baxa #21 Fluid Transfer Pump set to the female Luer fitting of an Ashton Pumpmatic pipet. Place the Ashton pipet into a bottle of CCM.
12. Place the Corning 5-Layer Cell stack on its side and fill the Cell stack with 750 mL of CCM, using a second bottle of CCM as needed (*see* **Note 8**).
13. Manipulate the Cell Stack as per the manufacturer's procedures; replace the filling cap with the original vent cap.
14. Incubate the cells at 37 °C with 5% CO_2 for four days.

3.2 Cell Expansion

1. Aspirate the spent CCM from the 5-Layer Cell Stack and discard.
2. Replace the cap on the Cell Stack with a Corning filling cap with female Luer and connect the cap to the male Luer fitting of a Baxa #21 Fluid Transfer Pump set.
3. Connect the remaining end of the Baxa #21 Fluid Transfer Pump set to the female Luer fitting of Ashton Pumpmatic pipet. Place the Ashton pipet into a bottle of DPBS.
4. Pump 200 mL of DPBS into the 5-Layer Cell Stack.
5. Manipulate the Cell Stack as per the manufacturer's procedures to distribute the DPBS over the entire surface area.
6. Aspirate the DPBS from the 5-Layer Cell stack and discard.
7. Pipette 50 mL of TrypLE SELECT™ or Trypsin to the 5-Layer Cell Stack and incubate at 37 °C until the majority of the cells become dislodged or for up to 15 min.
8. Pipette 100 mL of CCM to the 5-Layer Cell Stack.
9. Replace the cap on the Cell Stack with a Corning filling cap with female Luer and connect the cap to the male Luer fitting of a Baxa #21 Fluid Transfer Pump set.
10. Connect the remaining end of the Baxa #21 Fluid Transfer Pump set to the female Luer fitting of Ashton Pumpmatic pipet. Place the Ashton pipet into an empty 500 mL bottle.
11. Pump the cell suspension from the 5-Layer Cell Stack into the bottle.
12. Count the cells and determine their viability and concentration.
13. Add the volume of cell suspension necessary to seed 3.18E8 viable cells into each of the three Corning 10-Layer Cell Stacks.
14. Replace the cap on the Cell Stack with a new Corning filling cap with female Luer and connect the Baxa #21 pump set to a 10 L bag of CCM.

15. Place the Corning 10-Layer Cell stack on its side and fill the Cell stack with 1500 mL of CCM. Repeat until all three 10-Layer Cell Stacks have been filled.
16. Incubate the cells at 37 °C with 5% CO_2 for four days.

3.3 Transient Transfection

1. Aspirate the spent CCM from each 10-Layer Cell Stack and discard.
2. Replace the cap on the Cell Stack with a Corning filling cap with female Luer and connect the cap to the male Luer fitting of a Baxa #21 Fluid Transfer Pump set.
3. Connect the remaining end of the Baxa #21 Fluid Transfer Pump set to the female Luer fitting of Ashton Pumpmatic pipet. Place the Ashton pipet into a bottle of DPBS.
4. Pump 500 mL of DPBS into each 10-Layer Cell Stack.
5. Manipulate the Cell Stack as per the manufacturer's procedures to distribute the DPBS over the entire surface area.
6. Aspirate the DPBS from each 10-Layer Cell stack and discard.
7. Add 100 mL of TrypLE SELECT™ or Trypsin into each 10 Layer Cell Stack and incubate at 37 °C until the majority of the cells become dislodged or for up to 15 min.
8. Add 150 mL of CCM into each 10-Layer Cell Stack.
9. Replace the cap on the Cell Stack with a Corning filling cap with Luer and place the Ashton pipet into a sterile 1 L bottle. Pump the cell suspension from each 10 Layer Cell Stack into the bottle.
10. Count the cells and determine their viability and concentration.
11. Using an Ashton pipet, and a Baxa #21 pump set connected to a 10 L bag of CCM previously prepared, add 1300 mL of CCM into each of three roller bottles.
12. Add the volume of cell suspension necessary to seed 1.7E9 viable cells into each of the three Roller Bottles prefilled with CCM.
13. Add extra CCM to bring the total volume to at least 1415 mL, if necessary.
14. IMPORTANT: Periodically swirl the roller bottles during the remainder of this section to prevent the cells from clumping.
15. Label a 500 mL sterile bottle "Transfection Mixture" and add 131 mL of 2× HEPES Buffered Saline. Set aside.
16. Label a 250 mL sterile bottle "DNA/H2O/$CaCl_2$" and add the appropriate volume of plasmids and add sterile water for a final volume of 117.9 mL (*see* **Note 9**). Swirl to mix well.

17. Add 13.1 mL of 2.5 M $CaCl_2$ to the container labeled "DNA/H2O/$CaCl_2$" (*see* **Note 10**). Swirl to mix well.
18. Using a 100 mL pipette, transfer the contents of the flask labeled "DNA/H2O/$CaCl_2$" to the bottle labeled "Transfection Mixture" while gently swirling.
19. Incubate for 20 min at room temperature.
20. Transfer 85 mL of the "Transfection Mixture" to each of the three roller bottles containing cells and CCM.
21. Without delay, transfer the contents of each bottle into a corresponding 10-Layer Cell Stack. Ensure the culture volume is level across each Cell Stack.
22. Incubate at 37 °C with 5% CO_2 for 16–19 h.
23. Prepare a mixture of 666.8 μL of Benzonase (50 U/mL) and 6 mL of DMEM that has not been supplemented with FBS. Sterilize using a 0.22 μm PVDF syringe filter.
24. Aspirate the transfection media from the Cell Stacks.
25. Add 2 mL of the filtered Benzonase mixture to each Cell Stack.
26. Add 10 mL of 1 M $MgCl_2$ to each Cell Stack.
27. Using a #21 fluid transfer set, Baxa pump, and filling cap with Luer, add 1000 mL of pre-warmed CCM into each Cell Stack.
28. Incubate at 37 °C with 5% CO_2 for 22–26 h.

3.4 Harvest and Filtration

1. Using a Baxa pump and a #21 fluid transfer set, transfer the supernatant from each Cell Stack to a corresponding 1 L bioprocess the bag.
2. Refeed the Cell Stacks by pumping back in 1000 mL of fresh CCM.
3. Incubate the Cell Stacks at 37 °C with 5% CO_2 for an additional 22–26 h.
4. Using the Baxa pump, a #21 fluid transfer set, and a modified leukocyte reduction filter (LRF), filter each Cell Stack harvest into a single 5 L bag.
5. Store the bag at 4 °C until harvest 2 has been collected.
6. Repeat **step 1** for harvest 2.
7. Repeat **step 4** for harvest 2.
8. Remove harvest 1 from 4 °C and using a pump set, filter through a 0.45 μm Millipore Gamma Gold filter and into a 10 L bioprocess bag.
9. After all of harvest 1 has been filtered, filter harvest 2 into the same 10 L bioprocess bag.

3.5 Mustang Q Purification (See Note 11)

1. Organize all the filtration components in the biosafety cabinet and assemble according to the schematic in Fig. 1. Use sterile forceps to place the silicone gaskets between the filter capsule (MQ Filter set 1A) and the two connectors (MQ Filter set 1B and 1C). Secure all barb connections with zip ties.
2. Cap all open lines as indicated in Fig. 1 and connect one pressure transducer ("P" in the schematic) to the Luer connector on the inlet end of the filter. Cap the open side of the pressure transducer with a male Luer cap.
3. Close all vents and attach the MQ filter capsule to a ring stand with the clamps.
4. Connect the pressure transducer to the KrosFlo Pilot pump and tare the pressure transducer.
5. Remove the plunger from a 60 mL syringe and connect to the line attached to the top capsule vent of the MQ filter. Keep the syringe in an upright position (use zip tie, if necessary).
6. Connect a 1 Liter Bioprocess Bag to the line attached to the lower capsule vent (bottom) of the MQ filter capsule. This will be used during draining of the MQ capsule filter.

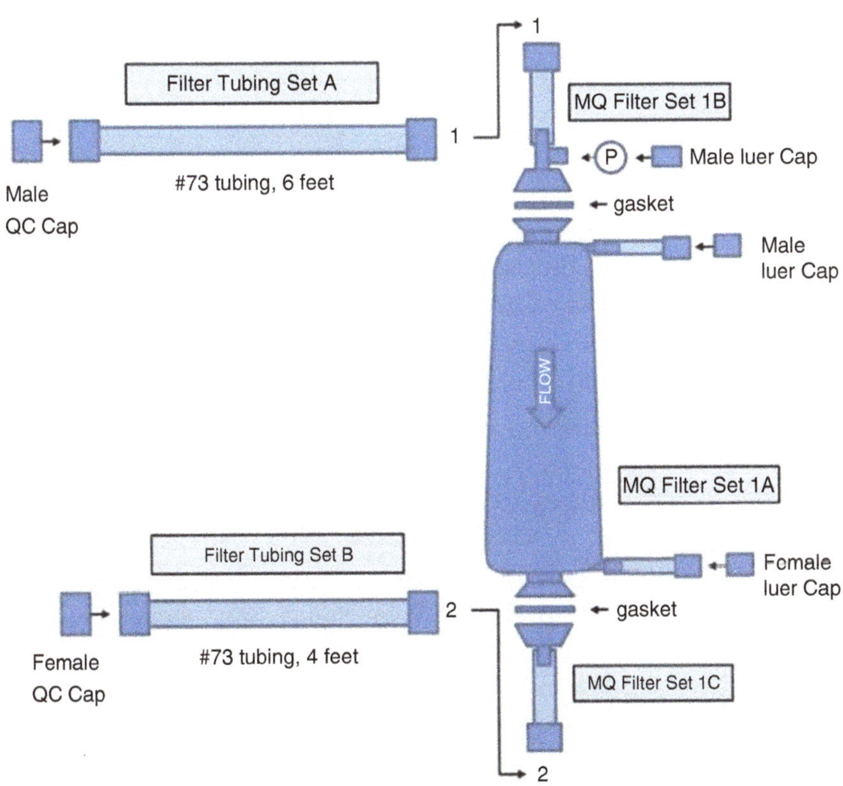

Fig. 1 Assembly schematic of the Mustang Q capsule and associated tubing. Remove components from their sterile wrappings and assemble in a biosafety cabinet as shown

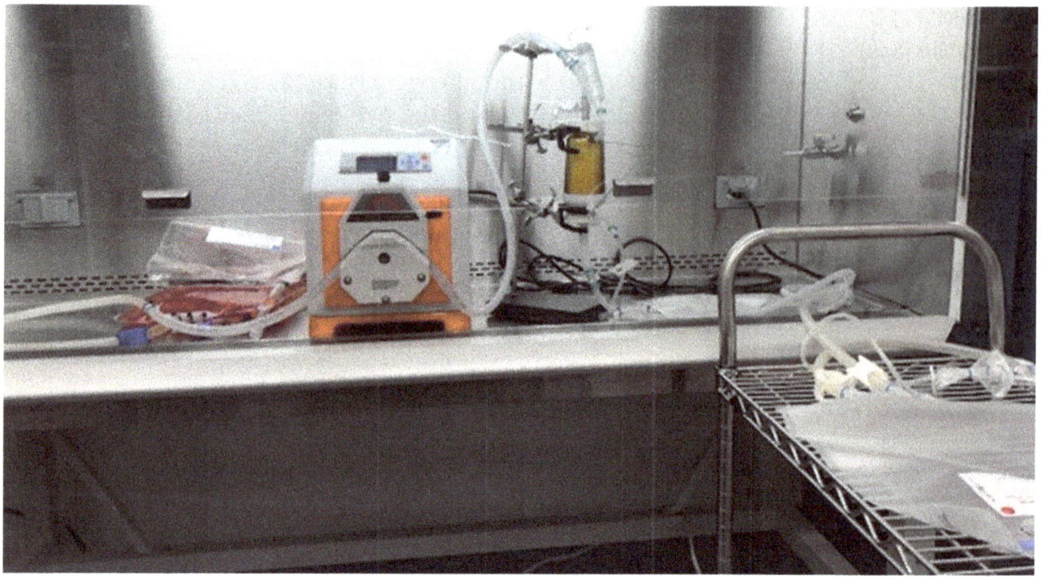

Fig. 2 Photo of the completely assembled Mustang Q capsule with pump and bioprocessing bag attachments

7. Connect a 1 Liter Bioprocess bag to the outlet line of the MQ Filter Capsule (Filter tubing set B). This and subsequent outlet bags may be located outside the biosafety cabinet as shown in a photo of the completed setup in Fig. 2.
8. Place the supply line (Filter tubing set A) into the KrosFlo Pilot pump.
9. Connect the bag of 1 M NaOH to the inlet line of the Mustang Q Capsule filter.
10. Close off the lower capsule vent using a hemostat (or slide clamp from the attached bag).
11. Close the outlet line (Mustang Filter set 1C) with a hemostat.
12. Leave open the upper capsule vent (60 mL syringe).
13. Pump the buffer from the bag into the capsule filter between 200 to 300 mL/min until most of the air has been displaced from the capsule filter and it is nearly full. Stop the pump.
14. Close the upper capsule vent with a hemostat and remove the hemostat from the outlet line. Verify that the fluid path is clear and start the pump.
15. Increase the pump speed until the maximum speed listed in Table 1 has been achieved. Pump the entire contents of the bag through the Mustang Q capsule.
16. Turn off the pump when most of the buffer has disappeared from the capsule.

Table 1
Mustang Q buffer information and order of use

Order of use	Buffer	Buffer volume (mL)	Speed (mL/min)
1	1 M NaOH	600	600
2	1 M NaCl	600	600
3	Equilibration buffer	2400	600
4	Filtered vector supernatant	6000	600
5	Wash buffer	2400	600
6	Elution buffer	1200	600

17. Facilitate the draining of any remaining liquid from the capsule into the waste bags by temporarily opening the top and bottom capsule vents (*see* **Note 12**).
18. Remove the empty bag of buffer from the inlet line and replace with the next buffer in the series as noted in Table 1.
19. Disconnect the outlet line waste bag, if necessary, and attach an appropriately sized replacement.
20. Repeat **steps 10** through **19** for each buffer. IMPORTANT: Be sure to attach the collection buffer (instead of a waste bag) to the outlet line when the elution buffer is being pumped through the column.
21. Immediately proceed with concentration by TFF.

3.6 Tangential Flow Filtration (See Note 13)

1. Assemble the TFF column with the tubing sets according to the schematic in Fig. 3. A photo of the complete system is shown in Fig. 4.
2. TFF Tubing Sets A and C connect the Universal Spike #29 in the processing bag to the pressure transducers at the top outlet port and bottom inlet port of the TFF, respectively.
3. TFF Tubing Sets B and D are attached to the top permeate port with pressure transducer and the bottom permeate port of the TFF, respectively.
4. The Baxa #21 attaches to a female Luer on one of the integrated tubing ports of the processing bag. The unused ports can be closed with the roller clamps.
5. Attach an empty waste bag to the end of the permeate line and place on the scale. Tare the scale.
6. Connect the pressure transducer and scale data wires to the KrosFlo miniKros Pilot pump.
7. Connect the KrosFlo miniKros Pilot pump to a laptop or tablet running KF Comm.

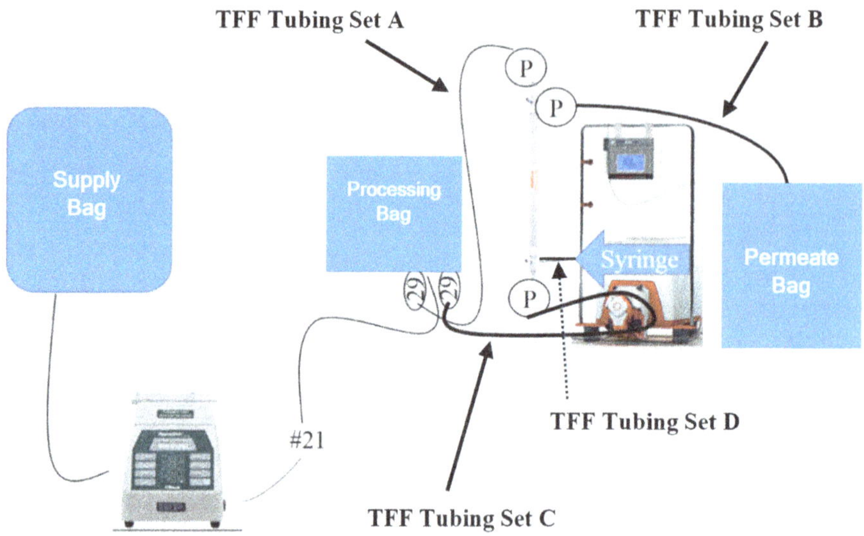

Fig. 3 Assembly schematic of the TFF system, pumps, accessories, and the associated tubing. Only the TFF, support stands, processing bag, and KrosFlo miniKros Pilot pump are setup in the biosafety cabinet. However, any open connections should be made in the biosafety cabinet. The "P" denotes the locations of pressure transducers. The "29" denotes the locations of the Universal Spike #29 s. The "#21" refers to the Baxa fluid transfer set location which passes through the Baxa pump

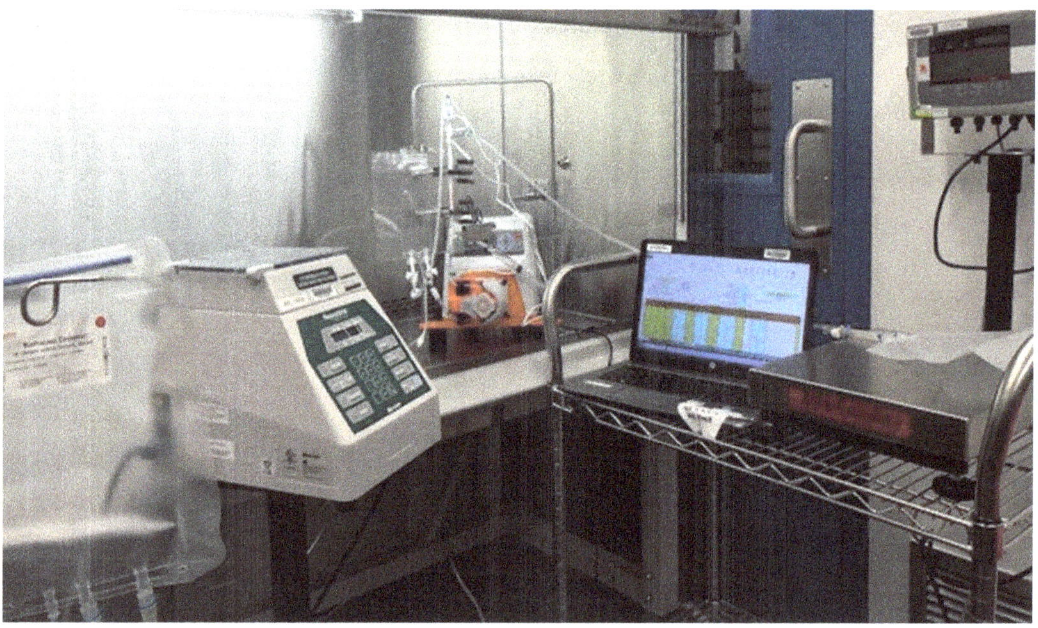

Fig. 4 Photo of the completely assembled TFF system. Note that much of the system is located outside of the biosafety cabinet

8. Connect the 20% ethanol bag to the end of TFF tubing set A in the "Supply bag" position indicated on Fig. 3.
9. Using the Baxa pump, move the entire volume into the "Processing Bag."
10. Begin pumping the 20% ethanol through the column using the KrosFlow pump at 195 mL/min.
11. As the ethanol breaks through the column and starts exiting the permeate and returning through the retentate line, clamp down the flow restrictor on the retentate line to achieve a transmembrane pressure (TMP) of 5 psi (*see* **Note 14**).
12. Run the system until most of the ethanol has moved through the line into the waste bag. Do NOT allow the column to run dry.
13. Switch out the now empty 20% ethanol bag with the bag of sterile H_2O and use the Baxa pump to move it into the processing bag.
14. Pump the H_2O through the column at 195 mL/min and adjust the flow restrictor to ensure a TMP of 5 psi.
15. Pump until all the water is out and air fills the line from the processing bag to the column.
16. Once the water has gone through the system, clamp the retentate and permeate line and slow the pump to 0.1 L/min (*see* **Note 15**).
17. Start the pump, allow inlet pressure to build to 5–7 psi, and stop the pump.
18. Allow the system to sit for 1 min and observe the inlet pressure. If the change is less than 0.5 psi/min proceed with the protocol. If not, connect another TFF column.
19. Switch out the now empty H_2O bag with the bag of selected TFF equilibration buffer (350 mL) and use the Baxa pump to move as much as possible into the processing bag. As the volume of H_2O in the processing bag runs allows, use the Baxa pump to add the remaining TFF equilibration buffer.
20. Set the KrosFlow pump to 195 ml/min and keep the TMP at 5 psi. Pump until most of the buffer is gone but do not allow the column to run dry.
21. Stop the pump and switch out the now empty TFF equilibration buffer bag with the bag of diluted MQ eluate and pump as much as possible into the processing bag. As the volume of H_2O in the processing bag runs allows, use the Baxa pump to add the remaining TFF equilibration buffer.
22. Run the KrosFlow pump at 195 mL/min and initiate the KF Comm software to track the pressures, weights, pump speed, and the calculated TFF shear. The shear should be kept within

$3700–4000\ s^{-1}$. This can be controlled by decreasing or increasing the pump speed. It is also important to maintain a TMP of 5 psi.

23. Concentrate the volume down as low as possible without allowing air into the system.
24. Stop the pump and switch out the now empty MQ eluate bag with the bag of diafiltration buffer and pump into the processing bag.
25. Run the KrosFlow pump at 195 mL/min and monitor the TFF conditions as described in **step 19** with the KF Comm software. Keep the shear within $3700–4000\ s^{-1}$ and TMP at 5 psi.
26. Concentrate the volume down as low as possible without allowing air into the system. Stop if excessive foaming is observed. The final volume should be around 13 mL.
27. Stop the pump and end data collection.
28. Clamp the permeate line with a hemostat and using the attached 10 mL syringe, push 5 mL of air into the TFF module. Re-clamp the syringe line.
29. Change the KrosFlow pump speed to 80 mL/min and allow the system to recirculate the product for 3 min.
30. After stopping the recirculation, lay the process bag on the biosafety cabinet surface and do not allow the product to reenter the system. If necessary, reverse the pump to get excess product out of the TFF column.
31. Collect the product from the processing bag using a 30 mL syringe.
32. Aliquot the required volume into the desired containers. We typically use 1–2 mL cryovials.

4 Notes

1. Use the 500 mL bottle size for cell culture and 10 L bag for vector production media changes. The use of antibiotics is optional. It is advisable to qualify each lot of FBS for lentiviral vector production at a small, T-flask scale prior to use in larger scales.
2. We obtain this reagent as a custom-made product in 50 mL aliquots. Other concentrations may be used but we recommend the solution be sterile and has an endotoxin level ≤ 4 EU/mL.
3. It is recommended the plasmids be made commercially to ensure a high quality preparation. If done in house, it is

recommended that plasmid isolation kits which eliminate (or drastically reduce) endotoxin are used.

4. All buffers are prepared in a certified biosafety cabinet and then 0.22 μm filtered. The buffers should be made in or transferred to appropriately sized labtainers or bioprocess bags to allow the fluids to be pumped. Filter system components are also assembled in a certified biosafety cabinet and then packaged and terminally sterilized by appropriate methods. Do not sterilize the assembled Mustang Q. Sterile the tubing sets described in Subheading 2 independently.
5. Other peristaltic pumps may be utilized as long the indicated tubing is supported. A separate pressure monitoring unit will be required if this capability is not integrated into the pump.
6. The equilibration and diafiltration buffer are the same composition but are split between two bioprocess bags. The composition varies with the intended application but phenol red free is recommended since it can become concentrated with TFF. Commercial preparations such as X-VIVO™ 10 or "custom" buffered solutions containing human serum albumin are typical examples. Note that the diafiltration buffer volume is typically 1/7th of the diluted MQ eluate volume.
7. The transient transfection should be timed to be performed on Tuesday afternoon if avoiding weekend work is desired. In this case, initiate cell thawing twelve days prior to the planned transfection date (a Thursday).
8. We typically use the volume markings on the Cell Stack to measure input volumes. However, if greater accuracy is desired, the Cell Stack may be placed on an appropriately sized scale (5 kg or higher limit) and use 1 g = 1 mL as a measure of added fluid.
9. Calculate the volumes of each plasmid needed based on optimization studies and stock concentration and add sterile water to a final volume of 117.9 mL.
10. The volume of $CaCl_2$ is calculated such that the final concentration is 125 mM when taking into account the 2× HEPES buffered saline combined volume.
11. Assemble the Mustang Q filter system within a biosafety cabinet to maintain sterility.
12. Complete **steps 17–19** as quickly as possible so the capsule filter surface does not dry out.
13. By utilizing KF Comm software it is possible to have exacting control over pressures and flow rates to ensure the vector is under the same conditions through each iteration of the process and that the cGMP process mirrors the small-scale work for a given vector. There are unavoidable product losses

associated with TFF, but those can be limited by completing the process in a timely and proficient manner.

14. The back pressure of the ethanol is not strong enough to generate a high TMP, but it is important to get as close as possible to 5 psi without going over. The retentate flow restrictor is the white clamp located on the Universal Spike #29.
15. The TFF column is integrity tested by the manufacturer; but in order to ensure that the integrity of the column remains after wetting and rinsing the column, an additional check is required.

Acknowledgments

Our facility is generously supported by the Cincinnati Children's Hospital Research Foundation. Most importantly, we would like to acknowledge the hard work and dedication of the entire staff of the Translational Core Laboratories at Cincinnati Children's Hospital; without it, our contributions to the gene therapy field would not be possible.

References

1. Otto-Wilhelm M, Hebben M, Bovolenta C (2016) Production of lentiviral vectors. Mol Ther Methods Clin Dev 3:16017. https://doi.org/10.1038/mtm.2016.17
2. Coffin J, Hughes S, Varmus H (eds) (1997) Retroviruses. Cold Spring Harbor Laboratory Press, New York
3. Rosenberg S, Aebersold P, Cornetta K et al (1990) Gene transfer into humans--immunotherapy of patients with advanced melanoma, using tumor-infiltrating lymphocytes modified by retroviral gene transduction. N Engl J Med 323:570–578
4. Blaese R, Culver K, Miller A et al (1995) T lymphocyte-directed gene therapy for ADA–SCID: initial trial results after 4 years. Science 270:475–480
5. Cavazzana-Calvo M, Hacein-Bey S, de Saint Basile G et al (2000) Gene therapy of human severe combined immunodeficiency (SCID)-X1 disease. Science 288:669–672
6. Kohn D, Weinberg K, Nolta J et al (1995) Engraftment of gene-modified umbilical cord blood cells in neonates with adenosine deaminase deficiency. Nat Med 1:1017–1023
7. Aiuti A, Slavin S, Aker M et al (2002) Correction of ADA-SCID by stem cell gene therapy combined with nonmyeloablative conditioning. Science 296:2410–2413. https://doi.org/10.1126/science.1070104
8. Hacein-Bey-Abina S, Von Kalle C, Schmidt M et al (2003) LMO2-associated clonal T cell proliferation in two patients after gene therapy for SCID-X1. Science 302:415–419. https://doi.org/10.1126/science.1088547
9. Bushman F (2007) Retroviral integration and human gene therapy. J Clin Invest 117 (8):2083–2086
10. Schröder A, Shinn P, Chen H et al (2002) HIV-1 integration in the human genome favors active genes and local hotspots. Cell 110(4):521–529
11. Mitchell R, Beitzel B, Schroder A et al (2004) Retroviral DNA integration: ASLV, HIV, and MLV show distinct target site preferences. PLoS Biol 2(8):E234. https://doi.org/10.1371/journal.pbio.0020234
12. Schambach A, Swaney W, van der Loo J (2009) Design and production of retro- and lentiviral vectors for gene expression in hematopoietic cells. Methods Mol Biol 506:191–205. https://doi.org/10.1007/978-1-59745-409-4_14
13. Modlich U, Bohne J, Schmidt M et al (2006) Cell-culture assays reveal the importance of retroviral vector design for insertional

genotoxicity. Blood 108:2545–2553. https://doi.org/10.1182/blood-2005-08-024976
14. Dull T, Zufferey R, Kelly M et al (1998) A third-generation lentivirus vector with a conditional packaging system. J Virol 72 (11):8463–8471
15. Gama-Norton L, Botezatu L, Herrmann S et al (2011) Lentivirus production is influenced by SV40 large T-antigen and chromosomal integration of the vector in HEK293 cells. Hum Gene Ther 22:1269–1279. https://doi.org/10.1089/hum.2010.143
16. Slepushkin V, Chang N, Cohen R et al (2003) Large-scale purification of a lentiviral vector by size exclusion chromatography or Mustang Q ion exchange chromatography. Bioprocess J 2:89–95
17. Leath A, Cornetta K (2012) Developing novel lentiviral vectors into clinical products. Meth Enzymol 507:89–108. https://doi.org/10.1016/B978-0-12-386509-0.00005-3

Chapter 9

Current Use of Adenovirus Vectors and Their Production Methods

Ekramy E. Sayedahmed, Rashmi Kumari, and Suresh K. Mittal

Abstract

Various adenovirus (AdV) vector systems have proven to be lucrative options for gene delivery. They can serve as potential vaccine candidates for prevention of several common infectious diseases and hold the promise for gene therapy, especially for cancer. Several AdV vector-based therapies are currently at various stages of clinical trials worldwide, which make an immense interest of both the clinicians and researchers. Since these vectors are easy to manipulate, have broad tropism, and have the capability to yield high titers, this delivery system has a wide range of applications for different clinical settings. This chapter emphasizes on some of the current usages of AdV vectors and their production methods.

Key words Adenovirus vector, Gene delivery system, Gene therapy, Vector design, Vector production, Recombinant vaccines

1 Introduction

Adenoviruses (AdVs) are non-enveloped double-stranded DNA viruses containing genomes of approximately 26–48 kilobase pairs (kbp). Initially, a human AdV was first isolated in 1953 from the adenoid tissues [1] and hence was named AdV [2]. They are known to cause inapparent or symptomatic infections of the upper or lower respiratory tract, gastrointestinal tract, or eyes, which are usually self-limiting in healthy individuals. Although AdVs were known for a long time, their therapeutic potential as a gene delivery vehicle was realized only with the advent of recombinant DNA technology. With continual advancement in the biology of AdV, it became the first viral gene delivery vector to be used in humans. More than 60 types of human AdVs have been described, of which the vector backbone of human AdV type C5 has been used extensively for gene delivery [3].

Fredric P. Manfredsson and Matthew J. Benskey (eds.), *Viral Vectors for Gene Therapy: Methods and Protocols*, Methods in Molecular Biology, vol. 1937, https://doi.org/10.1007/978-1-4939-9065-8_9,

1.1 AdV Vectors: Pros and Cons

AdV vectors have several advantages which make them ideal for gene delivery. The AdV biology has been deciphered, which makes the molecular manipulation of its genome easier. Moreover, several AdVs have low or no virulence in humans and have high transduction efficiency for both replicating and nonreplicating cell types. The vector can also be grown and purified in very high titers and large quantities at a reasonable cost. Furthermore, AdV vectors possess the minimal risk of insertional mutagenesis because of their inability to integrate into the host genome. The transient transgene expression by AdV vectors have been harnessed for oncolytic therapy and also for expression of vaccine antigens [4–9].

However, the transient nature of transgene expression by AdV vectors sometimes limits their use where continuous transgene expression is necessary for a desired therapeutic effect. Apart from this limitation, AdV vectors are known to activate innate immunity, which can lead to severe toxicity at a very high vector dose. One such evidence is the death of a patient enrolled in the ornithine transcarbamylase (OTC) deficiency clinical trial due to high vector dosage leading to multiple organs failure [10].

Due to the high prevalence of human AdVs, nearly 80% of human population is exposed to one or more AdV types multiple times in their lives [11–13], thereby developing AdV neutralizing antibodies popularly known as "preexisting vector immunity" [14]. The issue of preexisting vector immunity can be rectified to some extent by increasing the vector dosage without increasing toxicity [13, 15, 16]. Alternatively, preexisting vector immunity can also be circumvented using nonhuman AdV vectors and heterologous prime-boost approaches. Innovation in vector engineering strategies and the use of different immunosuppressive agents can also be used to overcome some of these limitations in the existing vector systems [17–19].

1.2 Nonhuman AdV Vectors

Nonhuman AdV vector systems based on bovine AdV, simian AdV, porcine AdV, ovine AdV, canine AdV, avian AdV, and murine AdV [20, 21] were developed in search of safe and efficient gene delivery vehicles to overcome the shortcomings of human AdV vector systems, especially the concern of preexisting vector immunity. For example, bovine AdV vectors are not neutralized by human AdV-specific neutralizing antibodies, and the prevalence of bovine AdV cross-neutralizing antibodies was not detected in human serum samples [22, 23]. Moreover, various nonhuman AdV vectors use different receptors for internalization, thereby broadening the range of cell types that can be targeted [24, 25].

1.3 AdV Vectors: Usage and Current Status

Initially, when the therapeutic potential of AdV vectors was realized, they were evaluated for a broad range of medical conditions including genetic diseases and metabolic disorders. However, soon it was realized that transgene expression is usually for a short duration [26, 27]. This limits the use of AdV vectors to conditions where transient transgene expression is required for the desired effects, such as recombinant vaccines and cancer therapeutics.

1.3.1 AdV Vector-Based Vaccines for Infectious Diseases

With the advancement in the field of viral vectored vaccines, various AdV vectors have been tested in both preclinical and clinical studies [6, 28, 29] for different infectious diseases. This is due to the fact that AdV vector-based vaccines induce a balanced humoral and cell-mediated immune (CMI) responses [30, 31] by stimulating innate immunity through pathways that are both Toll-like receptor (TLR)-dependent and TLR-independent [32, 33]. AdV vectors expressing different antigens of influenza virus have been tested in different animal models and have shown high protection efficiency against homologous and heterologous influenza viruses [34–39]. A vaccine construct expressing hemagglutinin (HA) of an H5N1 influenza virus provided cross-protection in mice following challenge with different strains of highly pathogenic H5N1 influenza viruses [40]. Similarly, in a clinical trial, immunization with an AdV vector encoding the HA gene of influenza virus increased hemagglutination inhibition (HI) titers in more than 75% of the participants [41].

With time, several modifications were incorporated in the AdV vector system to overcome the existing limitations of preexisting vector immunity and inadequate antigen-specific immunogenicity. For this purpose, other AdV types from both human and nonhuman AdVs have been evaluated. An AdV35 vector-based HIV vaccine was assessed in a clinical trial and was found very effective and safe [42]. AdV26, another less common type of human AdV, has been recently evaluated for Ebola vaccine in a clinical trial and it elicited a favorable antibody response [43]. In addition to less prevalent human AdVs, several nonhuman AdVs, in particular chimpanzee AdV (ChAdV) vectors, have shown very encouraging results in clinical trials for malaria [44], leishmania [45], and Ebola [46]. Recently, ChAdV vector, ChAd3-EBOZ, encoding for the Ebola G glycoprotein gene of the Zaire strain showed robust antibody and T cell responses in Phase I and II clinical trials [47]. Another approach which has been adopted to improve transgene immunogenicity is the use of single cycle AdV vectors having the deletion of pIIIa protein-coding gene. A single cycle AdV vector encoding influenza HA was assessed for immunogenicity in both cotton rats and hamsters leading to enhanced immune responses at a low dosage [48].

1.3.2 Oncolytic AdV Vector-Based Therapies for Cancer

The oncolytic nature of AdV has been utilized to combat various forms of cancer. To achieve effective oncolysis, the virus should infect and replicate within the cancer cells. Most of human AdVs require Coxsackievirus and Adenovirus Receptor (CAR) for virus internalization, but in many forms of cancer, there is a marked downregulation or complete absence of CAR [49], the reason for a marked reduction in cell transduction with many AdV vectors. To increase the interaction between the virus and cancer cell surface molecules, an introduction of a motif, like RGD, in the knob region of AdV fiber improves the interaction with integrins which are expressed on the cancer cell surface [50]. In some cases, a complete swapping of fiber is done for its preferred interaction with a cell surface molecule such as desmoglein 2, which is expressed in large number on cancer cells [51–54]. It seems very assuring that only tumor cells can be lysed by these oncolytic AdV vectors since their replication competency is dependent on the presence of a specific tumor antigen. A successful oncolytic AdV therapy also requires some other vector modifications to overcome the immunological as well as structural barriers of the tumor microenvironment. Oncolytic AdV vector expressing relaxin facilitates better vector spread in the dense extracellular matrix (ECM) [55]. VCN-01 is another armed oncolytic AdV vector which expresses hyaluronidase and is currently being tested in Phase 1 clinical trials [56]. Recently, oncolytic AdV vector ONCOS 102 expressing GM-CSF demonstrated a potent therapeutic effect with minimal side effects [57]. Many other molecules like interferon alpha, tumor necrosis factor alpha, and other interleukins are also being investigated as delivery molecules with oncolytic AdV vectors. It seems that targeting the tumor microenvironment, in addition to the tumor cell lysis, is a better approach for cancer therapeutics using oncolytic AdV vectors.

1.4 AdV Vector Types

Several changes have been made in the AdV vector design methodology to improve vector recovery, transgene expression, and safety. AdV vectors can be broadly classified into three types based on the deletions of the viral genes.

1.4.1 First- and Second-Generation AdV Vectors

First-generation AdV vectors contain the deletion of the early (E) region 1 (E1) or E1 & E3 regions of the viral genome. Deletion of the E1 region results in a replication-incompetent vector, and it also serves the purpose of increasing the capacity of the foreign gene cassette for insertion [58]. E1-deleted AdV vectors can only be grown in a cell line (e.g., HEK 293) that constitutively expresses E1 proteins [59]. However, anchorage-dependent cell lines can be used only for the small-scale production of vector preparations. To achieve scalability and batch-to-batch consistency, a suspension cell culture bioreactor system is used with a variant of HEK 293 cell line capable of growing cultures in suspension without serum. A bioreactor with 10,000 L capacity is projected to yield 10^9–10^{10} viral particle/milliliter (VP/mL) [60]. However, the usage of HEK 293 cells can result in the production of contaminating replication-competent AdV due to homologous recombination. The PER.C6 [61] and SL0036 cell lines [62] have been developed with a minimal E1 region to eliminate the possibility of homologous recombination. The major drawback of the first-generation AdV vector system is high immunogenicity in the host, which raises safety concern in situations where a very high vector dose is required for desired effects.

Second-generation AdV vectors were created to minimize the shortcomings of first-generation vectors. Second-generation AdV vectors were designed with deletion of two more gene regions, E2 and/or E4, along with E1 and E3 deletions. The idea was to reduce the vector immunogenicity by minimizing the leaky expression of viral genes [63]. Apart from this, the deletion/s also increase/s the transgene carrying capacity of the vector. However, these multiple regions deleted vectors require an appropriate complimentary cell line for their propagation.

1.4.2 Third-Generation AdV Vectors

Third-generation AdV vectors include the helper-dependent vectors, also known as gutless vectors, which are designed by removing all the AdV genes. These vectors retain only the AdV packaging signal along with the inverted terminal repeat (ITR) sequences of the viral genome [64–67]. Due to the complete absence of protein-coding regions in the helper-dependent vectors, a significant reduction in vector immunogenicity with improved safety occurs when these vectors are used in patients. Moreover, the transgene carrying capacity of helper-dependent vector system can be up to 36 kbp. Production of these vectors requires a helper virus—a first-generation empty AdV vector. Initially, the low yield of helper-dependent vector and the contamination with the helper virus were two major concerns with this vector system [64]. Both of these concerns were addressed by replacing the helper virus with the AdV vector in which the packaging sequences are flanked with a site-specific recombination sequence, e.g., loxP. The helper-dependent vector is grown with this novel helper vector in a cell line that expresses an appropriate recombinase, e.g., Cre recombinase for loxP sites [67, 68]. The loxP-Cre recombinase or an equivalent system will result in the generation of novel helper virus genomes without the packaging sequences, thereby allowing efficient packaging of the helper-dependent vector genomes. Third-generation AdV vectors have shown promising results in different animal models with minimal adverse effects [69].

1.5 Construction of AdV Vectors

Several techniques were developed to construct AdV vectors which can be broadly divided into two approaches: (1) direct insertion of the foreign gene into the viral genomic DNA, in a plasmid form, using unique restriction enzymes [30, 70] and (2) recombination between two plasmids through homologous recombination either in bacteria or in a permissive cell line [71–75]. The two plasmids system includes a *genomic plasmid* that contains nearly the complete AdV genome with appropriate deletion/s, and a *shuttle plasmid* carrying the foreign gene cassette and AdV sequences that are essential for homologous recombination and generation of an infectious AdV vector. There are two commonly used recombination techniques for generating AdV vectors: (1) Homologous recombination in bacteria and (2) Cre/lox recombination in mammalian cells. The detailed protocols to create and purify AdV vectors using these two recombination approaches are described below.

2 Materials

2.1 Generation of AdV Vector by Homologous Recombination in Bacteria

1. Genomic plasmid (pAdV-ΔE1E3).
2. Shuttle plasmid containing the desired transgene (pAdV-shuttle-T).
3. *E. coli* BJ5183 strain: kept in aliquots at −80 °C.
4. *E. coli* DH5α strain: kept in aliquots at −80 °C.
5. BIO-RAD Gene Pulser II Electroporation System.
6. 0.2 cm electroporation cuvette.
7. Lysogeny broth or Luria-Bertani (LB) broth and LB agar.
8. Ampicillin or carbenicillin 50 mg/mL.
9. Miniprep plasmid kit (Mini Plus Plasmid DNA 250 Prp. #GF2002).
10. Enzymes- PacI, HindIII, and Antarctic phosphatase.
11. 1% TAE agarose (50 × 1% TAE: Tris 242 g, glacial acetic acid 57.1 mL, 0.5 M EDTA, pH 8.0100 mL, and MilliQ water to adjust the volume to 1 L).
12. GENECLEAN® III Kit (MP Biomedicals).
13. Maxi Fast Ion Plasmid Kit (IBI Scientific).
14. Minimum essential medium (MEM).
15. Opti-MEM (Gibco).
16. Fetal bovine serum (FBS).
17. Lipofectamine 2000 (Invitrogen).
18. Appropriate cell line (i.e., HEK 293) that supports the replication of the desired AdV vector.

2.2 Generation of AdV Vector Using Cre/loxP Recombination System

1. Genomic plasmid pAdVΔψ,E1,E3/loxP.
2. Shuttle plasmid pAdV-shuttle/loxP containing a transgene (pAdV-shuttle/loxP/T).
3. *E. coli* DH5α strain: kept in aliquots at −80 °C.
4. Lysogeny broth or Luria-Bertani (LB) broth and LB agar.
5. BIO-RAD Gene Pulser II Electroporation System.

6. 0.2 cm electroporation cuvette.
7. Miniprep plasmid kit (Mini Plus Plasmid DNA 250 Prp. #GF2002).
8. Ampicillin 50 mg/mL.
9. GENECLEAN® III Kit (MP Biomedicals).
10. Maxi Fast Ion Plasmid Kit (IBI Scientific).
11. 1% TAE agarose.
12. $CaCl_2$ 2.5 M.
13. Hepes buffer saline (HBS): Hepes 5 g/L, NaCl 8 g/L, KCl 0.37 g/L, $Na_2HPO_4.2H_2O$ 0.125 g/L, glucose 1 g/L; final pH 7.1.
14. Carrier DNA: salmon sperm DNA (SSDNA) 1 mg/mL.
15. Minimum essential medium (MEM).
16. Fetal bovine serum (FBS).
17. 293Cre4 cell line or any other appropriate cell line.

2.3 Purification of AdV Vectors

1. Minimum Essential Medium (MEM) Eagle (Corning).
2. Fetal bovine serum (FBS).
3. Gentamicin 50 mg/mL.
4. AdV vector permissive cell line, e.g., HEK 293 for human AdV vectors.
5. 10 × Phosphate buffer saline (PBS) pH 7.4: NaCl 80 g, KCl 2 g, Anhydrous Na_2HPO_4 14.4 g, KH_2PO_4 2.4 g, MilliQ H_2O 900 mL.
6. PBS^{++}: PBS containing 0.1% $MgCl_2$ & 0.1% $CaCl_2$.
7. Sterile cell lifter (scraper).
8. 500 mL centrifuge bottles.
9. 0.1 M Tris, pH 8.0.
10. Sterile 50 mL centrifuge tubes.
11. 5% Sodium deoxycholate solution.
12. Tissue homogenizer (THb Handheld Tissue Homogenizer, Omni International, Inc. #10046–846).
13. Saturated cesium chloride (CsCl) solution in 0.01 M Tris and 0.001 M EDTA.
14. Beckman Swinging rotors, SW 40 Ti and SW 55 Ti.
15. Beckman Coulter polyallomer tubes size 13 mL (#331374) and size 5 mL (#326819).
16. Plastic syringes (3 mL) and 18 G needles.
17. Dialysis tube (Thermo Scientific™ SnakeSkin™ 3.5 K MWCO Dialysis Tubing #PI68035).
18. Sterile glycerol.

3 Methods

3.1 Generation of AdV Vector by Homologous Recombination in Bacteria

For the construction of first-generation AdV vectors, the *genomic plasmid*, pAdV-ΔE1E3, comprises the entire AdV genome flanked with a unique restriction enzyme (e.g., PacI) and also contains appropriate E1 and E3 deletions (Fig. 1). The *shuttle plasmid*, pAdV-shuttle-T, contains both ends of the AdV genome. The E1 deletion replaced with an eukaryotic promoter (e.g., human cytomegalovirus (HCMV) promoter), multiple cloning sites (MCS) for foreign gene insertion, and a polyadenylation (poly A) signal (e.g., bovine growth hormone (BGH) polyA) followed by the AdV sequences for the protein IX gene for homologous recombination (Fig. 1). The shuttle plasmid also contains a unique restriction

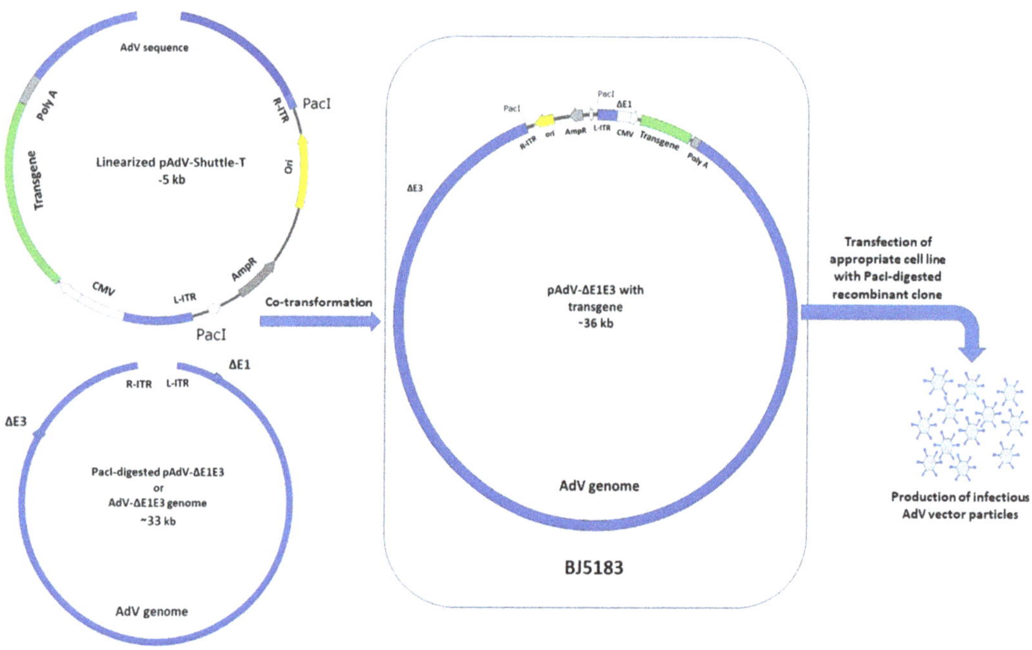

Fig. 1 Generation of AdV vector by homologous recombination in bacteria. AdV shuttle plasmid containing the desired transgene (pAdV-shuttle-T) is linearized with a unique restriction enzyme between the right and left ends of AdV genome. AdV genomic plasmid with a deletion in E1 and E3 regions (pAdV-ΔE1E3) is digested with a unique restriction enzyme to release AdV-ΔE1E3 sequences from the plasmid backbone. Alternatively, the genomic DNA extracted from the purified preparation of AdV-ΔE1E3 (empty vector) can be used. Co-transformation of *E. coli* BJ5183 with the linearized pAdV-shuttle-T and AdV-ΔE1E3 will result in the generation of recombinant clones (pAdV-ΔE1E3/T) containing the entire sequences of AdV-ΔE1E3 and the transgene cassette by homologous recombination. Transfection of an appropriate cell line expressing AdV E1 with pAdV-ΔE1E3/T digested with an appropriate restriction enzyme (i.e., PacI) will generate an infectious AdV vector expressing the desired transgene. AdV, adenovirus; R-ITR, right inverted terminal repeat of AdV genome; L-ITR, left inverted terminal repeat of AdV genome; CMV, cytomegalovirus promoter; PolyA, polyadenylation signal; AmpR, ampicillin resistance gene; Ori, bacterial origin of replication; ΔE1, deletion of AdV E1 region; ΔE3, deletion of AdV E3 region; BJ5183, *E. coli* strain for homologous recombination

enzyme site (e.g., PacI) at both ends of AdV sequences, and another unique restriction enzyme site (e.g., HindIII) in between the left and right AdV genome sequences. The genomic DNA is released from pAdV-ΔE1E3 by PacI digestion followed by gel purification. Alternatively, genomic DNA extracted from a purified preparation of AdV-ΔE1E3 (empty vector) can be used. The HindIII-linearized shuttle plasmid containing a transgene (pAdV-shuttle-T) and the AdV genomic DNA are used for co-transformation of *Escherichia coli* (*E. coli*) BJ5183. This bacterial strain is recBC sbc positive and thus is capable of homologous recombination (*see* **Note 1**). The desired recombination will produce pAdV-ΔE1E3/T plasmid carrying the AdV genome with a foreign insert in the E1 region (Fig. 1). The transfection of an appropriate cell line (HEK 293) with PacI-digested pAdV-ΔE1E3/T will result in the generation of infectious AdV vector carrying the desired transgene.

3.1.1 Preparation of Plasmid DNA for Bacterial Transformation

1. Digest 5 μg of genomic plasmid pAdV-ΔE1E3 with PacI to release AdV-ΔE1E3/T sequences from the plasmid backbone.
2. Digest 5 μg of the pAdV-shuttle-T plasmid containing the transgene with HindIII to linearize and then dephosphorylate it using Antarctic phosphatase (*see* **Note 2**).
3. Purify AdV-ΔE1E3 and the linearized pAdV-shuttle-T by running into 1% TAE agarose gel, collecting the appropriate bands and then recovering the DNA using the GENECLEAN III Kit following the manufacturer's instructions (*see* **Note 3**).

3.1.2 Transformation of E. coli BJ5183 for Homologous Recombination and Selection of Positive Clones

1. Mix the gel purified AdV-ΔE1E3 DNA (insert) to the linearized dephosphorylated pAdV-shuttle-T DNA (vector) at a 2:1 molar ratio (insert:vector). For example, 0.1 μg linearized and dephosphorylated pAdV-shuttle-T DNA should be combined with 1 μg of Ad-ΔE1E3 DNA.
2. Co-transform the *E. coli* BJ5183 with the insert, and vector DNA mixture using the blow described electro-transformation protocol.
3. Thaw a BJ5183 bacterial stock at room temperature and place the tube on ice.
4. In a cold 1.5 mL polypropylene tube, mix 50 μL of BJ5183 bacteria with 1–4 μL of the insert and vector DNA mixture.
5. Mix well and let sit on ice ~1 min.
6. Set the Gene Pulser apparatus at 25 μLF, 2.25 kV, and pulse controller to 200 Ω.
7. Transfer the mixture of bacteria and DNA to a cold, 0.2 cm electroporation cuvette, and shake the suspension to the bottom of the cuvette.

8. Place the cuvette in a shielded safety chamber, slide it into the chamber until the cuvette is seated between the electrode contacts at the base of the chamber.
9. Pulse once at the above settings. This setting will produce a pulse with a time constant of 4.5 to 5 ms. The field strength will be 12.5 kV/cm.
10. Remove the cuvette from the chamber and immediately add 1 mL of cold LB medium to the cuvette and resuspend the transformed BJ5183 bacteria with a Pasteur pipette. The rapid addition of LB medium after the pulse is essential for maximizing the recovery of transformants.
11. Transfer the transformed bacteria from the cuvette to a 15 mL tube and incubate for 15 min at 37 °C in a shaking incubator.
12. Centrifuge the LB medium containing bacteria at 1000 × *g* for 5 min, discard 750 μL of the supernatant, and then resuspend the bacterial pellet in the remaining 250 μL of LB medium.
13. Spread the transformed bacteria onto LB agar plates containing ampicillin or carbenicillin as a selection antibiotic and incubate the plates at 37 °C for 18 h.
14. Using standard culturing techniques, pick small colonies and grow them in LB broth containing the appropriate antibiotic at 37 °C overnight (*see* **Note 4**).
15. Extract DNA from miniprep cultures using the plasmid isolation kit according to the manufacturer's instructions. Digest the miniprep DNA samples with appropriate restriction enzyme/s to confirm the positive recombinant clones.
16. Since *E. coli* BJ5183 is not a high copy number strain, one or more positive recombinant DNA clones should be used to electro-transform *E. coli* Dh5α (as described above) to obtain a high yield of recombinant plasmid DNA.
17. Following electro-transformation, incubate the transformed bacteria at 37 °C for 1 h for the plasmid containing the ampicillin-resistant gene or 2 h for the plasmid containing the kanamycin-resistant gene. The tubes should be shaking at 225 rpm during this incubation to improve the recovery of transformants.
18. Spread suitable volume (50 μL) of the LB containing transformed Dh5α on LB agar containing the selective antibiotic (ampicillin). Incubate overnight at 37 °C for growing transformed colonies.
19. Pick up 5 colonies and incubate them separately in 2.5 mL of LB medium containing ampicillin overnight in a shaking incubator.

20. Extract the plasmid DNA using a miniprep kit. Digest the extracted DNA with appropriate restriction enzyme/s and run on the agarose gel to confirm the constructs.
21. Use the transformed bacteria representing the right constructs and grow them for large-scale plasmid DNA preparations in 200 mL of LB medium with an antibiotic in a 1 L flasks. Grow the clones in a 37 °C shaking incubator for overnight (16 h).
22. Extract the plasmid DNA using a maxiprep kit. Determine the plasmid DNA concentration spectrophotometrically at 260 nm.

3.1.3 Generation of Infectious AdV Vector Containing the Desired Transgene by Transfection of an Appropriate Cell Line with the Recombinant DNA Clones

1. Digest 50–100 μg of recombinant DNA clones with PacI to release the AdV genome containing the transgene cassette.
2. Clean the digested DNA by phenol/chloroform/isoamyl alcohol (25:24:1) solution, precipitate the DNA with absolute alcohol, and resuspend it in sterile water (*see* **Note 5**).
3. Grow the appropriate cells (HEK 293) in 4.5 mL of complete growth medium [minimum essential medium (MEM) supplemented with 10% fetal bovine serum and gentamycin (50 μg/mL)] overnight in four 60 mm tissue culture plates. Make sure the cells will not attain more than 70% confluency at the time of transfection.
4. Change the media to Opti-MEM (4.5 mL/plate) 2 h before transfection and keep the plates in a CO_2 incubator at 37 °C.
5. In four 1.5 mL sterile tubes, add 5 μg of PacI-digested recombinant plasmid DNA and 150 μL Opti-MEM into each tube.
6. In another four 1.5 mL sterile tubes, dilute 15 μL of lipofectamine 2000 into 150 μL Opti-MEM in each tube.
7. Incubate all tubes at room temperature for 20–30 min.
8. From each tube containing Opti-MEM/DNA, add the diluted DNA drop by drop into one of the tubes containing Opti-MEM/lipofectamine. Repeat this process to all four tubes.
9. Incubate the four tubes containing DNA/lipofectamine mixture at room temperature for 20 min.
10. Increase the volume of DNA/lipofectamine mixture in each tube to 500 μL with Opti-MEM.
11. Add the content of each tube (500 μL) drop by drop onto the cells of one of the 60-mm plates. Repeat this process for all four plates.
12. Incubate the four plates in a CO_2 incubator at 37 °C for 5–6 h, change the medium to MEM containing 5% fetal bovine serum and continue the incubation. Change the medium every 3–4 days.

13. The visible cytopathic effects (CPE) due to the generation of infectious AdV vector will be observed any time after 10 days post-transfection (*see* **Note 6**).
14. Once the CPE spread to approximately 50% of the cells in the culture plate, collect the cells with the medium in a sterile 15 mL centrifuge tube. A cell lifter can be used to remove the cells from the plate.
15. Centrifuge at 2000 × *g* for 10 min at 4 °C to pellet the infected cells, discard the medium, and keep approximately 1 mL of the medium to resuspend the pelleted cells.
16. Freeze and thaw the cell suspension 3 times. Centrifuge at 2000 × *g* for 10 min at 4 °C to pellet the cell debris and collect the supernatant. Use 0.5 mL of the supernatant to infect a 100 mm culture plate containing appropriate cells (e.g., HEK 293) at approximately 75% confluency.
17. Incubate the plate in a CO_2 incubator at 37 °C for approximately 42–48 h. Collect the infected cells when CPE is about 80–90%. Follow the freezing and thawing process again and this time infect 150 mm culture plate containing appropriate cells (e.g., HEK 293) at about 90% confluency. Incubate the plate in a CO_2 incubator at 37 °C for approximately 48 h.
18. Collect the infected cells when CPE is approximately 90–100%, centrifuge at 2000 × *g* for 10 min at 4 °C to pellet the infected cells and discard the supernatant. Resuspend the infected cell pellet in 1 mL of PBS++ containing 10% glycerol. Freeze and thaw 3 times and store this crude virus stock at −80 °C. This virus stock is used to make a large stock of the crude virus, which eventually will be used for growing a large scale of AdV vector for purification (*see* **Note 7**).
19. For downstream processing, please refer to Subheading 3.3.

3.2 Generation of AdV Vector Using Cre/loxP Recombination System

In this method, the genomic plasmid pAdVΔψ,E1,E3/loxP contains the entire AdV genome except for the packaging signal (ψ) sequences, E1, and E3 regions, and a loxP site is added just after the E1 deletion for recombination (Fig. 2). The shuttle plasmid, pAdV-shuttle/loxP, carries both ends of the AdV genome, the ψ sequences, the E1 region deletion is replaced with the HCMV promoter, multiple cloning sites, and the BGH poly A (Fig. 2). Just after the BGH ploy A, a loxP site is added. To obtain human AdV vectors by Cre/loxP recombination, HEK 293 cell line carrying the Cre recombinase gene (293Cre4) can be transfected with the genomic plasmid and the shuttle plasmid containing a transgene cassette [75, 76].

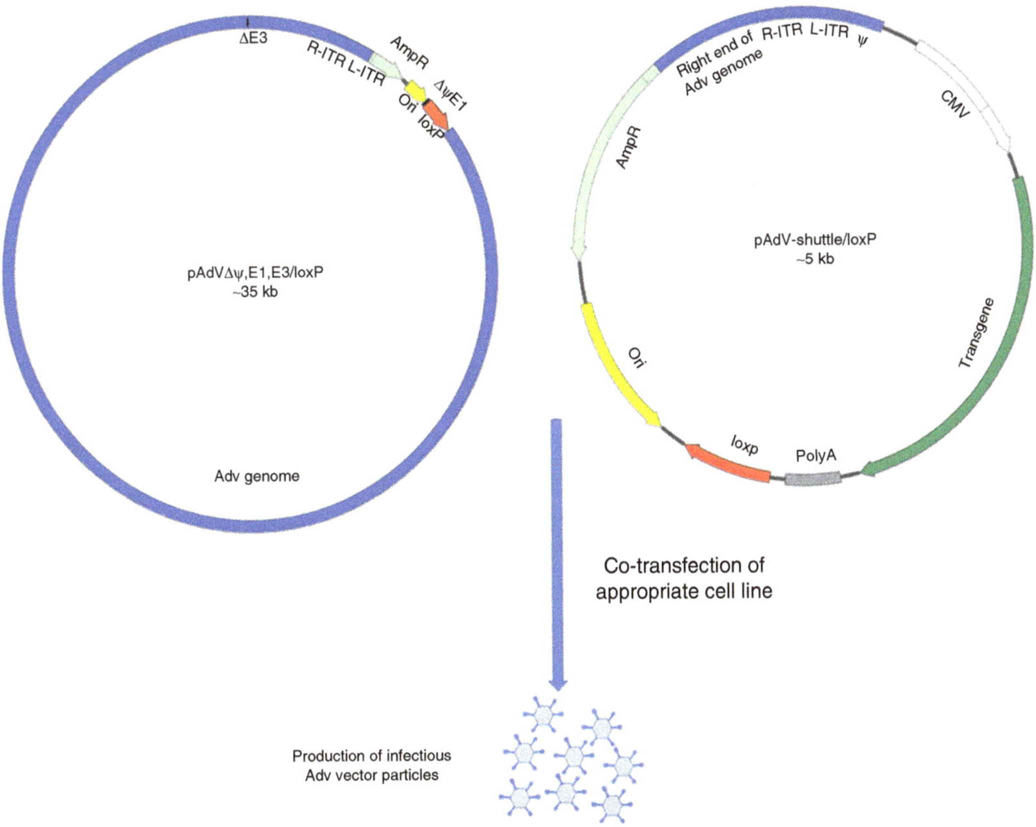

Fig. 2 Generation of AdV vector by Cre/loxP recombination system. Co-transfection of an appropriate cell line expressing Cre recombinase and AdV E1 (e.g., 293Cre4) with pAdVΔψ,E1,E3/loxP (genomic plasmid of AdV Cre/loxP recombination system) and pAdV-shuttle/loxP/T (AdV shuttle plasmid for Cre/loxP recombinase system containing the desired transgene) will generate an infectious AdV vector expressing the desired transgene. AdV, adenovirus; R-ITR, right inverted terminal repeat of AdV genome; L-ITR, left inverted terminal repeat of AdV genome; CMV, cytomegalovirus promoter; PolyA, polyadenylation signal; AmpR, ampicillin resistance gene; Ori, bacterial origin of replication; ΔE1, deletion of AdV E1 region; ΔE3, deletion of AdV E3 region; Δψ, deletion of AdV packaging sequences; loxP, 34 bp long (5'ATAACTTCGTATAATGTATGCTATAC-GAAGTTAT3') site-specific recombination sequences derived from the P1 bacteriophage

3.2.1 Co-transfection of 293Cre4 to Recover Infectious AdV Vector

1. Purify the genomic (pAdVΔψ,E1,E3/loxP) and shuttle plasmid containing the transgene (pAdV-shuttle/loxP/T) by maxiprep using a kit.
2. Grow 293Cre4 (or appropriate cells) overnight in 60 mm tissue culture plates making sure that the cells will attain confluency of approximately 70–90% at the time of transfection.
3. Change the medium with 4.5 mL of MEM containing 10% FBS 2 h before transfection.

4. 2 mL of HBS containing 20 μg SSDNA (10 μg/mL HBS) is sheared by vortex for 1 min, and 0.5 mL of this solution is aliquoted into four 1.5 mL sterile tubes.
5. Add 10 μg of pAdVΔψ,E1,E3/loxP and 5 μg of pAdV-shuttle/loxP/T to each 1.5 mL tube from **step 4**. Incubate HBS and plasmid DNA mixture for 30–45 min at room temperature (*see* **Note 8**).
6. Add 25 μL of 2.5 M $CaCl_2$ into each 1.5 mL tube and incubate them for 20–30 min at room temperature (*see* **Note 9**).
7. Add the content of each 1.5 mL tube drop by drop onto 293Cre4 cells in one of the 60-mm plates and repeat the same procedures in the other 3 plates.
8. Incubate the plates at 37 °C in a CO_2 incubator for 4–5 h, and then change the medium with 5 mL of MEM containing 5% FBS.
9. The media should be changed every 3–4 days, and the CPE depicting the generation of infectious AdV vector should appear any time after 7 days post-transfection.
10. After the spread of the CPE to at least 50% of the plate, collect the cells and the medium and prepare the crude virus stock as mentioned above (Subheading 3.1.3).
11. For downstream processing, please refer to Subheading 3.3.

3.3 Purification of AdV Vectors

Purified stocks of AdV vectors are required for all preclinical and clinical studies to evaluate the efficacy of the desired AdV vectors. Cesium chloride (CsCl) density-gradient centrifugation technique is the most common and reliable method for AdV vector purification for preclinical studies in animal models.

3.3.1 Purification of a Large Stock of AdV Vector

1. Split HEK 293 cells into thirty 150 mm tissue culture plates in MEM containing 10% FBS and incubate them overnight at 37 °C in a CO_2 incubator (*see* **Note 10**).
2. When the cells are approximately 90% confluent (usually after 24 h incubation), infect the cells at a multiplicity of infection (MOI) of 5 plaque forming units (PFU) of the desired AdV vector after diluting the crude vector stock in PBS^{++} (*see* **Note 11**).
3. For infection, remove the medium from each plate and add 3 mL of PBS^{++} containing the appropriate amount of vector.
4. Incubate the infected plates at 37 °C in a CO_2 incubator for 30 min and then add the 25 mL of maintenance medium (MEM + 2% FBS) per plate.
5. Incubate the infected plates at 37 °C in a CO_2 incubator until complete CPE is observed. The CPE is characterized by cell rounding and detachment and is usually achieved within 48 h post-infection.

6. Harvest the infected cells and the culture medium using a cell lifter and collect into a sterile 500 mL centrifuge bottle. Centrifuge the contents at 2000 × *g* for 10 min at 4 °C to pellet the infected cells. Discard the supernatant in another container and autoclave the discarded supernatant.
7. Resuspend the pellet in 15 mL of 0.1 M Tris, pH 8.0 solution and transfer it to a 50 mL tube.
8. Add 1.5 mL of 5% sodium deoxycholate solution, vortex for 30 s (the content will be more viscous due to cell lysis and the release of cellular DNA), and incubate the tube on ice for 30 min and vortex every 10 min.
9. Use a tissue homogenizer to homogenize the mixture three times at the medium speed for 10 s and two times at high speed for 5 s to shear the DNA. This step should be done in a hood, and the sample should be kept on ice during processing.

3.3.2 AdV Vector Purification by CsCl Density Gradient Centrifugation

1. Add 0.58 mL saturated CsC1 solution for each 1 mL of the homogenized vector suspension (Subheading 3.3.1, **step 9**).
2. Mix well and distribute the mixture into two 13 mL Beckman polyallomer tubes and spin in SW40 Ti rotor for 16 h at 4 °C at 35,000 rpm (21,787 × *g*).
3. Collect the white vector bands (Fig. 3a) by piercing the tube wall just below the vector band with a needle attached to a 3 mL syringe. Pool the virus bands into a 5 mL Beckman polyallomer tube (use saturated CsC1 0.58 mL for each 1 mL of 0.1 M tris PH 8 for a balance tube) (*see* **Note 12**).
4. Centrifuge the pooled vector in SW 55 Ti rotor at 35,000 rpm at 4 °C for 16 h. Collect the white vector band from the 5 mL Beckman polyallomer tube (Fig. 3b) and keep it at 4 °C till dialysis (do not keep the vector with CsCl for longer than 1 day at 4 °C).
5. Prepare the dialysis tube by boiling in 500 mL of 1 mM EDTA, pH 8.0, for 10 min. After cooling of the dialysis tube, clip the tube from one end and transfer the purified vector in CsCl solution to the tube and clip the other end.
6. Dialyze the vector in 1 L PBS for 2 h twice and in 1 L PBS^{++} for 2 h once at 4 °C. After dialysis, collect the purified AdV vector from the dialysis tube into a sterile tube in a biosafety cabinet.
7. Add 1/10 volume of sterile glycerol in the purified AdV vector preparation, mix, make aliquots (0.5–1 mL each), label the vials and store at −80 °C.
8. Titration of the AdV vector can be done by the plaque assay in a suitable cell line. The vector particle count can be done by measuring OD at 260 nm using a spectrophotometer [77–79].

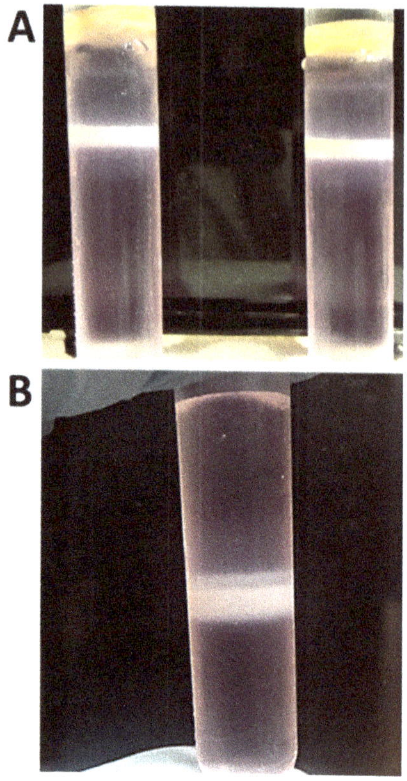

Fig. 3 Purification of AdV vector by CsCl density-gradient ultracentrifugation. (**a**) After the first round of CsCl density-gradient ultracentrifugation, the whitish color vector band is collected from each 13 mL tubes and pooled. (**b**) The pooled vector preparation is subjected to a second round of CsCl density-gradient ultracentrifugation in a 5 mL tube, and the whitish color vector band is collected for further processing

4 Notes

1. AdV vector can be modified at any part of its genome by bacterial homologous recombination.
2. Dephosphorylation of the vector plasmid for recombination in bacteria is essential to decrease the shuttle vector colonies during selection.
3. If DNA fragment/s will be used for cloning or transfection, electrophoresis of DNA in an agarose gel containing ethidium bromide should be done in the dark to avoid the DNA damage by the light.
4. Colonies with the right recombinant plasmid are usually of small size following bacterial recombination in *E. coli* BJ5183.

5. All DNA samples for transfection should be dissolved in sterile water under sterile conditions.
6. The adenovirus CPE will be visible under a light microscope with rounded darker cells, which later detached from the plate and float in the medium.
7. A good AdV crude stock should have approximately 10^{10} infectious virus particles/mL and 5–10 μL of this stock will be enough to infect one 150 mm plate with 90% cell confluency to get complete CPE within 48 h.
8. The Cre/loxP transfection method differs from lab to lab in the ratio of the big plasmid (genomic) and the small plasmid (shuttle). We get the best results with 10 μg of the genomic plasmid +5 μg of the shuttle plasmid.
9. After adding 2.5 M $CaCl_2$ into the HBS/DNA mixture, fine precipitates are expected, making the solution somewhat whitish. If coarse white precipitates are formed, it may adversely impact the transfection efficiency.
10. The split ratio of HEK293 cells is 1:3 and will take 3–4 days to be confluent. Ten 150 mm plates with confluent HEK293 cells will be needed to seed thirty 150 mm plates, which will become 90% confluency in 3–4 days.
11. For infecting thirty 150 mm culture plates with the crude Adv stock, use 300–600 μL of the crude virus stock in 90 mL PBS+ +, mix and use 3 mL for each plate.
12. After ultracentrifugation of the AdV vector homogenate in CsCl density-gradient, the vector band appears as a white band in the ultracentrifuge tube. To collect the vector band, a 3 mL syringe and 18 G needle are used to puncture the tube at an oblique angle 5 mm below the band. The band will be sucked slowly into the syringe.

Acknowledgments

This work was supported by the Public Health Service grant—AI059374 from the National Institute of Allergy and Infectious Diseases, and the Hatch and Animal Health funds.

References

1. Rowe WP, Huebner RJ, Gilmore LK et al (1953) Isolation of a cytopathogenic agent from human adenoids undergoing spontaneous degeneration in tissue culture. Proc Soc Exp Biol Med 84(3):570–573
2. Enders JF, Bell JA, Dingle JH et al (1956) Adenoviruses: group name proposed for new respiratory-tract viruses. Science 124 (3212):119–120
3. Davison AJ, Benko M, Harrach B (2003) Genetic content and evolution of adenoviruses. J Gen Virol 84(Pt 11):2895–2908. https://doi.org/10.1099/vir.0.19497-0

4. Ewer KJ, Lambe T, Rollier CS et al (2016) Viral vectors as vaccine platforms: from immunogenicity to impact. Curr Opin Immunol 41:47–54. https://doi.org/10.1016/j.coi.2016.05.014
5. Su C (2011) Adenovirus-based gene therapy for cancer. In: Xu K (ed) Viral gene therapy. InTech, Rijeka, p 06. https://doi.org/10.5772/19757
6. Pesonen S, Kangasniemi L, Hemminki A (2011) Oncolytic adenoviruses for the treatment of human cancer: focus on translational and clinical data. Mol Pharm 8(1):12–28. https://doi.org/10.1021/mp100219n
7. Tatsis N, Ertl HC (2004) Adenoviruses as vaccine vectors. Mol Ther 10(4):616–629. https://doi.org/10.1016/j.ymthe.2004.07.013
8. Vemula SV, Amen O, Katz JM et al (2013) Beta-defensin 2 enhances immunogenicity and protection of an adenovirus-based H5N1 influenza vaccine at an early time. Virus Res 178(2):398–403. https://doi.org/10.1016/j.virusres.2013.09.013
9. Sharma A, Tandon M, Bangari DS et al (2009) Adenoviral vector-based strategies for cancer therapy. Curr Drug ther 4(2):117–138
10. Raper SE, Chirmule N, Lee FS et al (2003) Fatal systemic inflammatory response syndrome in a ornithine transcarbamylase deficient patient following adenoviral gene transfer. Mol Genet Metab 80(1–2):148–158
11. Harvey BG, Worgall S, Ely S et al (1999) Cellular immune responses of healthy individuals to intradermal administration of an E1-E3-adenovirus gene transfer vector. Hum Gene Ther 10(17):2823–2837. https://doi.org/10.1089/10430349950016555
12. Zaiss AK, Machado HB, Herschman HR (2009) The influence of innate and pre-existing immunity on adenovirus therapy. J Cell Biochem 108(4):778–790. https://doi.org/10.1002/jcb.22328
13. Pandey A, Singh N, Vemula SV et al (2012) Impact of preexisting adenovirus vector immunity on immunogenicity and protection conferred with an adenovirus-based H5N1 influenza vaccine. PLoS One 7(3):e33428. https://doi.org/10.1371/journal.pone.0033428
14. Mast TC, Kierstead L, Gupta SB et al (2010) International epidemiology of human pre-existing adenovirus (Ad) type-5, type-6, type-26 and type-36 neutralizing antibodies: correlates of high Ad5 titers and implications for potential HIV vaccine trials. Vaccine 28 (4):950–957. https://doi.org/10.1016/j.vaccine.2009.10.145
15. Pratt WD, Wang D, Nichols DK et al (2010) Protection of nonhuman primates against two species of Ebola virus infection with a single complex adenovirus vector. Clin Vaccine Immunol 17(4):572–581. https://doi.org/10.1128/cvi.00467-09
16. Wold WS, Toth K (2013) Adenovirus vectors for gene therapy, vaccination and cancer gene therapy. Curr Gene Ther 13(6):421–433
17. Seregin SS, Amalfitano A (2010) Improving adenovirus based gene transfer: strategies to accomplish immune evasion. Viruses 2 (9):2013–2036. https://doi.org/10.3390/v2092013
18. Kay MA, Meuse L, Gown AM et al (1997) Transient immunomodulation with anti-CD40 ligand antibody and CTLA4Ig enhances persistence and secondary adenovirus-mediated gene transfer into mouse liver. Proc Natl Acad Sci U S A 94(9):4686–4691
19. Engelhardt JF, Ye X, Doranz B et al (1994) Ablation of E2A in recombinant adenoviruses improves transgene persistence and decreases inflammatory response in mouse liver. Proc Natl Acad Sci U S A 91(13):6196–6200
20. Bangari DS, Mittal SK (2006) Development of nonhuman adenoviruses as vaccine vectors. Vaccine 24(7):849–862. https://doi.org/10.1016/j.vaccine.2005.08.101
21. Mittal SK, Ahi YS, Vemula SV (2016) 19 - xenogenic adenoviral vectors A2 - curiel. In: David T (ed) Adenoviral vectors for gene therapy, 2nd edn. Academic Press, San Diego, pp 495–528. https://doi.org/10.1016/B978-0-12-800276-6.00019-X
22. Singh N, Pandey A, Jayashankar L et al (2008) Bovine adenoviral vector-based H5N1 influenza vaccine overcomes exceptionally high levels of pre-existing immunity against human adenovirus. Mol Ther 16(5):965–971. https://doi.org/10.1038/mt.2008.12
23. Moffatt S, Hays J, HogenEsch H et al (2000) Circumvention of vector-specific neutralizing antibody response by alternating use of human and non-human adenoviruses: implications in gene therapy. Virology 272 (1):159–167. https://doi.org/10.1006/viro.2000.0350
24. Li X, Bangari DS, Sharma A et al (2009) Bovine adenovirus serotype 3 utilizes sialic acid as a cellular receptor for virus entry. Virology 392(2):162–168. https://doi.org/10.1016/j.virol.2009.06.029

25. Bangari DS, Sharma A, Mittal SK (2005) Bovine adenovirus type 3 internalization is independent of primary receptors of human adenovirus type 5 and porcine adenovirus type 3. Biochem Biophys Res Commun 331 (4):1478–1484. https://doi.org/10.1016/j.bbrc.2005.04.058
26. Ehrhardt A, Xu H, Kay MA (2003) Episomal Persistence of Recombinant Adenoviral Vector Genomes during the Cell Cycle In Vivo. J Virol 77(13):7689–7695. https://doi.org/10.1128/jvi.77.13.7689-7695.2003
27. Seiler MP, Cerullo V, Lee B (2007) Immune response to helper dependent adenoviral mediated liver gene therapy: challenges and prospects. Curr Gene Ther 7(5):297–305
28. Gene Therapy Clinical Trials Worldwide (2017). Available from; http://www.abedia.com/wiley/vectors.php). http://www.abedia.com/wiley/vectors.php
29. Smaill F, Jeyanathan M, Smieja M et al (2013) A human type 5 adenovirus-based tuberculosis vaccine induces robust T cell responses in humans despite preexisting anti-adenovirus immunity. Sci Transl Med 5(205):205ra134. https://doi.org/10.1126/scitranslmed.3006843
30. Vemula SV, Mittal SK (2010) Production of adenovirus vectors and their use as a delivery system for influenza vaccines. Expert Opin Biol Ther 10(10):1469–1487. https://doi.org/10.1517/14712598.2010.519332
31. Ahi YS, Bangari DS, Mittal SK (2011) Adenoviral vector immunity: its implications and circumvention strategies. Curr Gene Ther 11 (4):307–320
32. Zhu J, Huang X, Yang Y (2007) Innate immune response to adenoviral vectors is mediated by both Toll-like receptor-dependent and -independent pathways. J Virol 81 (7):3170–3180. https://doi.org/10.1128/jvi.02192-06
33. Sharma A, Tandon M, Ahi YS et al (2010) Evaluation of cross-reactive cell-mediated immune responses among human, bovine and porcine adenoviruses. Gene Ther 17 (5):634–642. https://doi.org/10.1038/gt.2010.1
34. Hassan AO, Amen O, Sayedahmed EE et al (2017) Adenovirus vector-based multi-epitope vaccine provides partial protection against H5, H7, and H9 avian influenza viruses. PLoS One 12(10):e0186244. https://doi.org/10.1371/journal.pone.0186244
35. Kim EH, Han GY, Nguyen H (2017) An adenovirus-vectored influenza vaccine induces durable cross-protective hemagglutinin stalk antibody responses in mice. Viruses 9(8): E234. https://doi.org/10.3390/v9080234
36. Kamlangdee A, Kingstad-Bakke B, Anderson TK et al (2014) Broad protection against avian influenza virus by using a modified vaccinia Ankara virus expressing a mosaic hemagglutinin gene. J Virol 88(22):13300–13309. https://doi.org/10.1128/jvi.01532-14
37. Hoelscher MA, Singh N, Garg S et al (2008) A broadly protective vaccine against globally dispersed clade 1 and clade 2 H5N1 influenza viruses. J Infect Dis 197(8):1185–1188. https://doi.org/10.1086/529522
38. Cao W, Liepkalns JS, Hassan AO et al (2016) A highly immunogenic vaccine against A/H7N9 influenza virus. Vaccine 34(6):744–749. https://doi.org/10.1016/j.vaccine.2015.12.062
39. de Vries RD, Rimmelzwaan GF (2016) Viral vector-based influenza vaccines. Hum Vaccin Immunother 12(11):2881–2901. https://doi.org/10.1080/21645515.2016.1210729
40. Hoelscher MA, Garg S, Bangari DS et al (2006) Development of adenoviral-vector-based pandemic influenza vaccine against antigenically distinct human H5N1 strains in mice. Lancet 367(9509):475–481. https://doi.org/10.1016/s0140-6736(06)68076-8
41. Van Kampen KR, Shi Z, Gao P et al (2005) Safety and immunogenicity of adenovirus-vectored nasal and epicutaneous influenza vaccines in humans. Vaccine 23(8):1029–1036. https://doi.org/10.1016/j.vaccine.2004.07.043
42. Fuchs JD, Bart PA, Frahm N et al (2015) Safety and immunogenicity of a recombinant adenovirus serotype 35-vectored HIV-1 vaccine in adenovirus serotype 5 seronegative and seropositive individuals. J AIDS Clin Res 6(5):461. https://doi.org/10.4172/2155-6113.1000461
43. Milligan ID, Gibani MM, Sewell R et al (2016) Safety and immunogenicity of novel adenovirus type 26- and modified vaccinia ankara-vectored ebola vaccines: a randomized clinical trial. JAMA 315(15):1610–1623. https://doi.org/10.1001/jama.2016.4218
44. O'Hara GA, Duncan CJ, Ewer KJ et al (2012) Clinical assessment of a recombinant simian adenovirus ChAd63: a potent new vaccine vector. J Infect Dis 205(5):772–781. https://doi.org/10.1093/infdis/jir850
45. Osman M, Mistry A, Keding A et al (2017) A third generation vaccine for human visceral leishmaniasis and post kala azar dermal leishmaniasis: first-in-human trial of ChAd63-KH.

PLoS Negl Trop Dis 11(5):e0005527. https://doi.org/10.1371/journal.pntd.0005527
46. Ledgerwood JE, DeZure AD, Stanley DA et al (2017) Chimpanzee adenovirus vector ebola vaccine. N Engl J Med 376(10):928–938. https://doi.org/10.1056/NEJMoa1410863
47. Tapia MD, Sow SO, Lyke KE et al (2016) Use of ChAd3-EBO-Z Ebola virus vaccine in Malian and US adults, and boosting of Malian adults with MVA-BN-Filo: a phase 1, single-blind, randomised trial, a phase 1b, open-label and double-blind, dose-escalation trial, and a nested, randomised, double-blind, placebo-controlled trial. Lancet Infect Dis 16(1):31–42. https://doi.org/10.1016/s1473-3099(15)00362-x
48. Crosby CM, Matchett WE, Anguiano-Zarate SS et al (2017) Replicating single-cycle adenovirus vectors generate amplified influenza vaccine responses. J Virol 91(2):e00720. https://doi.org/10.1128/jvi.00720-16
49. Li Y, Pong RC, Bergelson JM et al (1999) Loss of adenoviral receptor expression in human bladder cancer cells: a potential impact on the efficacy of gene therapy. Cancer Res 59(2):325–330
50. Shen YH, Yang F, Wang H et al (2016) Arg-Gly-Asp (RGD)-modified E1A/E1B double mutant adenovirus enhances antitumor activity in prostate cancer cells in vitro and in mice. PLoS One 11(1):e0147173. https://doi.org/10.1371/journal.pone.0147173
51. Wang H, Li ZY, Liu Y et al (2011) Desmoglein 2 is a receptor for adenovirus serotypes 3, 7, 11 and 14. Nat Med 17(1):96–104. https://doi.org/10.1038/nm.2270
52. Biedermann K, Vogelsang H, Becker I et al (2005) Desmoglein 2 is expressed abnormally rather than mutated in familial and sporadic gastric cancer. J Pathol 207(2):199–206. https://doi.org/10.1002/path.1821
53. Harada H, Iwatsuki K, Ohtsuka M et al (1996) Abnormal desmoglein expression by squamous cell carcinoma cells. Acta Derm Venereol 76(6):417–420
54. Schmitt CJ, Franke WW, Goerdt S et al (2007) Homo- and heterotypic cell contacts in malignant melanoma cells and desmoglein 2 as a novel solitary surface glycoprotein. J Invest Dermatol 127(9):2191–2206. https://doi.org/10.1038/sj.jid.5700849
55. Lee SY, Park HR, Rhee J et al (2013) Therapeutic effect of oncolytic adenovirus expressing relaxin in radioresistant oral squamous cell carcinoma. Oncol Res 20(9):419–425. https://doi.org/10.3727/096504013x13657689383139
56. Vera B, Martinez-Velez N, Xipell E et al (2016) Characterization of the Antiglioma Effect of the Oncolytic Adenovirus VCN-01. PLoS One 11(1):e0147211. https://doi.org/10.1371/journal.pone.0147211
57. Ranki T, Pesonen S, Hemminki A et al (2016) Phase I study with ONCOS-102 for the treatment of solid tumors - an evaluation of clinical response and exploratory analyses of immune markers. J Immunother Cancer 4:17. https://doi.org/10.1186/s40425-016-0121-5
58. Danthinne X, Imperiale MJ (2000) Production of first generation adenovirus vectors: a review. Gene Ther 7(20):1707–1714. https://doi.org/10.1038/sj.gt.3301301
59. Graham FL, Smiley J, Russell WC et al (1977) Characteristics of a human cell line transformed by DNA from human adenovirus type 5. J Gen Virol 36(1):59–74. https://doi.org/10.1099/0022-1317-36-1-59
60. Kamen A, Henry O (2004) Development and optimization of an adenovirus production process. J Gene Med 6(Suppl 1):S184–S192. https://doi.org/10.1002/jgm.503
61. Fallaux FJ, Bout A, van der Velde I et al (1998) New helper cells and matched early region 1-deleted adenovirus vectors prevent generation of replication-competent adenoviruses. Hum Gene Ther 9(13):1909–1917. https://doi.org/10.1089/hum.1998.9.13-1909
62. Howe JA, Pelka P, Antelman D et al (2006) Matching complementing functions of transformed cells with stable expression of selected viral genes for production of E1-deleted adenovirus vectors. Virology 345(1):220–230. https://doi.org/10.1016/j.virol.2005.09.029
63. Wen S, Schneider DB, Driscoll RM et al (2000) Second-generation adenoviral vectors do not prevent rapid loss of transgene expression and vector DNA from the arterial wall. Arterioscler Thromb Vasc Biol 20(6):1452–1458
64. Mitani K, Graham FL, Caskey CT et al (1995) Rescue, propagation, and partial purification of a helper virus-dependent adenovirus vector. Proc Natl Acad Sci U S A 92(9):3854–3858
65. Parks R, Evelegh C, Graham F (1999) Use of helper-dependent adenoviral vectors of alternative serotypes permits repeat vector administration. Gene Ther 6(9):1565–1573. https://doi.org/10.1038/sj.gt.3300995
66. Alba R, Bosch A, Chillon M (2005) Gutless adenovirus: last-generation adenovirus for gene therapy. Gene Ther 12(Suppl 1):S18–S27. https://doi.org/10.1038/sj.gt.3302612

67. Parks R, Chen L, Anton M et al (1996) A helper-dependent adenovirus vector system: Removal of helper virus by Cre-mediated excision of the viral packaging signal. Proc Natl Acad Sci U S A 93(24):13565–13570
68. Hardy S, Kitamura M, Harris-Stansil T et al (1997) Construction of adenovirus vectors through Cre-lox recombination. J Virol 71 (3):1842–1849
69. Vetrini F, Ng P (2010) Gene therapy with helper-dependent adenoviral vectors: current advances and future perspectives. Viruses 2 (9):1886–1917. https://doi.org/10.3390/v2091886
70. Stow ND (1981) Cloning of a DNA fragment from the left-hand terminus of the adenovirus type 2 genome and its use in site-directed mutagenesis. J Virol 37(1):171–180
71. He TC, Zhou S, da Costa LT et al (1998) A simplified system for generating recombinant adenoviruses. Proc Natl Acad Sci U S A 95 (5):2509–2514
72. Chartier C, Degryse E, Gantzer M et al (1996) Efficient generation of recombinant adenovirus vectors by homologous recombination in Escherichia coli. J Virol 70(7):4805–4810
73. Wu C, Lei X, Wang J et al (2011) Generation of a replication-deficient recombinant human adenovirus type 35 vector using bacteria-mediated homologous recombination. J Virol Methods 177(1):55–63. https://doi.org/10.1016/j.jviromet.2011.06.016
74. Luo J, Deng ZL, Luo X et al (2007) A protocol for rapid generation of recombinant adenoviruses using the AdEasy system. Nat Protoc 2 (5):1236–1247. https://doi.org/10.1038/nprot.2007.135
75. Ng P, Parks RJ, Cummings DT et al (1999) A high-efficiency Cre/loxP-based system for construction of adenoviral vectors. Hum Gene Ther 10(16):2667–2672. https://doi.org/10.1089/10430349950016708
76. Chen L, Anton M, Graham FL (1996) Production and characterization of human 293 cell lines expressing the site-specific recombinase Cre. Somat Cell Mol Genet 22(6):477–488
77. Mittereder N, March KL, Trapnell BC (1996) Evaluation of the concentration and bioactivity of adenovirus vectors for gene therapy. J Virol 70(11):7498–7509
78. Maizel JV Jr, White DO, Scharff MD (1968) The polypeptides of adenovirus. I. Evidence for multiple protein components in the virion and a comparison of types 2, 7A, and 12. Virology 36(1):115–125
79. Rux JJ, Burnett RM (2007) Large-scale purification and crystallization of adenovirus hexon. In: Wold WSM, Tollefson AE (eds) Adenovirus methods and protocols, Ad proteins, RNA lifecycle, host interactions, and phylogenetics, vol 2. Humana Press, Totowa, NJ, pp 231–250. https://doi.org/10.1007/978-1-59745-277-9_17

Chapter 10

Construction of Oncolytic Herpes Simplex Virus with Therapeutic Genes of Interest

Andranik Kahramanian, Toshihiko Kuroda, and Hiroaki Wakimoto

Abstract

Herpes simplex virus (HSV) is one of the most extensively studied oncolytic virus platforms. The recent FDA approval of talimogene laherparepvec (T-VEC) has been accelerating translational research of oncolytic HSV (oHSV) as a promising therapeutic for refractory cancers such as glioblastoma, the deadliest primary malignancy in the brain. The large genome size of HSV readily allows arming of oHSV by incorporating therapeutic transgenes within the virus, as exemplified by T-VEC carrying GM-CSF, thereby enhancing the anticancer activity of oHSV. Here we describe a bacterial artificial chromosome-based method for construction of an oHSV expressing a transgene, which we routinely use in the laboratory to create a number of different recombinant oHSV bearing either therapeutic or reporter genes.

Key words Herpes simplex virus (HSV), Glioblastoma, Bacterial artificial chromosome, Cre recombination, Flp recombination

1 Introduction

Oncolytic virus therapy is increasingly being utilized as a treatment modality against cancer, fueled by the recent FDA approval of T-VEC for metastatic melanoma [1]. The use of oncolytic virus as a gene delivery vehicle and subsequent expression of a therapeutic gene in situ at the tumor site has been shown to enhance the anticancer potency of virus therapy [2, 3]. For this reason, the ability to genetically engineer a virus armed with a specific therapeutic gene is a crucial skill that may aid cancer researchers. In this chapter, we will focus on genetically engineered oncolytic herpes simplex virus (oHSV), one of the most extensively studied oncolytic virus platforms. We will first overview strategies of arming oHSV, and then provide detailed methods for constructing an armed oHSV using the flip-flop HSV-bacterial artificial chromosome (HSV-BAC) system that is reliable and time and labor efficient [4, 5].

Fredric P. Manfredsson and Matthew J. Benskey (eds.), *Viral Vectors for Gene Therapy: Methods and Protocols*, Methods in Molecular Biology, vol. 1937, https://doi.org/10.1007/978-1-4939-9065-8_10,

The following introduction will provide a context for the translational relevance of armed oHSV in the treatment of the malignant brain cancer, glioblastoma (GBM). Based on distinct aspects of the pathophysiology of GBM that transgenes in oHSV are designed to target, armed oHSV can be conceptually divided into subcategories, including extracellular matrix modification and anti-angiogenesis, immunogenicity, cell death pathways, and prodrug conversion (in situ chemotherapy).

1.1 Tumor Extracellular Matrix Modification

One of the main issues hindering the effectiveness of oncolytic virus therapy is limited virus distribution throughout tumors, and the extracellular matrix (ECM) of GBM represents one such barrier to efficient virus spread. To address the tumor ECM, Dmitrieva and colleagues exploited a naturally occurring bacterial enzyme, Chondroitinase ABC (Chase ABC), that removes the sugar side chains from proteoglycans found within the ECM [6]. A novel virus known as OV-Chase was created by inserting recombinant Chase-ABC into the HSV genome under an IE4/5 promoter. In an orthotopic xenograft mouse model, OV-Chase demonstrated enhanced spread throughout intracranial gliomas, without causing dispersal of glioma cells, and the median survival of OV-Chase-treated mice was significantly longer than mice receiving control virus. Collagenase and hyaluronidase are among potential transgene candidates for a similar strategy targeting tumor ECM to facilitate therapeutic virus to spread better [7, 8].

1.2 Anti-Angiogenesis

The tumor-associated vasculature not only supports GBM growth but also affects the entry and infiltration of immune cells within the tumor immune microenvironment. G47Δ-mAngio is an oHSV armed with angiostatin, a proteolytic fragment of plasminogen and an inhibitor of angiogenesis [9]. The clinical application of angiostatin is limited by its short half-life, an issue that could be overcome with local delivery of this protein using G47Δ-mAngio. This study demonstrated that intratumoral injection of G47Δ-mAngio into flank U87 GBM xenografts induced a large anti-angiogenic effect, and decreased intratumoral levels of VEGF. Furthermore, a combination of G47Δ-mAngio with an FDA-approved VEGF inhibitor, Bevacizumab, mediated a decrease in tumor vascularity, greater virus spread and enhanced tumor cell lysis, resulting in a decrease in tumor size and increased animal survival. Subsequently, a combination of two armed oHSV, G47Δ-mIL12 and G47Δ-mAngio, was shown to have a pronounced anti-angiogenic effect in orthotopic GBM xenograft models, revealing an additive effect of expressing two distinct anti-angiogenic proteins, IL12 and angiostatin, in the context of oHSV [10].

The gene coding for an anti-angiogenic proteolytic fragment, vasculostatin, was cloned into the rHSVQ backbone, an attenuated version of HSV-1, to create RAMBO [11]. RAMBO demonstrated a strong antitumor effect, but only 20% of tumor-bearing mice displayed a treatment response. To improve these results, the same group cloned vasculostatin into the rQnestin34.5 viral backbone, an HSV-1 that expresses γ34.5 under the control of nestin promoter [12]. This new virus, 34.5ENVE, demonstrated a greater antitumor effect compared to RAMBO, significantly prolonged survival and increased areas of necrosis of intracranial gliomas. Similar approaches introducing anti-angiogenic oHSV have been also reported with similar results [13, 14].

1.3 Immune Modulation

Immunosuppression is now recognized as a crucial component in the tumor immune microenvironment of GBM. Since oHSV infection of tumors triggers immune and inflammatory reactions, its manipulation to augment an antitumor immune response is a rational strategy to boost oHSV efficacy. M002 and G47Δ-mIL12 are armed with IL-12, a cytokine that is known to promote a Th1 pro-inflammatory antitumor immune response by directly acting on T effector cells, Natural Killer cells, and Natural Killer T cells [15, 16]. Saha and colleagues tested G47Δ-IL12 in combination with anti-CTLA-4 and anti-PD-1 immune checkpoint antibodies [17]. This triple combination therapy was associated with an influx of antitumor M1 polarized macrophages and an increased T effector/T regulatory cell ratio in tumor sections. This triple therapy showed potent curative activity in immunocompetent mice bearing 005 GBM stem cell brain tumors, which was mediated by CD4+ T cells, CD8+ T cells, and macrophages. Together, this study demonstrated that targeting the local and systemic immune-suppressive state by a combination of IL12-armed oHSV and immune checkpoint inhibitors can lead to profound efficacy, notwithstanding a poorly immunogenic GBM. Additionally, this study underscores the need for a combination strategy to effectively treat GBM. Other immune modulatory cytokines that have been used to arm oHSV include mIL4 [18, 19] and Flt3L [20].

1.4 Cell Death Pathways

Tumor cells are known to have aberrant cell growth and death pathways. Tamura and colleagues exploited these pathways in GBM by arming an oncolytic HSV with TNF-related Apoptosis Inducing Ligand (TRAIL), called oHSV-TRAIL [21]. This oncolytic virus induced secretion of TRAIL upon infection, leading to increased apoptosis and more potent killing of TRAIL resistant GBM cells compared to control oHSV. oHSV-TRAIL was also able to alter cell signaling pathways, downregulating ERK-MAPK signaling, and upregulating JNK signaling, resulting in inhibition of GBM invasion and significant prolongation of survival times in mice. oHSV-TRAIL was thus able to target both cell proliferation signaling and death pathways in GBM. oHSV-TRAIL was further shown to have activity in orthotopic GBM models based on chemoresistant GBM stem cells [22].

1.5 Prodrug Conversion

Prodrug converting enzymes convert non-active prodrugs to active drugs. Virus-induced tumor expression of prodrug converting enzymes, in conjunction with systemically administered prodrugs, allows local delivery of traditional chemotherapeutics, thereby bypassing liver metabolism reducing systemic toxicity. rRp450 is an oHSV armed with the enzyme CYP2B that bioactivates Cyclophosphamide [23]. The same group further created MGH2, an oHSV carrying the genes encoding CYP2B and shiCE [24]. shiCE is a secreted form of human intestinal carboxyl esterase that activates Irinotecan, a Topoisomerase I inhibitor. Having two prodrug converting enzymes, MGH2 was designed to be used in conjunction with Cyclophosphamide and Irinotecan. This therapeutic strategy displayed enhanced tumor cell death, as well as chemosensitization, without affecting viral replication. Importantly a significant increase in survival was noted using an orthotopic xenograft model in mice.

In line with a prodrug conversion strategy, another oHSV was created to express Cytosine Deaminase (CD), an enzyme that is responsible for the conversion of nontoxic 5-fluorocytosine (5-FC) to the highly toxic, chemotherapeutic agent, 5-fluorouracil (5-FU) [25]. 5-FU disrupts RNA metabolism and blocks DNA synthesis of cancer cells, but lacks clinical application in GBM due to its toxicity with systemic delivery. This virus could solve the problem of toxicity by producing 5-FU only within the tumor microenvironment. The virus, M012, was able to effectively convert 5-FC to 5-FU in mice, resulting in reduced growth rates of subcutaneous neuroblastoma tumors, and did not show neurotoxicity upon intracerebral inoculation in HSV-susceptible mice. Of note, 5-FU is also clinically known to be a potent radiosensitizer, and thus may be synergistic with radiotherapy that is part of the current standard GBM treatment regimen.

1.6 The Flip-Flop BAC System

The method described below is based on the Bacterial Artificial Chromosome (BAC) system that is combined with dual site-specific DNA recombinase systems, Cre/loxP and FLP/FRT [4, 5]. BAC is a large circular plasmid that is able to replicate within bacteria as an artificial genome.

More specifically, the following method will describe the use of a shuttle vector containing an expression cassette of a therapeutic transgene, and a lacZ open reading frame flanked by a loxP site as well as a stuffer sequence flanked by an FRT site. This plasmid is recombined into an HSV-BAC (Fig. 1), which is then selected to grow in bacteria with two antibiotics, chloramphenicol and kanamycin. The recombinant HSV DNA is then co-transfected with Flp recombinase into mammalian Vero cells to remove the BAC genome and stuffer sequences, allowing reduction in DNA size and production of the recombinant, replication-competent oHSV. Vero cells are used as they lack interferon signaling, allowing the production of a high viral titer.

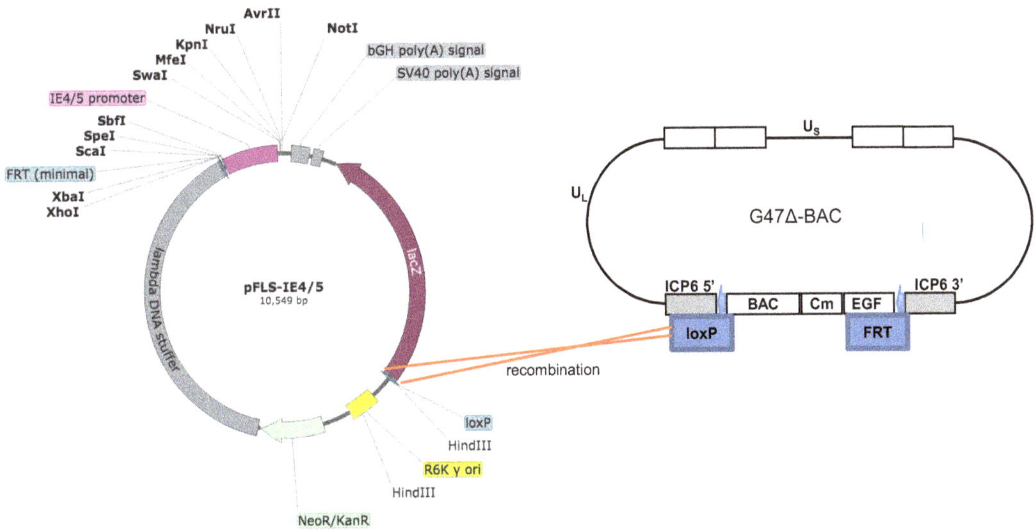

Fig. 1 Schematic showing Cre-loxP recombination between pFLS-IE4/5 plasmid DNA and G47Δ-BAC DNA. BAC, plasmid F-derived sequences. Cm, Chloramphenicol resistance gene

This system is advantageous as the researcher can genetically modify and insert a therapeutic transgene of choice within the ICP6 locus of a widely studied HSV-1 genome. Additionally, the BAC system allows simultaneous construction of multiple oncolytic viruses armed with different transgenes in a relatively short period of time (about 2 months), allowing direct comparison of these newly constructed viruses. We have utilized this method to construct a number of recombinant oHSV that bear therapeutic or reporter transgenes [9, 13, 14, 16, 21, 26].

2 Materials

2.1 DNA

1. Shuttle vector plasmid, pFLS-IE4/5.
2. G47Δ-BAC.
3. pCAGS-FLPe.

2.2 Cells

1. ONE Shot TOP10 chemically competent *E. coli* (Thermo Fisher).
2. ElectroMAX DH10B cells (Thermo Fisher).
3. Vero (Cercopithecus aethiops epithelial kidney cells, ATCC).

2.3 Media and Culture Related

1. LB and AB agar.
2. SOC.
3. DMEM supplemented with 10% calf serum.
4. Phosphate-buffered saline (PBS).

5. Trypsin (0.05%)–EDTA (0.53 mM).
6. OPTI-MEM (Thermo Fisher).

2.4 Reagents

1. TE (Tris10mM, EDTA 1 mM) buffer with DNase-free RNase (50 μg/mL).
2. Cre recombinase (New England Biolab).
3. 10× Cre buffer (New England Biolab).
4. Ethachinmate (Wako).
5. 100% and 70–75% Ethanol.
6. Kanamycin.
7. Chloramphenicol.
8. HindIII Restriction enzyme.
9. Agarose gel (0.6% in Tris-acetate-EDTA) with ethidium bromide (0.5 μg/mL).
10. High molecular weight (above 10 kb) and regular range (0.5–10 kb) DNA molecular weight ladders.
11. Solution 1: 50 mM Glucose, 10 mM EDTA, 25 mM Tris-HCl, pH = 8.0.
12. Solution 2: 0.2 N NaOH, 1% SDS.
13. Solution 3: 5 M Potassium acetate, 2.5 M acetic acid, pH = 4.8.
14. 2-propanol.
15. dimethyl formamide.
16. Lipofectamine LTX with Plus reagent (Thermo Fisher).
17. Dry ice in 100% ethanol.

3 Methods

3.1 Cloning of Gene of Interest into the Shuttle Vector Plasmid, pFLS-IE4/5

The first step is to clone a therapeutic gene (cDNA) of interest into an appropriate shuttle vector plasmid. We have created numerous recombinant oHSV using the pFLS-IE4/5 shuttle vector plasmid (Fig. 1), in which the IE(immediate early)4/5 promotor of HSV drives expression of the therapeutic gene following oHSV infection. The pFLS-IE4/5 shuttle vector has multiple cloning sites that are unique: SwaI, MfeI, KpnI, NruI, AvrII, NotI, for facilitating convenient insertion of a gene of interest that may be derived from PCR or plasmid DNA.

1. Use conventional molecular cloning methodology to clone your gene of interest to the multi-cloning sites of pFLS-IE4/5. We use Top10 competent *Escherichia Coli*. pFLS-IE4/5 has kanamycin resistance gene and LB agar plates

should contain kanamycin at 20 μg/mL. Use restriction digest of miniprep DNA of clones and agarose electrophoresis to identify proper clones.

2. The miniprep DNA, extracted by either a commercial kit or alkaline lysis method, dissolved in TE buffer containing DNase-free RNaseA (50 μg/mL) is ready for use in the next recombination step.

3.2 Cre-mediated DNA Recombination, and Isolation of Proper Recombinants

Construction of virus entails two main steps. The first step involves Cre recombinase-mediated site-specific recombination to integrate the shuttle vector containing your therapeutic gene of interest within the HSV BAC. The BAC contains a loxP site, that is targeted by Cre recombinase, allowing shuttle vector integration in the ICP6 locus of HSV genome to generate recombinant BAC. We routinely use a HSV BAC clone that carries the genomic DNA of oHSV G47Δ to create G47Δ-based recombinant oHSV. To obtain oHSV recombinant that has no transgene (we call this G47Δ-empty), the shuttle plasmid without transgene insert should be utilized.

1. Prepare solution as follows in a 1.5 mL microcentrifuge tube.
   - G47Δ-BAC DNA: 0.5–1 μg.
   - Shuttle vector plasmid containing gene of interest: 50–200 ng.
   - 10× Cre buffer: 1 μL.
   - Cre recombinase: 1 μL

   Add water to achieve a final mixture volume of 10 μL and mix by finger tapping.
2. Incubate for 30 mins at 37 °C for Cre reaction to proceed.
3. Inactivate Cre recombinase for 10 min at 70 °C and cool down to room temperature for 10 min.
4. Ethanol precipitate DNA using Ethachinmate (*see* **Note 1**), and wash with 75% ethanol. After brief air-drying, dissolve precipitated DNA in 5 μL of water.
5. Thaw ElectroMAX DH10B *E. coli* on ice and aliquot 20 μL in a precooled electroporation cuvette with 1 mm gap. Add 1 μL of DNA (from **step 4**) and mix gently by finger tapping.
6. Perform electroporation of the cells using the following setting: Voltage, 1.5 kV; Resistance, 200 Ohms; and Capacitance, 25 uF.
7. Add 500 μL of SOC medium and transfer with *E. coli* to a microcentrifuge tube.
8. Vigorously shake the tube in a shaking incubator for 60 min at 37 °C.
9. Plate 100 μL and 400 μL of *E. coli* to two LB/Kanamycin (20 μg/mL)/Chloramphenicol (15 μg/mL) plates.

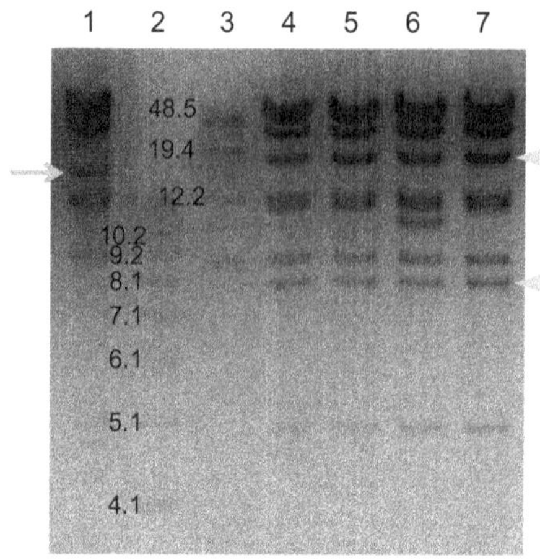

Fig. 2 Example of restriction analysis of recombinant BAC clones after HindIII digestion. Lane 1, HindIII digested G47Δ-BAC DNA; 2, molecular weight ladder; 3, high molecular weight ladder; 4–7, HindIII digested recombinant BAC DNA clones. Arrow, 15 kb band. Arrowheads, 18 and 8 kb bands. Three of four clones (lanes 4, 5, and 7) show the expected digestion pattern, indicative of the occurrence of proper Cre recombination. Numbers denote size of DNA markers in kilo base pairs

10. Incubate the plates overnight in a 37 °C incubator.
11. Pick up 6–10 colonies, inoculate in 10 mL of LB/Kanamycin (20 μg/mL)/Chloramphenicol (15 μg/mL) in centrifuge tubes, and culture overnight in a shaking incubator at 37 °C.
12. For each clone, perform DNA extraction using miniprep (*see* **Note 2**).
13. Digest 10 μL of DNA with HindIII at 37 °C for 2 h in a 25 μL reaction mix. In parallel, digest 0.5–1 μg of G47Δ-BAC DNA with HindIII.
14. Electrophorese these samples in a 0.6% agarose gel with ethidium bromide at 20 V overnight (12–16 h). Include run of 2 DNA molecular weight ladders: high molecular weight ladders (>10 kb) and ladders that cover 0.5–10 kb (Fig. 2).
15. Take picture under UV and analyze the gel. Cre recombination causes loss of the 15 kb fragment that includes the loxP site of G47Δ-BAC DNA. Two new fragments, 8 and 0.5 kb in size, should instead appear upon recombination, although the latter is difficult to observe. Additionally, one or multiple bands should appear, and their numbers and sizes are dependent on whether the insert of the shuttle plasmid contains HindIII sites; e.g., if the insert does not have any HindIII sites, a single band of (6.5+ the entire size of the shuttle plasmid) kb is seen (Fig. 2), and this fragment will be split into two if the insert has one HindIII site (*see* **Note 3**).

3.3 BAC DNA and Flp Plasmid Co-transfection to Generate Replication-Competent Recombinant oHSV

The second step of the process involves excision of the unnecessary BAC sequence by FLPe recombinase in mammalian cells in culture. The BAC and eGFP sequences are flanked by FRT sites that are the target of excision. This results in reduction of the DNA size of oversized BAC recombinants, allowing a successfully recombined oHSV to be competent in replication and progeny production. The replication-competent recombinant oHSV does not express eGFP and does express lacZ reporter, which can be conveniently used for histochemical staining (i.e., X-gal staining) to identify plaques and infected cells.

1. **Day 1:** plate Vero cells at 40,000 cells per well in 500 μL of DMEM supplemented with 10% Calf Serum to 24-well plates. Prepare four wells for each BAC clone.
2. **Day 2:** Vero cells should be 70% confluence.
3. Prepare following reaction mixture for transfection of four wells. Label as Tube 1.
   - HSV-BAC recombinant DNA: 1200–1600 ng.
   - pCAGS-FLPe: 120 ng.
   - Plus reagent: 9.6 μL.
   - OPTI-MEM: 120 μL, total about 140 μL

   Mix by tapping gently. Do not pipet up and down.
4. Prepare following reaction mixture for transfection of four wells. Label as Tube 2.
   - OPTI-MEM: 120 μL.
   - Lipofectamine LTX: 9.6 μL.
5. Mix Tube 1 and Tube 2 gently.
6. Incubate at room temperature for 5 min.
7. Add 60 μL of this solution dropwise per well.
8. **Day 3:** Check GFP expression for assessing efficiency of BAC transfection.
9. **Days 5–7:** Under microscope, look for the formation of plaques with no GFP (*see* **Note 4**).
10. Scrape cells and medium when cytopathic effects are spread throughout the well.
11. Freeze/Thaw three times with dry ice/ethanol and 37 °C water bath to release virus from cells. Spin at a low speed (300–400 × *g*) for 5 min to pellet cell debris, and collect supernatant containing oHSV in a new tube.
12. Titrate virus in the sample (*see* **Note 5**).

3.4 Virus Plaque Isolation and Amplification

Crude virus prep that is obtained above should be considered a mixture of viruses with presumably different genetic backgrounds due to some uncertainty associated with Flp recombination. Therefore, it is critical to isolate a single virus based on a single plaque, which should be followed by amplification and characterization.

1. **Day 1:** Plate 3000 Vero cells per well in a 96 well plate.
2. **Day 2**: Inoculate Vero cells with 0.3 virus particle per well according to the titer calculated above.
3. **Days 3–4**: Observe each well, and identify wells that initially contain a single plaque. Mark these wells for follow-up (*see* **Note 6**).
4. **Days 4–5:** Follow up marked wells, and select those where plaque shows continuous growth.
5. **Days 5–6:** Harvest both cells and supernatant, titer and repeat this process, i.e., secondary plaque isolation, to ensure isolation of purely single recombinant oHSV. Resulting recombinant oHSV clones will be amplified separately in Vero cells and subject to characterization studies including confirmation of transgene expression.

4 Notes

1. Ethachinmate helps visualization of DNA precipitation.
2. We use the classic alkaline lysis method for DNA miniprep. Solution 1: 50 mM Glucose, 10 mM EDTA, 25 mM Tris–HCl, pH = 8.0; Solution 2: 0.2 N NaOH, 1% SDS; Solution 3: 5 M Potassium acetate, 2.5 M acetic acid, pH = 4.8. Bacteria will be pelleted and 250 μL of Solution 1, 2 and 3 is used for each sample to separate a crude plasmid DNA fraction, followed by precipitation of DNA with 600 μL of 2-propanol. DNA will be washed with 75% ethanol and briefly air-dried before dissolved in TE with RNase. It is a good practice to make bacterial stock by freezing a small part of *E. coli* culture with 8% dimethyl formamide.
3. The Cre recombination is a robust and efficient process, as we typically observe 70–80% of clones are proper recombinants exhibiting an expected digestion pattern.
4. Replication-competent oHSV should have lost eGFP.
5. Use oHSV titration method of choice. We routinely use plaques formation in Vero cells in conjunction with human IgG that neutralizes HSV, and staining plaques with X-gal (5-bromo-4-chloro-3-indolyl-D-galactopyranoside).
6. If titration and dilution of virus are correct, many wells should have no plaques growing.

References

1. Ott PA, Hodi FS (2016) Talimogene laherparepvec for the treatment of advanced melanoma. Clin Cancer Res 22(13):3127–3131. https://doi.org/10.1158/1078-0432.CCR-15-2709
2. Peters C, Rabkin SD (2015) Designing herpes viruses as oncolytics. Mol Ther Oncolytics 2:15010. https://doi.org/10.1038/mto.2015.10
3. Kaur B, Cripe TP, Chiocca EA (2009) "Buy one get one free": armed viruses for the treatment of cancer cells and their microenvironment. Curr Gene Ther 9(5):341–355
4. Kuroda T, Martuza RL, Todo T et al (2006) Flip-Flop HSV-BAC: bacterial artificial chromosome based system for rapid generation of recombinant herpes simplex virus vectors using two independent site-specific recombinases. BMC Biotechnol 6:40. https://doi.org/10.1186/1472-6750-6-40
5. Kuroda T, Rabkin SD, Martuza RL (2006) Effective treatment of tumors with strong beta-catenin/T-cell factor activity by transcriptionally targeted oncolytic herpes simplex virus vector. Cancer Res 66(20):10127–10135. https://doi.org/10.1158/0008-5472.CAN-06-2744
6. Dmitrieva N, Yu L, Viapiano M et al (2011) Chondroitinase ABC I-mediated enhancement of oncolytic virus spread and antitumor efficacy. Clin Cancer Res 17(6):1362–1372. https://doi.org/10.1158/1078-0432.CCR-10-2213
7. Guedan S, Rojas JJ, Gros A et al (2010) Hyaluronidase expression by an oncolytic adenovirus enhances its intratumoral spread and suppresses tumor growth. Mol Ther 18 (7):1275–1283. https://doi.org/10.1038/mt.2010.79
8. McKee TD, Grandi P, Mok W et al (2006) Degradation of fibrillar collagen in a human melanoma xenograft improves the efficacy of an oncolytic herpes simplex virus vector. Cancer Res 66(5):2509–2513. https://doi.org/10.1158/0008-5472.CAN-05-2242
9. Zhang W, Fulci G, Buhrman JS et al (2012) Bevacizumab with angiostatin-armed oHSV increases antiangiogenesis and decreases bevacizumab-induced invasion in U87 glioma. Mol Ther 20(1):37–45. https://doi.org/10.1038/mt.2011.187
10. Zhang W, Fulci G, Wakimoto H et al (2013) Combination of oncolytic herpes simplex viruses armed with angiostatin and IL-12 enhances antitumor efficacy in human glioblastoma models. Neoplasia 15(6):591–599
11. Hardcastle J, Kurozumi K, Dmitrieva N et al (2010) Enhanced antitumor efficacy of vasculostatin (Vstat120) expressing oncolytic HSV-1. Mol Ther 18(2):285–294. https://doi.org/10.1038/mt.2009.232
12. Yoo JY, Haseley A, Bratasz A et al (2012) Antitumor efficacy of 34.5ENVE: a transcriptionally retargeted and "Vstat120"-expressing oncolytic virus. Mol Ther 20(2):287–297. https://doi.org/10.1038/mt.2011.208
13. Liu TC, Zhang T, Fukuhara H et al (2006) Dominant-negative fibroblast growth factor receptor expression enhances antitumoral potency of oncolytic herpes simplex virus in neural tumors. Clin Cancer Res 12 (22):6791–6799. https://doi.org/10.1158/1078-0432.CCR-06-0263
14. Liu TC, Zhang T, Fukuhara H et al (2006) Oncolytic HSV armed with platelet factor 4, an antiangiogenic agent, shows enhanced efficacy. Mol Ther 14(6):789–797. https://doi.org/10.1016/j.ymthe.2006.07.011
15. Cheema TA, Wakimoto H, Fecci PE et al (2013) Multifaceted oncolytic virus therapy for glioblastoma in an immunocompetent cancer stem cell model. Proc Natl Acad Sci U S A 110(29):12006–12011. https://doi.org/10.1073/pnas.1307935110
16. Markert JM, Cody JJ, Parker JN et al (2012) Preclinical evaluation of a genetically engineered herpes simplex virus expressing interleukin-12. J Virol 86(9):5304–5313. https://doi.org/10.1128/JVI.06998-11
17. Saha D, Martuza RL, Rabkin SD (2017) Macrophage polarization contributes to glioblastoma eradication by combination immunovirotherapy and immune checkpoint blockade. Cancer Cell 32(2):253–267 e255. https://doi.org/10.1016/j.ccell.2017.07.006
18. Terada K, Wakimoto H, Tyminski E et al (2006) Development of a rapid method to generate multiple oncolytic HSV vectors and their in vivo evaluation using syngeneic mouse tumor models. Gene Ther 13(8):705–714. https://doi.org/10.1038/sj.gt.3302717
19. Andreansky S, He B, van Cott J et al (1998) Treatment of intracranial gliomas in immunocompetent mice using herpes simplex viruses that express murine interleukins. Gene Ther 5 (1):121–130. https://doi.org/10.1038/sj.gt.3300550
20. Barnard Z, Wakimoto H, Zaupa C et al (2012) Expression of FMS-like tyrosine kinase 3 ligand

by oncolytic herpes simplex virus type I prolongs survival in mice bearing established syngeneic intracranial malignant glioma. Neurosurgery 71(3):741–748.; discussion 748. https://doi.org/10.1227/NEU.0b013e318260fd73

21. Tamura K, Mawaribuchi S, Yoshimoto S et al (2010) Tumor necrosis factor-related apoptosis-inducing ligand 1 (TRAIL1) enhances the transition of red blood cells from the larval to adult type during metamorphosis in Xenopus. Blood 115(4):850–859. https://doi.org/10.1182/blood-2009-04-218966
22. Jahan N, Lee JM, Shah K et al (2017) Therapeutic targeting of chemoresistant and recurrent glioblastoma stem cells with a proapoptotic variant of oncolytic herpes simplex virus. Int J Cancer 141(8):1671–1681. https://doi.org/10.1002/ijc.30811
23. Chase M, Chung RY, Chiocca EA (1998) An oncolytic viral mutant that delivers the CYP2B1 transgene and augments cyclophosphamide chemotherapy. Nat Biotechnol 16 (5):444–448. https://doi.org/10.1038/nbt0598-444
24. Tyminski E, Leroy S, Terada K et al (2005) Brain tumor oncolysis with replication-conditional herpes simplex virus type 1 expressing the prodrug-activating genes, CYP2B1 and secreted human intestinal carboxylesterase, in combination with cyclophosphamide and irinotecan. Cancer Res 65(15):6850–6857. https://doi.org/10.1158/0008-5472.CAN-05-0154
25. Guffey MB, Parker JN, Luckett WS Jr et al (2007) Engineered herpes simplex virus expressing bacterial cytosine deaminase for experimental therapy of brain tumors. Cancer Gene Ther 14(1):45–56. https://doi.org/10.1038/sj.cgt.7700978
26. Sgubin D, Wakimoto H, Kanai R et al (2012) Oncolytic herpes simplex virus counteracts the hypoxia-induced modulation of glioblastoma stem-like cells. Stem Cells Transl Med 1 (4):322–332. https://doi.org/10.5966/sctm.2011-0035

Chapter 11

Poxviruses as Gene Therapy Vectors: Generating Poxviral Vectors Expressing Therapeutic Transgenes

Steven J. Conrad and Jia Liu

Abstract

Treatments with poxvirus vectors can have long-lasting immunological impact in the host, and thus they have been extensively studied to treat diseases and for vaccine development. More importantly, the oncolytic properties of poxviruses have led to their development as cancer therapeutics. Two poxviruses, vaccinia virus (VACV) and myxoma virus (MYXV), have been extensively studied as virotherapeutics with promising results. Vaccinia virus vectors have advanced to the clinic and have been tested as oncolytic therapeutics for several cancer types with successes in phase I/II clinical trials. In addition to oncolytic applications, MYXV has been explored for additional applications including immunotherapeutics, purging of cancer progenitor cells, and treatments for graft-versus-host diseases. These novel therapeutic applications have encouraged its advancement into clinical trials. To meet the demands of different treatment needs, VACV and MYXV can be genetically engineered to express therapeutic transgenes. The engineering process used in poxvirus vectors can be very different from that of other DNA virus vectors (e.g., the herpesviruses). This chapter is intended to serve as a guide to those wishing to engineer poxvirus vectors for therapeutic transgene expression and to produce viral preparations for preclinical studies.

Key words Poxvirus, Poxvirus genetic engineering, Vaccinia virus (VACV), Myxoma virus (MYXV), Infection/transfection, Transgene expression, Poxvirus transfer vector

1 Introduction

1.1 Overview of Recombinant Pox Viruses

This chapter provides general guidelines for researchers intending to engineer therapeutic recombinant poxviruses in a laboratory setting with the capacity to safely handle BSL-2 agents.

The United States Food and Drug Administration (FDA) considers any agent(s) which "...mediate their effects by transcription and/or translation of transferred genetic material and/or by integrating into the host genome and that are administered as nucleic acids, viruses, or genetically engineered microorganisms" [1] to be examples of gene therapeutic agents. By this definition, poxviruses have much to offer as delivery vectors for therapeutic genetic sequences. The poxviruses (family *Poxviridae*) are large, double-

Fredric P. Manfredsson and Matthew J. Benskey (eds.), *Viral Vectors for Gene Therapy: Methods and Protocols*, Methods in Molecular Biology, vol. 1937, https://doi.org/10.1007/978-1-4939-9065-8_11,

stranded DNA (dsDNA) viruses that have had an outsized impact upon human health and history, both positive and negative. Smallpox has been a scourge of humanity from prehistory until the close of the twentieth century, a highly contagious disease that killed between one-fourth and one-third of its victims and routinely left survivors with lifelong scarring [2, 3]. The causative agent of smallpox, the variola virus (VARV), was the impetus for Edward Jenner's early vaccination experiments with the cowpox virus (CPXV). Jenner's experiments and insight led to the practice of vaccination and, eventually, the global eradication of smallpox through a program of comprehensive human vaccination with the vaccinia virus (VACV) [2, 3]. Because of its great utility as a smallpox vaccine, the VACV is the prototype and the best characterized of all the poxviruses.

In addition to the VACV's well-known role as a vaccine against smallpox, the ease with which VACV and other poxviruses can be engineered to express foreign proteins has made them valuable therapeutic vectors. Until the discovery of the mimivirus in 2003 [4], the poxviruses were thought to be the largest of the viruses and were the first viruses to be observed directly by light microscopy. Poxviruses possess large genomes and are capable of integrating at least 25 kilobases of foreign DNA [5]. Two of the most common poxvirus vectors used in therapy and gene delivery are VACV and myxoma virus (MYXV). Both can be easily manipulated to integrate therapeutic transgenes into the viral genome [6]. Because their replication cycle occurs entirely within the cytoplasm [7], there is no integration of viral DNA into the host genome and no splicing of viral transcripts [8]. This property minimizes the potential for genetic impact upon hosts that receive them as therapeutic agents. Finally, these poxviruses can also be prepared at high titers [9, 10] to meet the demand of escalated doses for treatment in preclinical models and in human patients.

Poxviruses, especially VACV and MYXV, along with other viruses are currently being intensively investigated as anticancer therapeutic agents [11]. Originally oncolytic viruses were envisioned as replication-competent viruses that would preferentially infect and destroy cancerous cells via cytolytic activity. The theoretical advantage of utilizing replication-competent viruses is that they can produce progeny viruses to amplify the therapeutic outcome, all the while leaving noncancerous cells unscathed. While the promise of a single efficacious viral agent to treat cancer has not yet been achieved, we now realize that successful oncolytic viruses are capable of shaping the host immune response to tumor cells [12] and even to alter the tumor microenvironment to facilitate the action of cytotoxic T cells [13]. Viruses can be further armed [14] with therapeutic genes (transgenes) including cytokines and chemokines [15, 16], prodrug activators [17], tumor-associated antigens [18], inducers of apoptosis [19], and ligands of the toll-like receptors

(TLRs) [20, 21]. In addition, imaging enablers such as the sodium iodide symporter [22] and luciferase [23] can also be incorporated. More importantly, virotherapy can complement and even synergize with chemotherapy and immunotherapy [24–26]. Treatment benefits of some chemotherapeutic agents can be further improved by the prior or subsequent use of an oncolytic viral agent [27], while concurrent delivery of both virotherapy and chemotherapy [24, 27] or radiotherapy [28] can also be beneficial. Of the poxviruses, the VACV is by far the furthest advanced as an oncolytic virus, and several VACV-based oncolytic virus candidates have been tested in humans in Phase I/II clinical trials, including Pexa-Vec (JX-594, SillaJen, Inc.) [29, 30] and GLV-1h68 (GL-ONC1, Genelux Corporation) [17]. Pexa-Vec is a VACV modified by insertion of human granulocyte-monocyte colony stimulating factor (hGM-CSF) into the thymidine kinase (TK) gene [30, 31]. GLV-1h68 is a Lister strain of VACV modified via insertion of *Renilla* luciferase (RLuc-GFP) into the F14.5L ORF, the insertion of ß-galactosidase into the viral TK gene, and the insertion of ß-glucuronidase into the viral hemagglutinin ORF (*A56R*) [32].

Myxoma virus (MYXV) has recently been recognized for its oncolytic potential in treating cancerous conditions in preclinical models [33] and has proven to be a valuable candidate for virotherapy [11]. The discovery of MYXV oncolytic potential was serendipitous [34], as MYXV has a strict tropism for European rabbits and is not known to cause disease in any other organism [35]. Myxoma virus (genus *Leporipoxvirus*) is antigenically distinct from VACV (genus *Orthopoxvirus*), making it a viable alternative to VACV virotherapy; it is often an efficacy concern when patients have been previously vaccinated with VACV against smallpox, as the aging population is most likely to experience cancer. Despite its narrow host tropism for pathogenesis, MYXV productively infects neoplastic cells derived from humans [36, 37], mice [24, 27, 38, 39], canines [40], and felines [41]. Although not yet tested in the clinic, MYXV has shown promise in preclinical animal models of neoplasms that are particularly difficult to treat and cure, including pancreatic cancer [24], gliomas [36], medulloblastoma [42], and ovarian cancer [27]. In addition to conventional oncolytic applications, MYXV has been examined for its immunotherapeutic potential and effectiveness in combinatorial treatments with chemotherapy [24, 27]. It has also been shown that MYXV efficiently binds to and subsequently destroys $CD138^+$ cells derived from human multiple myeloma patients. The binding of MYXV to these multiple myeloma cells induced apoptosis efficiently via the extrinsic pathway [43]. This allows for a potential ex vivo therapy by efficiently purging neoplastic cells from autologous stem cell grafts before re-implantation into the host [44, 45].

Based on our understanding of poxvirus virology, we provide a general guideline for the engineering, isolation, and purification of recombinant poxviruses suitable for in vitro and preclinical testing.

Each protocol (in Subheading 3) refers to reagents, consumables (such as tissue culture vessels), and equipment necessary to fully execute the described procedure. These referenced materials are detailed in Subheading 2 entitled "Materials."

1.2 Genetic Engineering of Recombinant Poxviruses

During poxvirus infection, recombination occurs in the virus genome [6]. This is a universal characteristic of poxviruses that has shaped the development of the methods used to produce recombinant poxviruses. The activities of viral enzymes, including the 3′-to-5′ exonuclease activity of the virally encoded DNA polymerase [46], topoisomerase I [47–49] and Holliday junction resolvase [50] play roles in recombination during infection. The poxvirus infection that is needed to generate the recombinant virus can be from the destination parental poxvirus or a helper poxvirus [51]. Nevertheless, to engineer a recombinant poxvirus, the first step is to produce the transfer plasmids that contain the gene of interest with its expression driven by a poxvirus-specific promoter. In a transfer plasmid the expression cassette for the gene of interest is flanked by approximately 500–1000 base pairs (bp) of dsDNA homologous to the sequence in the parental virus genome. This design will lead to integration of the transgene into the poxvirus genome in a site-specific manner. Initial attempts to insert and subsequently express foreign genes in VACV made use of the VACV thymidine kinase gene as an insertion site [52] and inserted the herpesvirus thymidine kinase, providing its own selection mechanism [53]. In the VACV, the viral thymidine kinase gene [53] and the VACV growth factor gene [54] loci have both been successfully used as insertion sites for foreign genetic material without reducing infectivity. If a wild-type (WT) MYXV vector is desirable for the study, insertion at the locus between *M135R* and *M136R* gene can be considered, as it will not affect either viral infectivity in vitro or pathogenesis in rabbits [55]. The MYXV vector with a single gene knockout of either *M063R* [56, 57] or *M135R* [45] is also commonly used as the backbone vector for therapeutic testing in preclinical models. Because *M063R*-null [56] and *M135R*-null [58] MYXV do not cause pathogenic diseases in European rabbits, this accommodates the need for a replicating MYXV vector without the possibility of endangering other animals in the process. Other genes in MYXV, including *M153R* [59], *M062R* [27], and Serp2 [40], can also be individually deleted to enhance its immunotherapeutic effect. In MYXV deletion of *M062R* alone results in a replication-defective viral vector in cells from all species tested [60]. However, treatments using *M062R*-null MYXV can provide immunological benefits in preclinical models and in some cases it has surpassed the therapeutic effect of WT MYXV (unpublished data J Liu).

1.2.1 Construction of the Transfer Plasmid Vector

Once a poxvirus viral vector has been chosen for the study and the insertion site in the viral genomic backbone has been determined, the promoters to drive gene expression also need to be identified. The actual methods used to construct the transfer vector are not in any way specific to poxvirus biology, and researchers wishing to make transfer vectors have a number of options.

New DNA manipulation and cloning technologies, as well as inexpensive whole-gene synthesis [61] mean that any researcher wishing to insert novel sequences into poxvirus transfer vectors has a wealth of choices. A systematic survey of the history and current state of DNA manipulation and molecular cloning methodologies is beyond the scope of this publication, and many reviews exist to guide the research in their choice of cloning procedures. One of such guides and current, detailed advice can be found published by the iGEN/TU Eindhoven Community [62].

Gateway cloning [63, 64] has been extensively utilized to engineer poxvirus transfer plasmids [55, 65, 66] and is the methodology of choice in our laboratory. The Gateway cloning system allows for the simple, rapid, and highly efficient insertion of "att" bounded sequences into a destination vector using a recombinase derived from lambda phage. Smallwood et al. has provided an excellent and detailed protocol [67].

1.2.2 Promoters for Transgene Expression in the Recombinant Poxvirus

Poxvirus genes can be categorized into three temporal classes corresponding to the kinetics of their transcription: early, intermediate, or late [68, 69]. Viral gene transcription occurs in a cascade fashion. During the infection, each kinetic class of genes encodes factors necessary for the transcription of the subsequent temporal class [69]. Poxvirus-specific promoters are required to drive transgene expression in the context of a poxvirus infection. Early promoters operate before viral DNA replication, while intermediate and late promoters operate only after DNA replication [70]. Across the *Poxviridae*, promoters of a particular temporal class differ slightly but have highly conserved core regions [68]. The promoter structures of each temporal class have been well characterized in the VACV [71, 72]. Because transcription factors are highly conserved across the *Chordopoxvirinae*, promoters from one poxvirus can often be used to drive transcription during infections caused by poxviruses of different genera [68]. For example, the VACV p7.5 early/late promoter drives transcription in the fowlpox virus (FWPV, genus *Avipoxvirus*) [73], the VACV p11 late promoter drives transcription in MYXV (genus *Leporipoxvirus*) [55], and the synthetic early/late promoter (vvSyn E/L) drives transcription in the tanapox virus (TANV, genus *Yatapoxvirus*) [21]. Poxvirus promoters combining the temporal classes have been made by ligating regions from temporally distinct promoters and have been extensively utilized to drive robust transgene expression for both virology studies and therapeutic applications [55, 74].

The researcher wishing to express exogenous genes in the context of a poxvirus infection should be aware of the constraints imposed by each type of poxvirus promoter used. For example, to engineer poxviruses for high levels of transgene production (e.g., vaccine antigens [75]) it is appropriate to choose a promoter such as the early/late VACV p7.5 [76] or the vvSyn E/L [39, 74, 77] promoter to drive continuous expression throughout the course of an infection. In cases where the gene product must be expressed late in the replicative process due to toxicity or as a selection control for transgene expression only during a productive infection, a late promoter such as the poxvirus p11 late promoter [55] may be the appropriate choice. Table 1 shows several poxvirus promoters including both native and synthetic promoters from different temporal classes.

1.3 Engineering Recombinant Poxviruses

After construction of the transfer plasmid (or shuttle vector), recombinant poxvirus is generated via homologous recombination in poxvirus infected cells. Generally a length of 500–1000 bp on each flank of the insertion DNA fragment is sufficient to guide homologous recombination in a site-specific manner into the viral genome [6].

Utilizing a helper poxvirus such as fowlpox virus (FWPV) [83] or Shope fibroma virus (SFV) [84] to provide the necessary machinery for recombination has been shown to greatly improve the ease of downstream purification steps. In these modified protocols, instead of using live virus of the chosen poxvirus vector, purified poxvirus genomic DNA from the destination poxvirus vector is used along with the shuttle vector [83, 84]. In these systems significantly improved recombination efficiency between the transfer plasmid and the destination poxvirus vector has been reported. The marked improvement in abundance of recombinant virus is likely due to an effective inhibition of infection by the helper poxvirus in the subsequent purification process. By choosing a cell line for purification that is nonpermissive to the helper poxvirus but permissive to the chosen viral vector, one can significantly eliminate the presence of helper virus in subsequent steps. However, due to the need for additional reagents (live viruses such as FWPV or SFV) besides the chosen viral backbones (e.g., VACV or MYXV), in this chapter we will focus on the conventional methodology to generate recombinant poxviruses.

It is a common practice to transfect the vector into cells already hosting a poxvirus infection, typically referred to as an infection/transfection procedure [83]. This is usually accomplished by inoculating permissive cells with poxvirus (e.g., at a multiplicity of infection (MOI) of 1), followed by transfection of the transfer plasmid

Table 1
Promoter sequences can be used to drive transgene expression

Name	Kinetic class	Sequence	References
VACV TK	Early	GAATAAAGTGAACAATAATTAATTCTTTATTGTCATCATG	[78]
7.5 kd	Early	CGTAAAAGTAGAAAATATATTCTAATTTATTGCA	[71]
G8R	Intermediate	CATTTAACTTTAAATAATTTACAAAAATTTAAAATA	[79]
vvD6R	Late	ATATATGCTCATATATTTATAGAAGATATCACATATCTAAATG	[80]
vvP11	Late	GAATTTCATTTTGTTTTTTTCTATGCTATAAATG	[81]
vvSynE/L	Early/late	AAAAATTGAAATTTTATTTTTTTTTTTTGGAATATAAATA	[74]
pSynE/L	Early/late	AATTGGATCAGCTTTTTTTTTTTTTTTTTTGGCATATAAATAAGGTC-GAAGCTTGGTACCAAAAATTGAAAAACTATTCTAATTTATTGCACG	[82]

The transcription start sites are indicated by double underline where known. Start codons are shown highlighted in gray

1–2 h later. Because the majority of progeny virions will be the parental virus [85], it is necessary to apply a selection method to screen for recombinant viruses bearing the desired transgenes. Thus, a plaque or focus purification step is necessary to isolate the virus until a pure population of the recombinant virus is established.

There are many types of selection that allow the isolation of recombinant virus from the unaltered parental virus. Historically, selection was first conducted by probing of replica-plated plaques on nitrocellulose membranes followed by *in situ* hybridization to identify the successful recombinants [86]. Selection can be accomplished via biochemical means using selectable markers such as resistance gene to neomycin [86], hygromycin [87], or guanine phosphoriboxyltransferase (gpt) [88, 89]. However, the use of genotoxic selectable markers has been shown to cause secondary mutations elsewhere in the genome that are not immediately apparent without direct sequencing [83]. We thus advise caution if one of the above selection methods is used. Fluorescent reporters provide selection that can be used together with other selectable markers [90, 91] or used alone. Although we reference selection schemes for recombinant poxviruses based upon drugs or other biochemical properties, this protocol is written with fluorescence-based selection in mind.

1.4 Considerations for Clinical Application Using Poxvirus Vectors

When making recombinant viruses for purely research work it is usually acceptable to leave genes for selection proteins intact in the recombinant virus. If, however, a virus is intended for therapeutic use in a human clinical trial, a clinical-grade viral preparation must be made [92]. These recombinant viruses must also be free of expressible drug resistance proteins (such as the neomycin phosphotransferase used during neomycin selection), as there are potential safety concerns due to adventitious mutations in these viruses [93]. It is therefore necessary to excise genes encoding any drug resistance proteins used during the isolation of the virus. Such an excisable selection system was published by Rintoul et al. in 2011 [94]. The authors describe a selectable and excisable marker (SEM) system, which consists of a transfer plasmid, pSEM-1, which contains (a) a luciferase gene driven by a poxvirus early/late synthetic promoter and (b) a gene encoding yellow fluorescent protein (*yfp*) fused to the gene encoding the drug selection marker guanine phosphoribosyltransferase (*gpt*). The entire portion of pSEM-1 containing these two genes was flanked by sequence homologous to the viral thymidine kinase gene (dictating the site of insertion into the viral genome), while loxP sites flanked the portion containing the yfp-gpt gene. When the virus was inoculated onto U2OS cells, both YFP and luciferase activity were detected; when U2OS cells stably expressing the Cre recombinase were inoculated, only luciferase activity was detected. Thus, the *yfp-gpt* portion of the viral genome had been excised.

When live, replication-competent viruses (particularly VACV) are used as virotherapeutics, there exists a risk that productive viral infection within tumor cells in patients could spread to healthy tissue and cause adverse effects. To protect against this possibility, researchers should have access to antiviral drugs that can blunt any undesirable viral spread originating with the treated tumor tissue. VACV infection in research animals has been successfully treated with two investigational drugs, ST-246™ (SIGA Technologies, Inc) [95] and CMX-001™ (Chimerix, Inc) (Brincidofivir) [96, 97]. Although it has not been approved for use in poxvirus infections, Cidofovir (Vistide™ Gilead Sciences, Inc.) may also be used [98]. In one case, combinatorial treatment was provided to a pediatric patient with refractory atopic dermatitis who developed eczema vaccinatum from contact vaccinia infection and the patient recovered; the combinatorial treatment included intravenous vaccinia immune globulin, cidofovir, and ST-246 [99, 100]. ST-246 and CMX-001 are predicted to be able to control MYXV infection effectively. Although MYXV inoculation in a non-rabbit species is unlikely to cause infectious disease due to a strict species tropism, the availability of these drugs provides an extra measure of safety.

1.5 Summary

It is an exciting time to be in the laboratory working with poxviruses as they continue to be developed as virotherapeutic agents for the treatment of diseases, both in humans and in companion animals [101, 102]. Although VACV has held center stage in the ongoing oncolytic poxviruses story, MYXV is steadily establishing itself as a valuable addition to the virotherapeutic armamentarium. VACV and MYXV are, however, only the beginning of the poxvirus story. Within the *Chordopoxvirinae*, the International Committee on Taxonomy of Viruses (ICTV) currently lists 12 genera containing 40 recognized poxviruses. The therapeutic potential of most of these poxviruses remains to be characterized and developed. In addition to vector administration, we continue to discover novel gene functions encoded by poxviruses that modulate the cellular innate immune state, any of which could prove important for the development of new therapeutics. Genetic manipulation of poxvirus genome provides a powerful tool for future investigations.

2 Materials

This section contains a detailed list of cells, media, materials, consumables, apparatus, and equipment used in the subsequent protocols.

2.1 Cell Lines

1. BSC-40, ATCC (catalog CRL-2761). Derived from *Cercopithecus aethiops*, epithelial, adherent, from kidney. BSL-1.
2. CV1, ATCC (catalog CCL-70). Derived from *Cercopithecus aethiops*, fibroblast, adherent, from kidney. BSL-1.
3. Vero, ATCC (catalog CRL-812). Derived from *Cercopithecus aethiops*, epithelial, adherent, from kidney. BSL-1.

2.2 Tissue Culture Medium, Reagents, and Solutions

1. DMEM, with 4.5 g/L glucose and L-glutamine.
2. Penicillin/streptomycin (Pen Strep).
3. Fetal bovine serum (FBS).
4. Tris–HCl 10 mM, pH 8.0, (sterile).
5. Trypsin 10×, without phenol red.
6. Sucrose (D-Sucrose, saccharose), molecular biology grade. Individual solutions are made to 24, 28, 32, 36 and 40% W/V and filtered through a 0.22 μm membrane.
7. Low-melting-temperature agarose.
8. PBS, with calcium and magnesium, sterile.
9. ViaFect Transfection Reagent, Promega (catalog E4982).

2.3 Consumables

1. 150 mm cell culture dish, treated, DNase/RNase free, sterile.
2. 24 well tissue culture-treated plate, sterile.
3. 35 × 12 mm style tissue culture dish, tissue culture-treated, sterile.
4. 25 cm cell scraper, sterile, individually wrapped.
5. Centrifuge tube, 50 mL, conical, sterile.
6. Ultra-Clear Centrifuge Tubes 1 × 3½ in (25 × 89 mm).

2.4 Apparatus and Equipment

1. Dounce-style homogenizer, 40 mL, with "loose" pestle, Wheaton.
2. Dismembrator (sonicator), with inverted "cup" and sound-deadening enclosure.
3. Ultracentrifuge, from ThermoScientific/Sorvall, model WX-Ultra 80, used with swinging bucket rotor, model AH-629.

3 Methods

3.1 Infection/Transfection of Permissive Cells

1. Seed permissive cells (e.g., BSC-40) into a 35 mm tissue culture dish at a density that will produce 80–90% confluency on the next day.
2. Inoculation of poxviruses onto the monolayer:

(a) Use a poxvirus viral stock with sufficient plaque-forming units (pfu) to achieve an MOI of 1 in a minimal volume of growth medium (just sufficient to cover the bottom of the tissue culture dish). For a 35 mm culture dish 0.5 mL is sufficient. Allow viral attachment to the cells for 1 h, which can be performed at 4 °C. Every 10 min gently rock the plate to ensure complete exposure of the monolayer to the inoculum.

(b) After 1 h of incubation, remove the inoculum from the dish and wash three times with 1 mL of PBS each. Immediately add 2 mL of fresh growth medium to the cells.

3. Incubate the infected cell monolayer at 37 °C with 5% CO_2 for 1 h before transfection with the transfer vector.
4. Transfect the infected cells with complexes formed from the transfection reagent of choice (e.g., ViaFect, *see* above) and the transfer vector, according to the manufacturer's protocol. Remove the transfection complexes if so directed by the manufacturer's protocol, as prolonged exposure may be toxic and cause cell death. Allow the infection to proceed following transfection until a complete cytopathic effect (CPE) in all cells is observed (48–72 h post infection).
5. When the infection is suitably advanced, collect the cell lysate together with the culture medium by scraping the monolayer with a sterile cell scraper. At this point the crude viral preparation may be stored at −80 °C until further processing.

3.2 Selection and Purification of Recombinant Viruses

1. Before purification, subject the cell lysate to 3 cycles of freeze (−80 °C) -thaw (37 °C water bath) followed by sonication in an ice bath for 1 min using a cup sonicator (e.g., Fisher Scientific sonicator with the amplitude set to 50, *see* above, Subheading 2.4).
2. Purification of the recombinant virus is conducted on a monolayer of BSC-40 cells seeded in 35 mm dishes or in 6-well plates. Serially dilute the original cell lysate and transfer an appropriate amount of inoculum into each well for an incubation period of 1 h. During the 1-hour incubation gently rock the dish every 10 min to ensure that sufficient culture medium is covering the monolayer cells. After 1 h, aspirate the inoculum and overlay an appropriate volume of medium containing low-melting agarose (0.67% in normal growth medium). Allow 20 min for the agarose overlay to solidify at room temperature before transferring the plates into the incubator. An alternative is to use methylcellulose to replace low-melting agarose. You may plate one dish of cells at each dilution for this step. Once a suitable dilution is identified, more plates can be seeded until a successful isolation is achieved.

3. Once well-separated viral foci or plaques are detected, proceed to isolate the virus from them for further purification. For each round of purification, first identify a discrete, well-separated plaque of viral infection and isolate it for amplification and analysis (*see* **Note 1** for further explanation.)
4. Repeat the process of plaque (focus) purification until the resulting stock contains a uniform population of recombinant virus (*see* **Note 2**). We typically do a minimum of four rounds of plaque purification.

At each round of purification a sample of recombinant virus is isolated for viral DNA extraction and PCR verification for purity. For PCR verification, one can use a primer set that recognizes sequences surrounding the insertion site in the parental virus. It is also helpful to design a different set of primers to recognize sequences only present in the recombinant virus. Use viral DNA from your parental virus as a control. Once you confirm the purity of the recombinant virus, proceed with further confirmation by sequencing. We typically sequence the insert and several hundred bases surrounding the insertion site to ensure that the gene(s) of interest have integrated correctly and that no mutations have been introduced.

3.3 Poxvirus Amplification and Purification

Before testing in cells or animal models, it is necessary to produce a large quantity of poxvirus with high titer. Viral titer is usually expressed as plaque-forming units (pfu) per milliliter of virus preparation. Cell lines for the replication of VACV and other poxviruses include BSC-40 [103], CV-1 [104], and Vero [105]. These cells are obvious first choices for high-efficiency replication of poxviruses, and BSC-40 cells are most frequently used. A protocol for small- to medium-scale VACV and MYXV preparation is detailed below, which is modified from methods reported previously [106, 107].

1. Use cells that are fully permissive to the virus for amplification, such as BSC-40. Start with 20–40 tissue culture-treated dishes of size of 150 mm (each dish has 176.7 cm^2 growth area for adherent cells) and allow them to reach full confluence (approximately 2.0×10^7 cells/dish). Cells should not be over-confluent before infection.
2. Inoculate each dish with a low MOI (e.g., 0.001 for VACV).
   (a) For infection, prepare an inoculum in a low volume of growth medium per dish (e.g., no more than 10 mL) with an appropriate amount of virus. Replace the medium from each plate with the inoculum.
   (b) Incubate the inoculum at 37 °C for 1 h with occasional gentle rocking of the dish. This is to ensure sufficient

exposure of all cells to the inoculum and to avoid drying of the monolayer.

(c) Add an additional volume of complete growth medium to a total of 30 mL to each dish after 1 h of incubation.

3. Incubate and allow the infection to develop until cells reach full production capacity (e.g., 72 h for MYXV, and up to 5 days for VACV). Monitor the progress of the infection daily. For VACV stock preparation, monitor the cells until a CPE is apparent but cells have not started to lyse.
4. Harvest the infected cells by scraping the monolayer and transferring the scraped cells along with the medium into a 50 mL centrifuge tube. Recover as much of the cells as possible.
5. Centrifuge cells at 750 × *g* for 10 min at 4 °C. Carefully discard the supernatant and continue to collect scraped cells using the same 50 mL conical tubes. In each tube, a pellet typically contains 4–6 dishes of infected cells. For temporary storage, discard as much medium after centrifugation as possible and store the tubes at −80 °C until purification.
6. Resuspend briefly thawed cell pellets in ice-cold sterile 10 mM Tris–HCl, pH 8.0. Use a volume approximately 6 times the volume of the cell pellet. For example, cells from 5 of the 150 mm dishes make up approximately 1 mL of volume and thus 6 mL of 10 mM Tris–HCl pH 8.0 is needed.
7. Freeze (−80 °C) and thaw (37 °C water bath) the resuspended cell pellet two more times. Sonicate the suspension for one min in an ice bath-equipped cup sonicator (*see* Subheading 3.2, **step 1** and **Note 3**).
8. In a biosafety cabinet, transfer the sonicated cell pellet to a sterile 40 mL prechilled glass Dounce homogenizer (Wheaton) on ice and **use the loose pestle** for 20 strokes **slowly**. Collect the homogenized cell suspension into a 50 mL conical tube and centrifuge at 1500 × *g* for 15 min at 4 °C.
9. **Save the supernatant** in a fresh 50 mL conical tube on ice and resuspend the remaining cell pellet in 4 mL of ice-cold 10 mM Tris–HCl, pH 8.0. Repeat the homogenization step (**step 8**, above).
10. Centrifuge the homogenized pellet suspension at 1500 × *g* for 15 min at 4 °C. **Combine the supernatant** with that from **step 9** and discard the pellet.
11. Carefully layer the homogenate (no more than 20 mL of crude cell extract per centrifugation tube) over a 36% sucrose cushion in 10 mM Tris–HCl pH 8.0 (No less than 17 mL of volume) in an Ultra-Clear centrifuge tube.
12. Centrifuge at 44,500 × *g* for 80 min at 4 °C (*see* **Note 4**).

13. After centrifugation, the virus will form a tight pellet on the bottom of the centrifuge tube. Carefully discard the supernatant and resuspend the pellet in no more than 7 mL of ice-cold 10 mM Tris–HCl pH 8.0.
14. Sonicate the viral preparation 2 times for 60 s each in the ice bath using the settings in Subheading 3.2, **step 1**.
15. Load the sonicated viral preparation on top of the sucrose step-gradient, made with 6 mL each of 24, 28, 32, 36, and 40% sucrose that have been dissolved in 10 mM Tris--HCl, pH 8.0 (*see* **Note 5**). Continue with purification by centrifugation of the resuspension on the gradient for 40 min at 24,000 × *g* at 4 °C.
16. Centrifugation should result in 1 distinct band near the middle of the gradient. Carefully remove the top layers of gradient material. Collect the band with a pipette tip and discard the remaining liquid (*see* **Note 6**).
17. Resuspend the collected band in 10 mM Tris–HCl, pH 8.0 and further dilute the band to fill up the centrifugation tube.
18. Centrifuge the diluted band for 60 min at 31,500 × *g*, 4 °C.
19. Carefully discard the supernatant and resuspend the pellet in 1 mL of 10 mM Tris–HCl, pH 8.0.
20. Aliquot and store the purified virus preparation at −80 °C. Viral titer is evaluated by titration on BSC-40 cells. Proper execution of this protocol results in viral yields of 1–5 × 10^9 pfu/mL for MYXV, and approximately 1 × 10^{10} pfu/mL for replicating VACV.

For testing the recombinant virus in vitro and in animal models, stop at the **step 13**; virus preparation is of sufficient purity for this purpose. When generating recombinant virus to express cytokine or other therapeutic molecules, further purification through the remaining of the steps is necessary to further eliminate trace amount of transgene product which may remain with the virus pellet at **step 13**.

3.4 Determination of Viral Titer

1. If titration is conducted on a crude cell lysate which has not previously been processed, then freeze (at −80 °C) and thaw (37 °C water bath) the sample 3 times followed by sonication before proceeding to **step 2** for serial dilution of the virus stock. If a viral preparation is purified from a sucrose cushion (*see* Subheading 3.3, **step 13**) or sucrose gradient purification (*see* Subheading 3.3, **step 15**), sonicate the virus aliquot in an ice bath for 30 s before proceeding to **step 2**.
2. Make serial dilutions of your sample in PBS with Mg^{+2} and Ca^{+2} and 2% heat inactivated FBS (dilution medium). Use 1.5 mL Eppendorf tubes and make the initial 10^{-2} dilution

by adding 10 μL of sample to 990 μL dilution medium, and all subsequent serial 10^{-1} dilutions by adding 100 μL of diluted virus to 900 μL of dilution medium. For samples of unknown titer, we suggest dilutions up to 10^{-8} (*see* **Note 7**).

3. Use 24-well tissue culture-treated plates with a monolayer of permissive cells (e.g., BSC-40 cells) seeded the day before for titration. Remove the growth medium from each well, vortex the dilution tube before adding 200 μL of the dilution tube into three wells. Start with the 10^{-8} wells and proceed from the highest dilution to the lowest dilution (10^{-2}).
4. Place the dish(es) in a 37 °C with 5% CO_2 incubator for 60 min. Return every 10 min to gently rock the plate(s).
5. Prepare the overlay medium (as instructed in Subheading 3.2, **step 2** using low-melting agarose or methylcellulose) and warm it to 37 °C in a water bath.
6. After the 1-hour inoculation period, aspirate the inoculum off each well before adding 0.5 mL overlay medium.
7. Incubate the plate(s) for 2–3 days, or until plaques are clearly visible. For viruses expressing a fluorescent reporter gene, you will have to count the number of foci under a fluorescence microscope. Calculate the virus titer and express in units of pfu/mL.

4 Notes

1. The goal of this step is to find a dilution that produces well-separated regions of viral CPE so that you may isolate a pure clone of the virus. A suitable plaque can be marked using a sharpie under the microscope on the bottom of the dish. To identify a focus or plaque that expresses a fluorescent protein, simply search for and mark discrete foci of fluorescence on the bottom of the dish under a fluorescent microscope prior to isolating virus from plaques.
2. We recommend collecting viral DNA during each round of purification to check the purity. It is useful to pick many candidate plaques in the first round, then focus on your most promising samples in subsequent rounds. Each round is a potentially unique mixture or single clone of virus and should be stored at −80 °C until further purification or characterization.
3. We use a Fisher Scientific cup sonicator set to an amplitude of 50.
4. An ultracentrifuge and a rotor equipped with swinging buckets are needed.

5. After the sucrose solutions are made, they are sterilized by filtration through a 0.2 μm membrane using negative pressure. The sucrose gradient is set up the day before and left at 4 °C undisturbed overnight for the gradient formation.
6. Ten percent bleach or disinfectant is needed to disinfect the liquid before it is taken outside of the biosafety cabinet. Further disinfection (such as autoclaving) may be required before disposal.
7. Be sure to vortex the sample thoroughly before taking out a portion for the next dilution. You must use a new pipette tip each time to prevent carryover.

References

1. FDA guidance for industry: gene therapy clinical trials - observing subjects for delayed adverse events (2006)
2. Fenner F, Henderson D, Arita I et al (1988) Smallpox and its eradication. In: Smallpox and its eradication, vol 6. World Health Organization, Geneva
3. Stewart AJ, Devlin PM (2006) The history of the smallpox vaccine. J Infect 52(5):329–334. https://doi.org/10.1016/j.jinf.2005.07.021
4. La Scola B, Audic S, Robert C et al (2003) A giant virus in amoebae. Science 299 (5615):2033. https://doi.org/10.1126/science.1081867
5. Smith GL, Moss B (1983) Infectious poxvirus vectors have capacity for at least 25,000 base-pairs of foreign DNA. Gene 25(1):21–28. https://doi.org/10.1016/0378-1119(83)90163-4
6. Evans DH, Stuart D, McFadden G (1988) High levels of genetic recombination among cotransfected plasmid DNAs in poxvirus-infected mammalian cells. J Virol 62 (2):367–375
7. Schramm B, Locker JK (2005) Cytoplasmic organization of POXvirus DNA replication. Traffic 6(10):839–846. https://doi.org/10.1111/j.1600-0854.2005.00324.x
8. Yang Z, Martens CA, Bruno DP et al (2012) Pervasive initiation and 3′-end formation of poxvirus postreplicative RNAs. J Biol Chem 287(37):31050–31060. https://doi.org/10.1074/jbc.M112.390054
9. Kotwal GJ, Abrahams M-R (2004) Growing poxviruses and determining virus titer. In: Vaccinia virus and poxvirology. Springer, Berlin, pp 101–112
10. Smallwood SE, Rahman MM, Smith DW et al (2010) Myxoma virus: propagation, purification, quantification, and storage. Curr Protoc Microbiol Chapter 14:Unit 14A 11. https://doi.org/10.1002/9780471729259.mc14a01s17
11. Chan WM, Rahman MM, McFadden G (2013) Oncolytic myxoma virus: the path to clinic. Vaccine 31(39):4252–4258. https://doi.org/10.1016/j.vaccine.2013.05.056
12. Bell J (2014) Oncolytic viruses: immune or cytolytic therapy? Mol Ther 22 (7):1231–1232. https://doi.org/10.1038/mt.2014.94
13. Devaud C, John LB, Westwood JA et al (2013) Immune modulation of the tumor microenvironment for enhancing cancer immunotherapy. Oncoimmunology 2(8): e25961. https://doi.org/10.4161/onci.25961
14. Kirn DH, Thorne SH (2009) Targeted and armed oncolytic poxviruses: a novel multi-mechanistic therapeutic class for cancer. Nat Rev Cancer 9(1):64–71. https://doi.org/10.1038/nrc2545
15. Kirn DH, Wang Y, Le Boeuf F et al (2007) Targeting of interferon-beta to produce a specific, multi-mechanistic oncolytic vaccinia virus. PLoS Med 4(12):e353. https://doi.org/10.1371/journal.pmed.0040353
16. Li J, O'Malley M, Urban J et al (2011) Chemokine expression from oncolytic vaccinia virus enhances vaccine therapies of cancer. Mol Ther 19(4):650–657. https://doi.org/10.1038/mt.2010.312
17. Seubert CM, Stritzker J, Hess M et al (2011) Enhanced tumor therapy using vaccinia virus strain GLV-1h68 in combination with a beta-

galactosidase-activatable prodrug seco-analog of duocarmycin SA. Cancer Gene Ther 18 (1):42–52. https://doi.org/10.1038/cgt.2010.49
18. Zajac P, Oertli D, Marti W et al (2003) Phase I/II clinical trial of a nonreplicative vaccinia virus expressing multiple HLA-A0201-restricted tumor-associated epitopes and costimulatory molecules in metastatic melanoma patients. Hum Gene Ther 14 (16):1497–1510. https://doi.org/10.1089/104303403322495016
19. Loya SM, Zhang X (2015) Enhancing the bystander killing effect of an oncolytic HSV by arming it with a secretable apoptosis activator. Gene Ther 22(3):237–246. https://doi.org/10.1038/gt.2014.113
20. Rojas JJ, Sampath P, Bonilla B et al (2016) Manipulating TLR signaling increases the anti-tumor T cell response induced by viral cancer therapies. Cell Rep 15(2):264–273. https://doi.org/10.1016/j.celrep.2016.03.017
21. Conrad SJ, El-Aswad M, Kurban E et al (2015) Oncolytic tanapoxvirus expressing FliC causes regression of human colorectal cancer xenografts in nude mice. J Exp Clin Cancer Res 34(1):19. https://doi.org/10.1186/s13046-015-0131-z
22. Mansfield DC, Kyula JN, Rosenfelder N et al (2016) Oncolytic vaccinia virus as a vector for therapeutic sodium iodide symporter gene therapy in prostate cancer. Gene Ther 23 (4):357–368. https://doi.org/10.1038/gt.2016.5
23. Zhang Q, Yu YA, Wang E et al (2007) Eradication of solid human breast tumors in nude mice with an intravenously injected light-emitting oncolytic vaccinia virus. Cancer Res 67(20):10038–10046. https://doi.org/10.1158/0008-5472.can-07-0146
24. Wennier ST, Liu J, Li S et al (2012) Myxoma virus sensitizes cancer cells to gemcitabine and is an effective oncolytic virotherapeutic in models of disseminated pancreatic cancer. Mol Ther 20(4):759–768. https://doi.org/10.1038/mt.2011.293
25. Wennier ST, Liu J, McFadden G (2012) Bugs and drugs: oncolytic virotherapy in combination with chemotherapy. Curr Pharm Biotechnol 13(9):1817–1833
26. Yu F, Wang X, Guo ZS et al (2014) T-cell engager-armed oncolytic vaccinia virus significantly enhances antitumor therapy. Mol Ther 22(1):102–111. https://doi.org/10.1038/mt.2013.240
27. Nounamo B, Liem J, Cannon M et al (2017) Myxoma virus optimizes cisplatin for the treatment of ovarian cancer in vitro and in a syngeneic murine dissemination model. Mol Ther Oncolytics 6:90–99. https://doi.org/10.1016/j.omto.2017.08.002
28. Wilkinson MJ, Smith HG, McEntee G et al (2016) Oncolytic vaccinia virus combined with radiotherapy induces apoptotic cell death in sarcoma cells by down-regulating the inhibitors of apoptosis. Oncotarget 7 (49):81208–81222. https://doi.org/10.18632/oncotarget.12820
29. Cripe TP, Ngo MC, Geller JI et al (2015) Phase 1 study of intratumoral Pexa-Vec (JX-594), an oncolytic and immunotherapeutic vaccinia virus, in pediatric cancer patients. Mol Ther 23(3):602–608. https://doi.org/10.1038/mt.2014.243
30. Park BH, Hwang T, Liu TC et al (2008) Use of a targeted oncolytic poxvirus, JX-594, in patients with refractory primary or metastatic liver cancer: a phase I trial. Lancet Oncol 9 (6):533–542
31. Merrick AE, Ilett EJ, Melcher AA (2009) JX-594, a targeted oncolytic poxvirus for the treatment of cancer. Curr Opin Investig Drugs 10(12):1372–1382
32. Gentschev I, Muller M, Adelfinger M et al (2011) Efficient colonization and therapy of human hepatocellular carcinoma (HCC) using the oncolytic vaccinia virus strain GLV-1h68. PLoS One 6(7):e22069. https://doi.org/10.1371/journal.pone.0022069
33. Liu J, Wennier S, McFadden G (2010) The immunoregulatory properties of oncolytic myxoma virus and their implications in therapeutics. Microbes Infect 12 (14–15):1144–1152. https://doi.org/10.1016/j.micinf.2010.08.012
34. McFadden G (2015) The curious road from basic pathogen research to clinical translation. PLoS Pathog 11(6):e1004997. https://doi.org/10.1371/journal.ppat.1004997
35. McFadden G (2005) Poxvirus tropism. Nat Rev Microbiol 3(3):201–213. https://doi.org/10.1038/nrmicro1099
36. Lun XQ, Yang WQ, Alain T et al (2005) Myxoma virus is a novel oncolytic virus with significant antitumor activity against experimental human gliomas. Cancer Res 65 (21):9982–9990. https://doi.org/10.1158/0008-5472.Can-05-1201
37. Stanford MM, Barrett JW, Nazarian SH et al (2007) Oncolytic virotherapy synergism with

signaling inhibitors: Rapamycin increases myxoma virus tropism for human tumor cells. J Virol 81(3):1251–1260. https://doi.org/10.1128/JVI.01408-06
38. Stanford MM, Shaban M, Barrett JW et al (2008) Myxoma virus oncolysis of primary and metastatic B16F10 mouse tumors in vivo. Mol Ther 16(1):52–59. https://doi.org/10.1038/sj.mt.6300348
39. Tosic V, Thomas DL, Kranz DM et al (2014) Myxoma virus expressing a fusion protein of interleukin-15 (IL15) and IL15 receptor alpha has enhanced antitumor activity. PLoS One 9(10):e109801. https://doi.org/10.1371/journal.pone.0109801
40. Urbasic AS, Hynes S, Somrak A et al (2012) Oncolysis of canine tumor cells by myxoma virus lacking the serp2 gene. Am J Vet Res 73(8):1252–1261. https://doi.org/10.2460/ajvr.73.8.1252
41. MacNeill AL, Moldenhauer T, Doty R et al (2012) Myxoma virus induces apoptosis in cultured feline carcinoma cells. Res Vet Sci 93(2):1036–1038. https://doi.org/10.1016/j.rvsc.2011.10.016
42. Lun XQ, Zhou H, Alain T et al (2007) Targeting human medulloblastoma: oncolytic virotherapy with myxoma virus is enhanced by rapamycin. Cancer Res 67(18):8818–8827. https://doi.org/10.1158/0008-5472.Can-07-1214
43. Bartee MY, Dunlap KM, Bartee E (2016) Myxoma virus induces ligand independent extrinsic apoptosis in human myeloma cells. Clin Lymphoma Myeloma Leuk 16(4):203–212. https://doi.org/10.1016/j.clml.2015.12.005
44. Bartee E, Chan WM, Moreb JS et al (2012) Selective purging of human multiple myeloma cells from autologous stem cell transplantation grafts using oncolytic myxoma virus. Biol Blood Marrow Tr 18(10):1540–1551. https://doi.org/10.1016/j.bbmt.2012.04.004
45. Lilly CL, Villa NY, Lemos de Matos A et al (2017) Ex vivo oncolytic virotherapy with myxoma virus arms multiple allogeneic bone marrow transplant leukocytes to enhance graft versus tumor. Mol Ther Oncolytics 4:31–40. https://doi.org/10.1016/j.omto.2016.12.002
46. Gammon DB, Evans DH (2009) The 3′-to-5′ exonuclease activity of vaccinia virus DNA polymerase is essential and plays a role in promoting virus genetic recombination. J Virol 83(9):4236–4250. https://doi.org/10.1128/JVI.02255-08
47. Shuman S (1991) Site-specific interaction of vaccinia virus topoisomerase I with duplex DNA. Minimal DNA substrate for strand cleavage in vitro. J Biol Chem 266(17):11372–11379
48. Shuman S (1992) Two classes of DNA end-joining reactions catalyzed by vaccinia topoisomerase I. J Biol Chem 267(24):16755–16758
49. Petersen BO, Shuman S (1997) DNA strand transfer reactions catalyzed by vaccinia topoisomerase: hydrolysis and glycerololysis of the covalent protein-DNA intermediate. Nucleic Acids Res 25(11):2091–2097
50. Garcia AD, Otero J, Lebowitz J et al (2006) Quaternary structure and cleavage specificity of a poxvirus holliday junction resolvase. J Biol Chem 281(17):11618–11626. https://doi.org/10.1074/jbc.M600182200
51. Scheiflinger F, Dorner F, Falkner FG (1992) Construction of chimeric vaccinia viruses by molecular cloning and packaging. Proc Natl Acad Sci U S A 89(21):9977–9981
52. Panicali D, Paoletti E (1982) Construction of poxviruses as cloning vectors - insertion of the thymidine kinase gene from herpes-simplex virus into the DNA of infectious vaccinia virus. Proc Natl Acad Sci-Biol 79(16):4927–4931. https://doi.org/10.1073/pnas.79.16.4927
53. Mackett M, Smith GL, Moss B (1982) Vaccinia virus: a selectable eukaryotic cloning and expression vector. Proc Natl Acad Sci U S A 79(23):7415–7419
54. Moss B (1996) Genetically engineered poxviruses for recombinant gene expression, vaccination, and safety. Proc Natl Acad Sci U S A 93(21):11341–11348
55. Liu J, Wennier S, Reinhard M et al (2009) Myxoma virus expressing interleukin-15 fails to cause lethal myxomatosis in European rabbits. J Virol 83(11):5933–5938. https://doi.org/10.1128/JVI.00204-09
56. Barrett JW, Shun Chang C, Wang G et al (2007) Myxoma virus M063R is a host range gene essential for virus replication in rabbit cells. Virology 361(1):123–132. https://doi.org/10.1016/j.virol.2006.11.015
57. Bratke KA, McLysaght A, Rothenburg S (2013) A survey of host range genes in poxvirus genomes. Infect Genet Evol 14:406–425. https://doi.org/10.1016/j.meegid.2012.12.002
58. Barrett JW, Sypula J, Wang F et al (2007) M135R is a novel cell surface virulence factor

of myxoma virus. J Virol 81(1):106–114. https://doi.org/10.1128/JVI.01633-06
59. Ogbomo H, Zemp FJ, Lun X et al (2013) Myxoma virus infection promotes NK lysis of malignant gliomas in vitro and in vivo. PLoS One 8(6):e66825. https://doi.org/10.1371/journal.pone.0066825
60. Liu J, McFadden G (2015) SAMD9 is an innate antiviral host factor with stress response properties that can be antagonized by poxviruses. J Virol 89(3):1925–1931. https://doi.org/10.1128/JVI.02262-14
61. Stewart L, Burgin AB (2005) Whole gene synthesis: a gene-O-matic future. Front Drug Des Discov 1(1):297–341. https://doi.org/10.2174/1574088054583318
62. Eindhoven T (2015) The Cloning Guide.80
63. Katzen F (2007) Gateway((R)) recombinational cloning: a biological operating system. Expert Opin Drug Discov 2(4):571–589. https://doi.org/10.1517/17460441.2.4.571
64. Liang X, Peng L, Baek CH et al (2013) Single step BP/LR combined Gateway reactions. BioTechniques 55(5):265–268. https://doi.org/10.2144/000114101
65. Mohamed MR, Rahman MM, Lanchbury JS et al (2009) Proteomic screening of variola virus reveals a unique NF-kappaB inhibitor that is highly conserved among pathogenic orthopoxviruses. Proc Natl Acad Sci U S A 106(22):9045–9050. https://doi.org/10.1073/pnas.0900452106
66. Liu J, Wennier S, Zhang L et al (2011) M062 is a host range factor essential for myxoma virus pathogenesis and functions as an antagonist of host SAMD9 in human cells. J Virol 85(7):3270–3282. https://doi.org/10.1128/JVI.02243-10
67. Smallwood SE, Rahman MM, Werden SJ et al (2011) Production of Myxoma virus gateway entry and expression libraries and validation of viral protein expression. Curr Protoc Microbiol Chapter 14:Unit 14A 12. https://doi.org/10.1002/9780471729259.mc14a02s21
68. Broyles SS (2003) Vaccinia virus transcription. J Gen Virol 84(Pt 9):2293–2303. https://doi.org/10.1099/vir.0.18942-0
69. Yang Z, Maruri-Avidal L, Sisler J et al (2013) Cascade regulation of vaccinia virus gene expression is modulated by multistage promoters. Virology 447(1–2):213–220. https://doi.org/10.1016/j.virol.2013.09.007
70. Keck JG, Baldick CJ, Moss B (1990) Role of DNA replication in vaccinia virus gene expression: a naked template is required for transcription of three late trans-activator genes. Cell 61(5):801–809
71. Davison AJ, Moss B (1989) Structure of vaccinia virus early promoters. J Mol Biol 210(4):749–769
72. Davison AJ, Moss B (1989) Structure of vaccinia virus late promoters. J Mol Biol 210(4):771–784
73. Skinner MA, Laidlaw SM, Eldaghayes I et al (2005) Fowlpox virus as a recombinant vaccine vector for use in mammals and poultry. Expert Rev Vaccines 4(1):63–76. https://doi.org/10.1586/14760584.4.1.63
74. Chakrabarti S, Sisler JR, Moss B (1997) Compact, synthetic, vaccinia virus early/late promoter for protein expression. BioTechniques 23(6):1094–1097
75. Sanchez-Sampedro L, Perdiguero B, Mejias-Perez E et al (2015) The evolution of poxvirus vaccines. Viruses 7(4):1726–1803. https://doi.org/10.3390/v7041726
76. Cochran MA, Puckett C, Moss B (1985) In vitro mutagenesis of the promoter region for a vaccinia virus gene: evidence for tandem early and late regulatory signals. J Virol 54(1):30–37
77. Thomas DL, Doty R, Tosic V et al (2011) Myxoma virus combined with rapamycin treatment enhances adoptive T cell therapy for murine melanoma brain tumors. Cancer Immunol Immunother 60(10):1461–1472. https://doi.org/10.1007/s00262-011-1045-z
78. Weir JP, Moss B (1987) Determination of the promoter region of an early vaccinia virus gene encoding thymidine kinase. Virology 158(1):206–210. https://doi.org/10.1016/0042-6822(87)90254-6
79. Baldick CJ Jr, Moss B (1993) Characterization and temporal regulation of mRNAs encoded by vaccinia virus intermediate-stage genes. J Virol 67(6):3515–3527
80. Hagen CJ, Titong A, Sarnoski EA et al (2014) Antibiotic-dependent expression of early transcription factor subunits leads to stringent control of vaccinia virus replication. Virus Res 181:43–52. https://doi.org/10.1016/j.virusres.2013.12.033
81. Senkevich TG, Ward BM, Moss B (2004) Vaccinia virus A28L gene encodes an essential protein component of the virion membrane

with intramolecular disulfide bonds formed by the viral cytoplasmic redox pathway. J Virol 78(5):2348–2356
82. Hammond JM, Oke PG, Coupar BE (1997) A synthetic vaccinia virus promoter with enhanced early and late activity. J Virol Methods 66(1):135–138
83. Rice AD, Gray SA, Li Y et al (2011) An efficient method for generating poxvirus recombinants in the absence of selection. Viruses 3 (3):217–232. https://doi.org/10.3390/v3030217
84. Yao XD, Evans DH (2003) High-frequency genetic recombination and reactivation of orthopoxviruses from DNA fragments transfected into leporipoxvirus-infected cells. J Virol 77(13):7281–7290. https://doi.org/10.1128/Jvi.77.13.7281-7290.2003
85. Di Lullo G, Soprana E, Panigada M et al (2010) The combination of marker gene swapping and fluorescence-activated cell sorting improves the efficiency of recombinant modified vaccinia virus Ankara vaccine production for human use. J Virol Methods 163 (2):195–204. https://doi.org/10.1016/j.jviromet.2009.09.016
86. Franke CA, Rice CM, Strauss JH et al (1985) Neomycin resistance as a dominant selectable marker for selection and isolation of vaccinia virus recombinants. Mol Cell Biol 5 (8):1918–1924. https://doi.org/10.1128/Mcb.5.8.1918
87. Zhou J, Crawford L, Sun XY et al (1991) The hygromycin-resistance-encoding gene as a selection marker for vaccinia virus recombinants. Gene 107(2):307–312
88. Falkner FG, Moss B (1988) Escherichia coli gpt gene provides dominant selection for vaccinia virus open reading frame expression vectors. J Virol 62(6):1849–1854
89. Isaacs SN, Kotwal GJ, Moss B (1990) Reverse guanine phosphoribosyltransferase selection of recombinant vaccinia viruses. Virology 178(2):626–630
90. Lorenzo MM, Blasco R (1998) PCR-based method for the introduction of mutations in genes cloned and expressed in vaccinia virus. BioTechniques 24(2):308–313
91. Wong YC, Lin LCW, Melo-Silva CR et al (2011) Engineering recombinant poxviruses using a compact GFP-blasticidin resistance fusion gene for selection. J Virol Methods 171(1):295–298. https://doi.org/10.1016/j.jviromet.2010.11.003
92. Guo ZS, Liu Z, Sathaiah M et al (2017) Rapid generation of multiple loci-engineered marker-free poxvirus and characterization of a clinical-grade oncolytic vaccinia virus. Mol Ther Methods Clin Dev 7:112–122. https://doi.org/10.1016/j.omtm.2017.09.007
93. Dambach MJ, Trecki J, Martin N et al (2006) Oncolytic viruses derived from the γ34.5-deleted herpes simplex virus recombinant R3616 encode a truncated UL3 protein. Mol Ther 13(5):891–898
94. Rintoul JL, Wang J, Gammon DB et al (2011) A selectable and excisable marker system for the rapid creation of recombinant poxviruses. PLoS One 6(9):e24643. https://doi.org/10.1371/journal.pone.0024643
95. Grosenbach DW, Jordan R, Hruby DE (2011) Development of the small-molecule antiviral ST-246 as a smallpox therapeutic. Futur Virol 6(5):653–671. https://doi.org/10.2217/fvl.11.27
96. Olson VA, Smith SK, Foster S et al (2014) In vitro efficacy of brincidofovir against variola virus. Antimicrob Agents Chemother 58 (9):5570–5571. https://doi.org/10.1128/AAC.02814-14
97. Trost LC, Rose ML, Khouri J et al (2015) The efficacy and pharmacokinetics of brincidofovir for the treatment of lethal rabbitpox virus infection: a model of smallpox disease. Antivir Res 117:115–121. https://doi.org/10.1016/j.antiviral.2015.02.007
98. Guimaraes AP, de Souza FR, Oliveira AA et al (2015) Design of inhibitors of thymidylate kinase from Variola virus as new selective drugs against smallpox. Eur J Med Chem 91:72–90. https://doi.org/10.1016/j.ejmech.2014.09.099
99. Lederman ER, Davidson W, Groff HL et al (2012) Progressive vaccinia: case description and laboratory-guided therapy with vaccinia immune globulin, ST-246, and CMX001. J Infect Dis 206(9):1372–1385. https://doi.org/10.1093/infdis/jis510
100. Vora S, Damon I, Fulginiti V et al (2008) Severe eczema vaccinatum in a household contact of a smallpox vaccinee. Clin Infect Dis 46(10):1555–1561. https://doi.org/10.1086/587668
101. Adelfinger M, Bessler S, Frentzen A et al (2015) Preclinical testing oncolytic vaccinia virus strain GLV-5b451 expressing an anti-VEGF single-chain antibody for canine cancer therapy. Viruses 7(7):4075–4092. https://doi.org/10.3390/v7072811
102. Patil SS, Gentschev I, Nolte I et al (2012) Oncolytic virotherapy in veterinary medicine: current status and future prospects for canine patients. J Transl Med 10:3. https://doi.org/10.1186/1479-5876-10-3

103. Paszkowski P, Noyce RS, Evans DH (2016) Live-cell imaging of vaccinia virus recombination. PLoS Pathog 12(8):e1005824. https://doi.org/10.1371/journal.ppat.1005824
104. Villa NY, Bartee E, Mohamed MR et al (2010) Myxoma and vaccinia viruses exploit different mechanisms to enter and infect human cancer cells. Virology 401 (2):266–279. https://doi.org/10.1016/j.virol.2010.02.027
105. Fang Q, Yang L, Zhu W et al (2005) Host range, growth property, and virulence of the smallpox vaccine: vaccinia virus Tian Tan strain. Virology 335(2):242–251. https://doi.org/10.1016/j.virol.2005.02.014
106. Henderson B (1996) In: Ausubel F, Brent R, Kingston RE, Moore DD, Seidman JG, Smith JA, Struhl K (eds) Short protocols in molecular biology, 3rd edn. John Wiley and Sons, Chichester xxxii+ 700 pages,£ 6.00 (1995). Wiley Online Library
107. Liu J, Wennier S, Moussatche N et al (2012) Myxoma virus M064 is a novel member of the poxvirus C7L superfamily of host range factors that controls the kinetics of myxomatosis in European rabbits. J Virol 86 (9):5371–5375. https://doi.org/10.1128/jvi.06933-11

Part IV

Viral Vector Delivery

Chapter 12

AAV-Mediated Gene Delivery to the Mouse Liver

Sharon C. Cunningham and Ian E. Alexander

Abstract

The liver is an attractive target for gene therapy due to the high incidence of liver disease phenotypes. Adeno-associated viral vectors (AAV) are currently the most popular gene delivery system for targeting the liver, reflecting high transduction efficiency in vivo and the availability of a toolkit of multiple different capsids with high liver tropism. While AAV vectors confer stable gene transfer in the relatively quiescent adult liver, the predominantly episomal nature of AAV vector genomes results in less stable expression in the growing liver as a consequence of episome clearance during hepatocellular replication. This is an important consideration in experimental design involving young animals, particularly mice, where liver growth is rapid. Given the immense value of murine models for dissecting disease pathophysiology, experimental therapeutics and vector development, this technical manuscript focuses on AAV-mediated transduction of the mouse liver. Xenograft models, in which chimeric mouse-human livers can be established, are also amenable to AAV-mediated gene transfer and have proven to be powerful tools for in vivo selection and characterization of novel human-specific capsids. While yet to be confirmed, such models have the potential to more accurately predict transduction efficiency of clinical candidate vectors than nonhuman primate models.

Key words Mouse, Liver, Hepatocyte, Gene transfer, Adeno-associated virus, Viral vector, Metabolic, Hepatocellular, Transposon

1 Introduction

1.1 The Liver as a Target for Gene Transfer

Many genetic and acquired diseases involve the liver, either as a primary site of pathology or as a component of systemic disease. This reflects the many vital metabolic, biosynthetic, storage and detoxification functions performed by the liver, and renders the liver a highly attractive target for therapeutic gene transfer [1]. The liver can also be used as a site of ectopic expression of potentially therapeutic proteins, such as immunomodulatory molecules in the context of organ and tissue transplantation [2]. Among contemporary vector systems those based on recombinant Adeno-associated virus (rAAV) are the most widely used for liver-targeted gene transfer, reflecting the high efficiency of the AAV8 capsid in the murine liver observed in early preclinical studies, and

Fredric P. Manfredsson and Matthew J. Benskey (eds.), *Viral Vectors for Gene Therapy: Methods and Protocols*, Methods in Molecular Biology, vol. 1937, https://doi.org/10.1007/978-1-4939-9065-8_12,

therapeutically successful use in human clinical trials [3–6]. Capsid engineering strategies are progressively improving species- and tissue-specific tropism, particularly with a view to human therapy [7].

Mouse models are an invaluable tool for disease modeling, experimental therapeutics and the in vivo testing and development of gene transfer and genome editing technologies. They are relatively inexpensive to maintain, easy to handle, have large litters, and many disease models exist either through spontaneous mutation or through transgenic technology. Furthermore, humanized mouse models, such as the FRG mouse [8] in which chimeric mouse-human livers can be established, are excellent surrogate systems for direct studies in humans, and have recently been used successfully to develop novel AAV capsids with unprecedented tropism for primary human hepatocytes [7, 9].

1.2 Influence of Target Cell Proliferation State on AAV-Mediated Gene Transfer

While AAV vectors can integrate into the host genome [10], the majority of input vector genomes persist as transcriptionally active episomes that are relatively stable in tissues with low rates of cellular proliferation, such as the adult mouse liver. In contrast, episomal vector genomes are rapidly lost from more highly proliferative tissues, such as the growing mouse liver, by mechanisms that involve both dilution and degradation [11–13]. Since many disease phenotypes that are potentially amenable to gene therapy require therapeutic intervention early in life, both the proliferative state of the target cell and the level of gene modification required for therapeutic effect require consideration. For example, correction of intracellular (cell autonomous) defects such as ornithine transcarbamylase (OTC) deficiency will require a higher proportion of cells to be stably gene-modified than disorders involving secreted proteins, such as hemophilia B. In the OTC-deficient *spf*^{ash} mouse model, rAAV-mediated liver-targeted delivery of a therapeutic transgene, at a dose that transduces ~100% of hepatocytes, results in long-term phenotype correction in adult mice [14, 15]. Treatment in the newborn period, however, results in transient phenotype correction as a consequence of almost complete clearance of episomal vector genomes, with the small proportion of hepatocytes carrying stable integration events (~8%) being insufficient for disease control [16]. In the context of hemophilia B this level of stable gene transfer would almost certainly be therapeutic.

1.3 Overcoming the Effects of Hepatocellular Proliferation

We have explored a number of ways to overcome the effects of hepatocellular proliferation on AAV vector-mediated gene delivery. Vector readministration is one approach to counter growth-associated loss of transgene expression, provided the challenge of anti-vector humoral immunity is addressed. In a neonatal lethal mouse model of argininosuccinate synthetase (ASS) deficiency, vector delivery/redelivery prenatally and postnatally, with alternate capsids (and cross-fostering of vector-treated pups to vector-naive dams), resulted in mice surviving to adulthood [17]. Another strategy explored was the development of a hybrid rAAV/transposon system based on piggyBac (PB) [18]. The transposase coding sequence and the terminal inverted repeat (TIR) elements required for transposition were vectorized separately, with the TIR elements flanking the transgene cassette containing the gene of interest and the transposase delivered via a conventional rAAV. Using this system, phenotype correction and survival to adulthood was achieved with a single vector dose delivered prenatally in the same ASS mouse model.

1.4 Routes of Vector Delivery and Transgene Detection

The most efficient routes of delivery to the mouse liver are intraperitoneal (ip) and intravenous (iv), either systemically via the tail vein (tv) or directly to the liver via the portal vein (pv). There are advantages and disadvantages to each. Vector administered by ip injection is absorbed rapidly, primarily through the portal circulation, passing through the liver before reaching other organs [19]. The technique is relatively simple, and preferable when delivery to prenatal, newborn, and juvenile mice is required. However, there is a risk of inadvertent injection into the gut, abdominal fat, and subcutaneous tissues, and experimental design in terms of cohort size and downstream analyses should take this into account.

Direct iv delivery is of similar efficiency, but greater reliability [20]. Injection via the tv results in a rapid accumulation of high levels of vector particles in the liver, as the liver receives nearly 25% of cardiac output [21]. Portal vein injection delivers vector directly to the liver, but is technically more difficult, as a laparotomy is required.

Methods for assessing the success of gene delivery, level of transgene expression and functionality depend on the gene of interest and expected outcome (gene addition, gene knockdown, gene editing and associated phenotypic changes). Given the differences in efficiency and reproducibility between delivery techniques, it is important to correlate outcome not only with the input vector dose but also to the number of vector genomes reaching the target tissue. Analyses include detection of biochemical markers in blood and urine, quantitation of vector copies, determination of relative mRNA and protein expression in whole liver lysates, and immunohistochemistry/in situ activity assays on fixed liver sections.

2 Materials

2.1 Equipment

1. Needles and syringes (regular).
2. Hamilton syringe (gastight—1700 series) and 33G RN needles (25 mm).

3. Insulin syringes (27G needle).
4. Surgical clippers.
5. Sterile drapes.
6. Sterile gauze.
7. Surgical scissors, forceps, and needle clamps.
8. Sterile cotton swabs.
9. Sutures.
10. Infrared lamp.
11. Rodent restrainer.
12. Heating pad.

2.2 Solutions

1. Inhalant anesthetic (e.g., Isoflurane).
2. Injectable anesthetic (e.g., Xylazine/Ketamine).
3. Analgesic (e.g., Buprenorphine, Carprofen).
4. Antibiotic (e.g., Ampicillin).
5. 70% ethanol.
6. Sterile saline for injection.
7. Povidine-Iodine.
8. Purified rAAV diluted in sterile saline at the required dose/volume for injection (*see* **Note 1**).

3 Methods

3.1 Vector Delivery to the Quiescent (Adult) Mouse Liver

Stable and persistent transgene expression can be achieved in a quiescent adult mouse liver using a conventional rAAV. The dose required to transduce close to 100% of mouse hepatocytes using the AAV8 capsid is approximately 5×10^{10} vg/mouse.

3.1.1 Intraperitoneal Administration

1. Restrain the mouse by gently holding the loose skin at the back of the neck, with the abdomen facing upward.
2. Vector, diluted to the desired dose in sterile saline in a volume of 50–100 μL, is injected into the peritoneal cavity using an insulin syringe (29G needle).
3. The needle is inserted through the abdominal wall into the left lower quadrant of the abdomen at a 30° angle.
4. Inject the vector solution slowly and then withdraw the needle carefully.

3.1.2 Tail Vein Administration

1. Stimulate dilation of the tail vein by placing the mouse in a heated environment (28–30 °C), such as under an infrared lamp, for 10–15 min.

2. Place the mouse in a rodent restrainer.
3. Vector, diluted to the desired dose in sterile saline in a volume of 50–100 μL, is slowly injected into the tail vein using an insulin syringe (29G needle).

3.1.3 Portal Vein Administration

1. Portal vein injections require a laparotomy. Anesthetize mice using a combination of ketamine (100 mg/kg) and xylazine (10 mg/kg), delivered by ip injection. Buprenorphine (0.01 mg/kg) is given by subcutaneous injection.
2. Prepare the abdominal area for surgery in the following way. Shave the abdominal fur using surgical clippers and remove loose hair, then place the mouse on a sterile drape and heating pad. Swab the abdomen with Povidine-Iodine and then with 70% (v/v) ethanol to disinfect.
3. Make a ventral midline incision and then expose the portal vein by gently moving the gut and mesentery aside. Using sterile swabs, slightly elevate the portal vein for ease of access.
4. Vector, diluted to the desired dose in sterile saline in a volume of 50 μL, is slowly injected into the portal vein using a Hamilton syringe (33G needle). After withdrawing the needle, place pressure on the injection point with a sterile swab until hemostasis is achieved.
5. Close muscle and skin layers of the abdominal wall separately with interrupted sutures. Maintain mice on Carprofen in the drinking water (0.14 mL/250 mL) and inject with Ampicillin (5 mg/30 gm in 100 μL) subcutaneously for 3 days post-injection. Monitor for signs of ill health and weight loss.

3.2 Vector Delivery to a Growing Liver Undergoing Rapid Hepatocellular Proliferation

If only a small proportion of stably gene-modified cells are required, or transient gene expression is desirable, a conventional AAV system can be used. In the newborn mouse vector doses in the 5×10^{10} to 2.5×10^{11} vg/mouse range result in 2–8% of hepatocytes being stably gene-modified. To achieve a high proportion (~35–70%) of stably gene-modified hepatocytes in the growing mouse liver use the hybrid AAV/PB system, at a dose of 1×10^{11} vg/mouse for the AAV/transposon and 5×10^{10} vg/mouse for the AAV/PB [18].

3.2.1 Prenatal Delivery to the Fetus via the Intraperitoneal (Ip) Route

1. Fetuses are injected at stage E15, to allow time for the incision site to heal before parturition. Anesthetize pregnant females using isoflurane inhalation anesthetic (*see* **Note 2**) and administer buprenorphine (0.01 mg/kg) by subcutaneous injection.
2. Shave the abdomen using surgical clippers and remove loose hair, then place the mouse on a sterile drape and heating pad. Swab the abdomen with Povidine-Iodine and then with 70% (v/v) ethanol to disinfect.

3. Make a longitudinal incision (1 cm) along the midline to expose the gravid uterus, and place sterile saline-soaked gauze on the skin on either side of the incision. Gently lift one uterine horn out of the abdominal cavity onto the gauze using cotton swabs to minimize damage to the uterus and bruising of fetuses. Bathe internal organs with warm saline to prevent desiccation.
4. Vector, diluted in sterile saline to the required dose in a volume of 5 μL, is injected into the peritoneal cavity of each fetus in the vicinity of the liver, through the uterine wall, using a Hamilton syringe (33G needle).
5. After withdrawing the needle, hold a swab over the injection point for 10 s. After all fetuses in one uterine horn are injected, gently put it back into place, and carry out the same procedure on the other uterine horn.
6. Close the abdominal wall by suturing. Maintain the mice on Carprofen in the drinking water (0.14 mL/250 mL) and inject with Ampicillin (5 mg/30 gm in 100 μL) daily subcutaneously until parturition (designated day 0).

3.2.2 Postnatal Delivery to Newborn/Juvenile Mice via the Intraperitoneal (Ip) Route

1. Injections (ip) are carried out as described in Subheading 3.1.1, with a Hamilton syringe (33G needle) in mice up to 10–14 days of age, and an insulin syringe (29G needle) in older mice. Dilute vector in sterile saline to the required dose in a volume of 10–20 μL per mouse.

4 Notes

1. We routinely produce rAAVs in HEK 293 cells [22] by transfection using standard calcium phosphate/DNA coprecipitation methods, and vector particles are purified from the cell lysate using either CsCl or Iodixanol gradients [23].
2. Isoflurane inhalation was used to anesthetize the pregnant females in order to avoid possible needle injury to the fetus during administration of injectable anesthetics.

Acknowledgments

This work was supported by NHMRC grants APP1008021 and APP1065053 to I.E.A.

References

1. Alexander IE, Kok C, Dane AP et al (2012) Gene therapy for metabolic disorders: an overview with a focus on urea cycle disorders. J Inherit Metab Dis 35(4):641–645
2. Laurence JM, Allen RD, McCaughan GW et al (2009) Gene therapy in transplantation. Transplant Rev (Orlando) 23(3):159–170
3. Gao GP, Alvira MR, Wang L et al (2002) Novel adeno-associated viruses from rhesus monkeys as vectors for human gene therapy. Proc Natl Acad Sci U S A 99:11854–11859
4. Davidoff AM, Gray JT, Ng CY et al (2005) Comparison of the ability of adeno-associated viral vectors pseudotyped with serotype 2, 5, and 8 capsid proteins to mediate efficient transduction of the liver in murine and nonhuman primate models. Mol Ther 11:875–888
5. Nathwani AC, Gray JT, Ng CY et al (2006) Self-complementary adeno-associated virus vectors containing a novel liver-specific human factor IX expression cassette enable highly efficient transduction of murine and nonhuman primate liver. Blood 107:2653–2661
6. Nathwani AC, Reiss UM, Tuddenham EG et al (2014) Long-term safety and efficacy of factor IX gene therapy in haemophilia B. N Engl J Med 371(21):1994–2004
7. Lisowski L, Dane AP, Chu K et al (2014) Selection and evaluation of clinically relevant AAV variants in a xenograft liver model. Nature 506(7488):382–386
8. Azuma H, Paulk N, Ranade A et al (2007) Robust expansion of human hepatocytes in $Fah^{-/-}/Rag2^{-/-}/Il2rg^{-/-}$ mice. Nat Biotechnol 25:903–910
9. Paulk NK, Pekrun K, Zhu E et al (2018) Bioengineered AAV capsids with combined high human liver transduction *in vivo* and unique humoral seroreactivity. Mol Ther 26 (1):289–303
10. Nakai H, Montini E, Fuess S et al (2003) AAV serotype 2 vectors preferentially integrate into active genes in mice. Nat Genet 34 (2):297–302
11. Cunningham SC, Dane AP, Spinoulas A et al (2008) Gene delivery to the juvenile mouse liver using AAV2/8 vectors. Mol Ther 16 (6):1081–1088
12. Nakai H, Yant SR, Storm TA et al (2001) Extrachromosomal recombinant adeno-associated virus vector genomes are primarily responsible for stable liver transduction in vivo. J Virol 75:6969–6976
13. Wang Z, Zhu T, Qiao C et al (2005) Adeno-associated virus serotype 8 efficiency delivers genes to muscle and heart. Nat Biotechnol 23:321–328
14. Cunningham SC, Spinoulas A, Carpenter KH et al (2009) AAV2/8-mediated correction of OTC deficiency is robust in adult but not neonatal *spf^{ash}* mice. Mol Ther 17(8):1340–1346
15. Cunningham SC, Kok CY, Dane AP et al (2011) Induction and prevention of severe hyperammonemia in the *spf^{ash}* mouse model of ornithine transcarbamylase deficiency using shRNA and rAAV-mediated gene delivery. Mol Ther 19(5):854–859
16. Cunningham SC, Kok CY, Spinoulas A et al (2013) AAV-encoded OTC activity persisting to adulthood following delivery to newborn spfash mice is insufficient to prevent shRNA-induced hyperammonaemia. Gene Ther 20 (12):1184–1187
17. Kok CY, Cunningham SC, Carpenter KH et al (2013) Adeno-associated virus-mediated rescue of neonatal lethality in argininosuccinate synthetase-deficient mice. Mol Ther 21 (10):1823–1831
18. Cunningham SC, Siew SM, Hallwirth CV et al (2015) Modeling correction of severe urea cycle defects in the growing murine liver using a hybrid recombinant adeno-associated virus/piggyBac transposase gene delivery system. Hepatology 62:417–428
19. Lukas G, Brindle SD, Greengard P (1971) The route of absorption of intraperitoneally administered compounds. J Pharmacol Exp Ther 178(3):562–564
20. Dane AP, Wowro SJ, Cunningham SC et al (2013) Comparison of gene transfer to the murine liver following intraperitoneal and intraportal delivery of hepatotropic AAV pseudo-serotypes. Gene Ther 20(4):460–464
21. Vollmar B, Menger MD (2009) The hepatic microcirculation: mechanistic contributions and therapeutic targets in liver injury and repair. Physiol Rev 89:1269–1339
22. Graham FL, Smiley J, Russell WC et al (1977) Characteristics of a human cell line transformed by DNA from human adenovirus type 5. J Gen Virol 36:59–74
23. Grieger JC, Choi VW, Samulski RJ (2006) Production and characterization of adeno-associated viral vectors. Nat Protoc 1:1412–1428

Chapter 13

Surgical Methods for Inner Ear Gene Delivery in Neonatal Mouse

Kevin Isgrig and Wade W. Chien

Abstract

Inner ear gene therapy offers great potential as a treatment for hearing loss and dizziness. The surgical method used to deliver gene therapy into the inner ear is a critical step in determining the success of inner ear gene therapy. Here we describe two commonly used surgical methods for gene delivery in neonatal mouse inner ear: the round window approach and the posterior semicircular canal approach. Both of these approaches are effective at delivering gene therapy to the neonatal mouse inner ear.

Key words Gene therapy, Inner ear, Posterior semicircular canal, Round window, Hearing loss, Dizziness

1 Introduction

Inner ear gene therapy is an exciting area of investigation as a potential treatment for hearing loss and dizziness. Several studies have shown functional recovery of hearing and balance functions in mutant mice after inner ear gene therapy delivery [1–7]. One of the key factors in determining the success of inner ear gene therapy is the surgical approach used to access the inner ear. Ideally, the surgical approach would be easy to perform, the anatomic landmarks would be consistent and easy to identify, and the resulting transduction of targeted cell types would be high. In this article, we describe the steps involved with two commonly used surgical approaches for gene delivery in neonatal mouse inner ear. Both of these approaches are effective at delivering gene therapy to the neonatal mouse inner ear. The pros and cons of these two approaches are described.

Fredric P. Manfredsson and Matthew J. Benskey (eds.), *Viral Vectors for Gene Therapy: Methods and Protocols*, Methods in Molecular Biology, vol. 1937, https://doi.org/10.1007/978-1-4939-9065-8_13,

2 Materials

1. Micro-forceps (#5 and #55 Dumont).
2. Micro-scissors.
3. Heating pad.
4. Nanoliter 2000 micro-injector (Nanoliter2000, World Precision Instruments, Sarasota, FL).
5. Glass pipette.
6. 5–0 vicryl sutures.
7. Dissecting microscope or operating microscope.
8. Viral vectors.
9. Glass bead sterilizer.

3 Methods

All instruments should be sterilized by ethylene oxide in the beginning of the experiments. Instruments should be cleaned using bead sterilization in between animals.

3.1 Anesthesia

Hypothermia is used to anesthetize neonatal mice (*see* **Note 1**). This method is used to anesthetize neonatal mice and is approved by American Veterinary Medical Association [2, 3].

1. Prior to surgery, the mother is placed in a cage separate from the litter. During this time, the home cage containing the litter is placed on a recirculating heat pad to keep the mice warm.
2. The first mouse is placed in a heavy-duty glove finger and placed into a bucket of ice for ~2–5 min.
3. After the pup is anesthetized, it is transferred onto a large square commercial plastic freeze pack with a 4 × 4 gauze between the pup and the pack's surface.
4. A heavy-duty glove filled with crushed ice is then molded and placed over the pup. The pup is judged to be under hypothermic anesthesia by the complete lack of any response to various stimuli (including a firm toe pinch). Animals remain on the ice for the duration of the surgery (approximately 5–10 min).

3.2 Surgical Approaches (Fig. 1)

1. Once the animal is anesthetized, the postauricular area is prepped several times, first with iodine, then with alcohol wipes to sterilize the area.
2. A postauricular incision is made ~2 mm behind the pinna with micro-scissors, and the underlying soft tissues are retracted and

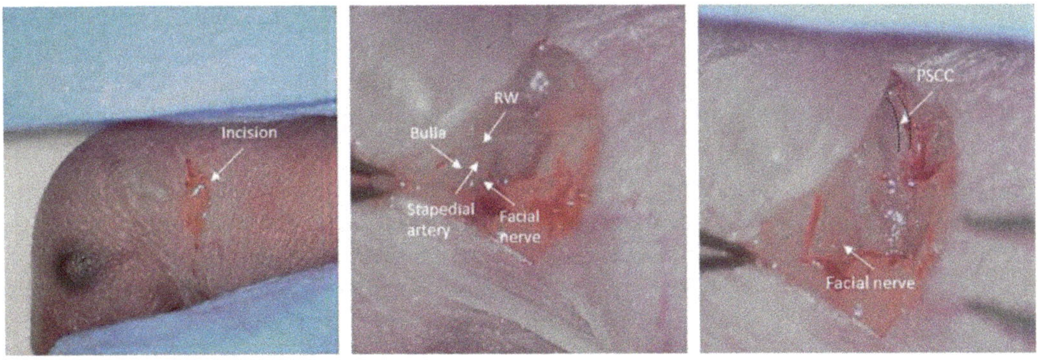

Fig. 1 Intraoperative images showing surgical access to the round window (RW) and posterior semicircular canal (PSCC) in a P0 mouse. The left ear is shown

divided using both #5 Dumont micro-forceps and micro-scissors.

3. The facial nerve is identified. The facial nerve is a key landmark as it lies lateral to the bulla, which is cartilaginous and semi-transparent at this age. The stapedial artery can often be seen through the bulla at this age, which is also a useful landmark.
4. For round window delivery, the round window niche is often visible through the semitransparent bulla, adjacent to the stapedial artery. If the round window niche is clearly visible, gene delivery can be performed through the bulla as described below. If the round window niche is not visible, then a #55 micro-forceps is used to peal the bulla open to visualize the round window *see* **Notes 2** and **3**).
5. For posterior semicircular canal delivery (*see* **Note 4**), the facial nerve is followed superiorly and posteriorly to locate the posterior semicircular canal (PSCC). The facial nerve runs medial to the sternocleidomastoid muscle, which should be divided with micro-scissors to expose the PSCC. The soft tissue overlying the posterior semicircular canal is removed with micro-scissors.
6. Solution containing viral vectors is loaded into a glass micropipette mounted onto a micro-injector (*see* **Note 5**).
7. For round window delivery, if the round window is clearly visible through the bulla, the glass micropipette can penetrate through the bulla and into the round window. If the round window is not clearly visible, it is advisable to gently remove the bulla wall over the round window, and the round window membrane can be penetrated with the glass micropipette.
8. For posterior semicircular canal delivery, the PSCC is cartilaginous at this age and it is penetrated using a glass micropipette.

9. Viral gene therapy is injected into the inner ear using the micro-injector system. Typically ~1 μL of viral gene therapy is injected into the neonatal mouse inner ear (*see* **Note 6**).
10. The glass micropipette is carefully withdrawn.

3.3 Post-operative Care

1. After surgery, mice are placed in a cage on a warming pad to maintain body temperature during recovery from anesthesia, with constant manual stimulation/rolling with warm thinly gloved human fingers. Warming is successful when the pup begins to breathe in approximately 5 min.
2. Typically it takes an additional 5 min of manual stimulation and warming before the pup is responsive.
3. Once the pups recover from anesthesia, they are placed in their home cage as a group (also set upon a warming pad) (*see* **Note 7**).
4. Prior to reintroducing the mother, each pup should be caressed with a cotton swab that has been exposed to the home cage bedding. The purpose of this is to have the mice smell as they did prior to surgery, which increases the likelihood of the mother re-accepting her litter post-surgery.
5. If possible, urine from the mother can be collected and rubbed on the pups using the cotton swabs to further decrease the likelihood of rejection.
6. Mineral oil is rubbed on the mother's nose using a cotton swab to desensitize her, which increases the likelihood of pup acceptance after surgery [8].

4 Notes

1. It is important to ensure that the total duration of hypothermia is kept below ~15 min to increase the pup survival.
2. In our experience, both the posterior semicircular canal approach and the round window approach are capable of delivering viral vectors to the mouse inner ear (Fig. 2). In the past, the round window approach was most commonly used to access the cochlea, whereas the posterior semicircular canal was typically used to access the vestibular organs [9]. We have shown that the posterior semicircular canal injections of viral gene therapy resulted in high efficiency of hair cell transduction in both the vestibular organs and the cochlea [5]. This may be due to the relatively large cochlear aqueduct in rodents, which is located at the cochlear base, adjacent to the round window. Therefore, viral vectors injected through the round window could be diverted into the cochlear aqueduct, reducing its ability to perfuse the entire inner ear [10].

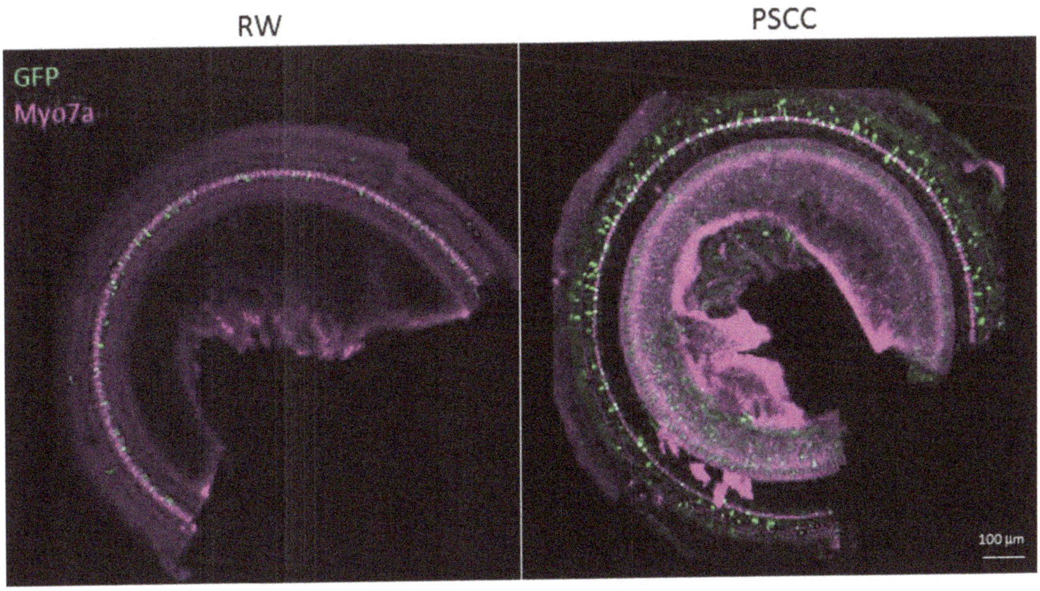

Fig. 2 Injection of AAV8-GFP into neonatal mouse inner ear through the round window and posterior semicircular canal resulted in GFP expression in cochlear inner hair cells. Hair cells are labeled with anti-Myo7a antibody (shown in magenta)

3. When the bulla is transparent, the round window can be easily located and viral vectors can be delivered without opening the bulla. However, the injection micropipette needs to penetrate through two barriers (bulla wall and the round window membrane), which can potentially make the gene delivery less consistent. If the bulla is not transparent, it will need to be opened in order to visualize the round window membrane directly, which increases the risk of hearing loss.
4. While our published data showed higher cochlear hair cell infection efficiency with the posterior semicircular canal approach compared to round window approach, other inner ear gene therapy studies have also achieved excellent results with the round window approach [3, 4].
5. When injecting the viral vectors into the mouse inner ear, it is important to ensure the tip of the micropipette is not blocked by debris and the viral vector solution flows properly into the inner ear.
6. In our lab, we typically inject a total of 1 μL of viral vector solution into the neonatal mouse inner ear. In our hands, this volume is sufficient to perfuse the entire mouse inner ear without causing noticeable damage. However, various volumes have been used in other studies for inner ear therapy [1, 3, 4].
7. After the surgeries are completed and the pups are placed back into the home cage with the mother, it is important to observe

their interactions for ~10–15 min to ensure the mother accepts the pups, in order to decrease the probability of cannibalism.

Acknowledgment

This work was supported by funds from the NIDCD Division of Intramural Research/NIH (DC000082-02 to W.W.C., as well as DC000081 to advanced imaging core). We are grateful for the NIDCD animal facility staff for caring for our animals. We would like to thank Dr. Lisa Cunningham and Dr. Nicole Schmitt for critiquing the manuscript.

Disclosures: The authors have no conflict of interest to disclose.

References

1. Emptoz A, Michel V, Lelli A et al (2017) Local gene therapy durably restores vestibular function in a mouse model of Usher syndrome type 1G. Proc Natl Acad Sci U S A 114:9695–9700
2. Askew C, Rochat C, Pan B et al (2015) Tmc gene therapy restores auditory function in deaf mice. Sci Transl Med 7:295ra108
3. Akil O, Seal RP, Burke K et al (2012) Restoration of hearing in the VGLUT3 knockout mouse using virally mediated gene therapy. Neuron 75:283–293
4. Pan B, Askew C, Galvin A et al (2017) Gene therapy restores auditory and vestibular function in a mouse model of Usher syndrome type 1c. Nat Biotechnol 35:264–272
5. Isgrig K, Shteamer JW, Belyantseva IA et al (2017) Gene therapy restores balance and auditory functions in a mouse model of Usher syndrome. Mol Ther 25:780–791
6. Lentz JJ, Jodelka FM, Hinrich AJ et al (2013) Rescue of hearing and vestibular function by antisense oligonucleotides in a mouse model of human deafness. Nat Med 19:345–350
7. Shibata SB, Ranum PT, Moteki H et al (2016) RNA interference prevents autosomal-dominant hearing loss. Am J Hum Genet 98:1101–1113
8. Van Sluyters RC, Obernier JA (2004) Guidelines for the care and use of mammals in neuroscience and behavioral research. Contemp Top Lab Anim 43:48
9. Chien WW, Monzack EL, McDougald DS et al (2015) Gene therapy for sensorineural hearing loss. Ear Hear 36:1–7
10. Salt AN, Gill RM, Hartsock JJ (2015) Perilymph kinetics of FITC-dextran reveals homeostasis dominated by the cochlear aqueduct and cerebrospinal fluid. J Assoc Res Otolaryngol 16:357–371

Chapter 14

Gene Transfer to Mouse Kidney In Vivo

C. J. Rocca and S. Cherqui

Abstract

Genetic nephropathies represent a challenging class of disorders to be treated by gene therapy. This is primarily due to the filtering properties of the kidney itself, which does not allow the vehicle carrying the transgene of interest to remain long enough in the organ to penetrate efficiently into the nephrotic cells. Also, the kidney has a complex anatomical structure composed of different cell types compartmentalized within isolated anatomic structures that limit their access. Here, we describe a simple surgical procedure to deliver recombinant adeno-associated virus (rAAV) to the whole kidney based on the hydraulic force of the retrograde renal vein injection. In its clinical form, this procedure would correspond to a renal venography where a catheter is threaded retrograde from the femoral vein under fluoroscopic guidance.

Key words Mouse, Kidney, Gene therapy, Genetic, Nephropathies, Recombinant adeno-associated virus, Renal vein injection

1 Introduction

1.1 General Introduction

In the past decades, a lot of effort has been dedicated to develop efficient and safe gene transfer technologies. Gene replacement therapy appears to be an attractive approach for monogenetic disorders. However, getting a gene expressed in a physiological manner, in a specific location without risk of oncogenic insertion in the genome, is not trivial. First of all, it requires a good vehicle. There are two ways to deliver a gene: (a) Nonviral approach: naked DNA is delivered to the tissue using physical or chemical methods to enhance gene delivery. This strategy presents with the advantage of having simple large-scale production and low host immunogenicity. However, there is low efficiency rate due to difficulty crossing physical body barriers. (b) The viral approach, on other hand, uses the infective properties of viruses to deliver a payload gene into cells and is consequently more efficient but with some risk of immunogenicity and/or insertional mutagenesis. However, genetic manipulations of the viral vectors considerably improve safety for in vivo usage. That said, it is crucial to opt for an adequate

Fredric P. Manfredsson and Matthew J. Benskey (eds.), *Viral Vectors for Gene Therapy: Methods and Protocols*, Methods in Molecular Biology, vol. 1937, https://doi.org/10.1007/978-1-4939-9065-8_14,

delivery method to achieve tissue-specific gene expression without off-site vector sequestration and viral inactivation. Furthermore, it is also important that the chosen delivery method does not artificially exacerbate the disease state by inducing, for instance, ischemia or traumatic injury of the organ. In this chapter, we describe an efficient and safe method for kidney-specific gene transfer in mice.

1.2 Kidney Gene Transfer

Inherited kidney diseases constitute at least 150 different disorders and they have an overall prevalence of about 60–80 cases per 100, 000 in Europe and the USA. At least 10% of adults and nearly all children who progress to renal-replacement therapy have an inherited kidney disease [1]. As with every other type of organ transplant, there are consequent disadvantages associated with renal allografts: (a) severe shortage of donor organs and average waiting time of 3–5 years [2]; (b) short average lifetime of renal allografts (10–15 years) [3]; (c) significant morbidity and mortality associated with transplant rejection [4]; and (d) patients have to take lifelong immunosuppressive drugs that lead to hypertension, increased cholesterol levels, tremors, headache, etc. For monogenic hereditary nephropathies, gene replacement therapy may represent an idealistic alternative. However, this organ is particularly difficult to target in comparison to others (brain, eye, or liver), because of its complex anatomical structure limiting the successful delivery of genetic material. So far, few studies have been performed to deliver genes into the kidney using nonviral and viral vectors, with different routes of injection (*see* Table 1). Recombinant adeno-associated viral (rAAV) seems to be the most promising vehicle for kidney-gene delivery, although the expression is always limited to a restricted part of the kidney. Therefore, rAAVs are the vectors of choice for targeted gene therapy because they overcome many of the problems associated with other vector systems. rAAVs are non-pathogenic and minimally immunogenic, can infect both dividing and nondividing cells, persist after infection, and have extensive cell and tissue tropism. Their main limitation is the size of the transgene capacity (<5 kb).

1.3 Summary

We published, in 2014, a comprehensive study for kidney-specific gene delivery using a procedure and vectors relevant for clinical application in mice [5]. We modified a technique that has first been described by Maruyama et al. in rats [4]. Postinjection, the levels of kidney transduction were investigated in both the cortex and medulla using two different reporter genes analyzed by both qualitative and quantitative assays. Our results showed that retrograde renal vein injection of rAAV9 represents a promising strategy for the treatment of kidney diseases. We also showed that we could significantly improve the specific expression of transgenes to the kidney by using a kidney-specific promoter. This is translationally relevant because there is already a clinical equivalent procedure that

Table 1
Examples of kidney gene delivery

Vehicle	Route of injection	Results	Paper
LacZ expression plasmid	Retrograde renal vein injection	Expression only in interstitial fibroblasts near proximal tubular cells	[7]
Adenoviral vector	– Intrarenal ureteral route. – Intrarenal arterial route.	– Expression in distal tubular and pyelic epithelial cells. – Expression in cortical interstitial cells.	[8]
rAAV2	Parenchymal injection	Low transgene expression in the tubular structures near the point of injection	[9]
rAAV2	Renal arterial injection	Limited transduction of the S3 segments of proximal tubular cells, straight segments of the proximal tubule descending into the outer medulla, for only 6 weeks. Significant inflammation and renal injury were noted and attributed to the procedure	[10]
rAAV serotype 1–5	Catheter inserted into the renal artery	Only rAAV2 could transduce the kidney but only the tubular epithelial cells	[11]
rAAV2, 8, and 9 mutants	Microinjection into the renal cortex	Only rAAV2 mutant led to a robust transgene expression in the distal tubular cells	[12]

is well established in humans: renal venography, a minimally invasive and readily performed outpatient procedure [6]. We already investigated the efficacy of this technique in a mouse model of Glycogen storage disease Ia (GSD-Ia), which improved the fasting blood glucose levels of the mice but also led to normal renal function for 52 weeks postinjection, last time-point tested (https://doi.org/10.1016/S1525-0016(16)33159-8). Here, we describe step by step how to perform this surgical procedure in mice, in order to efficiently and specifically deliver a gene to the kidney. We think that this procedure is scalable and could be performed in larger animals.

2 Materials

1. Mice: C57BL/6 mice aged 8–10 weeks purchased from Jackson Laboratories and bred continuously at UCSD vivarium.
2. rAAV-CMV-Luciferase or rAAV-P1-Luciferase (P1 is a parathyroid hormone receptor "kidney-specific" promoter) serotype 9 particles were produced by and purchased from the University of North Carolina Gene Therapy Center.

3. Phosphate Buffer Saline 1X.
4. BD insulin syringe with the BD Ultra-Fine™ needle 1 mL 31 G × 5/16 in. (8 mm).
5. Anesthetic: Isoflurane.
6. Complete rodent anesthesia system: anesthesia machine, vaporizer, non-rebreathing circuit and mask, evacuation cartridges, dual diverter valves, and induction chamber.
7. Small animal electric clipper.
8. Ethanol 70%.
9. Standard surgical instrument including:
    (a) Surgical scissors.
    (b) Retractor.
    (c) Forceps.
    (d) Vascular clamps 11 mm.
    (e) Clamp applicator.
    (f) Needle holder.
    (g) 5–0 absorbable vicryl suture.
    (h) 5–0 nonabsorbable surgical Ethilon suture.
10. Cotton Q-tip.
11. Tissue adhesive (Vetbond, 3 M).
12. Analgesic: Buprenorphine.
13. Heat pad.
14. IVIS Imaging System 200 Series (Caliper Life Sciences).
15. Luciferin.

3 Methods

During the whole procedure, the surgeon will wear scrubs, sterile latex gloves, Tyvek sleeves, surgical mask, and foot protection. All the surgical instruments will be sterilized with povidone iodine prior each surgery.

3.1 Virus Preparation

1. Dilute the rAAV stock solution in order to obtain 10E11 virus particles in a final volume of 100 μL of PBS-1X (*see* **Note 1**).

3.2 Mouse Preparation

1. Put the mouse in an induction box that delivers 2.5–3% isoflurane and also 5% oxygen by a precision vaporizer to put the mouse in a deeply sleeping state. Then, maintain the mouse in a deeply sleeping state throughout the whole duration of the procedure by using a nose cone associated with the vaporizer to continue delivering isoflurane (*see* **Note 2**).

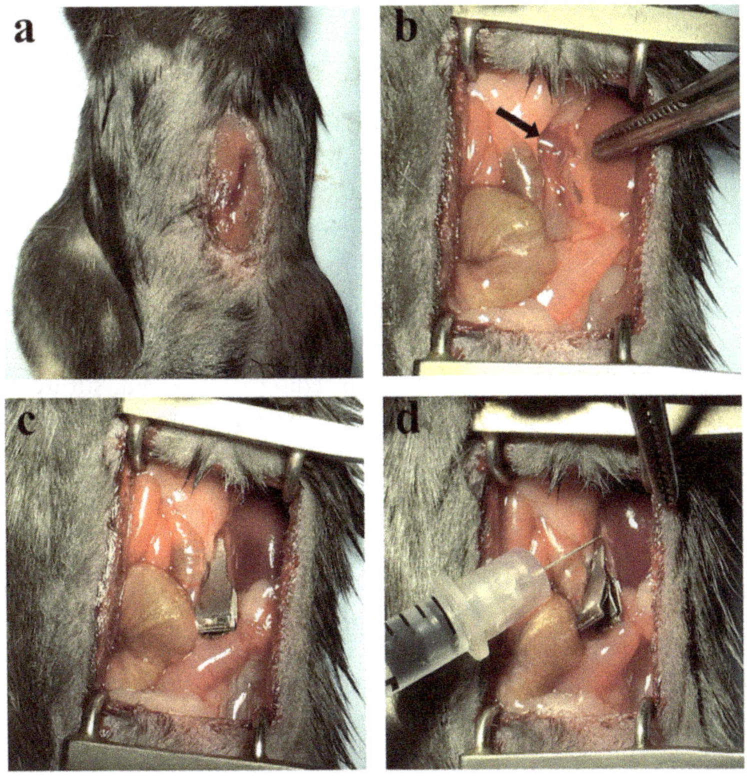

Fig. 1 Different steps of the retrograde renal vein injection. (**a**) Incision in the upper-left quadrant of the mouse abdomen. (**b**) Exposure of the kidney and identification of the renal vein (arrow). (**c**) Positioning of the clamp with the kidney appearing darker after a proper clamping. (**d**) Injection of the rAAV solution in the segment of the vein between the clamp and the kidney

2. Place the mouse on a clean blue pad. Remove hair on the left ventral part of the animal using an electric clipper and disinfect the skin with 70% ethanol.

3.3 Surgical Procedure

1. Perform a small incision in the median section of the abdomen on the left side of the anesthetized mice (Fig. 1a; *see* **Note 3**).
2. Expose the left kidney in order to see the renal vein. Use the retractor to gently push all the organs covering the kidney on each side (Fig. 1b).
3. Prepare the 31G needle syringe with 100 μL of rAAV particles (*see* **Note 4**).
4. Delicately, isolate the vein from the fatty tissue surrounding it using forceps with round points to avoid damaging the vein. Clamp the vein at a location close to the kidney to allow the base of the vein to inflate a little bit (Fig. 1c).
5. Pierce the inflated vein slowly and then quickly and forcefully inject the solution in order to create a hydraulic force that will

infiltrate the solution into all the different structures of the kidney (Fig. 1d; *see* **Note 5**).

6. Remove the needle while applying compression for 30 s with cotton Q tip for hemostasis.
7. Leave the vein clamp for 15 min to allow the rAAV particles to infect the renal cells and also prevent the rAAV from being flushed out through blood flow. Monitor the mouse for normal breathing, gagging, and for any bleeding (*see* **Note 6**).
8. Gently remove the clamp and continue to monitor the mouse for any bleeding for a couple of minutes (*see* **Note 7**).
9. If no bleeding is observed, suture the peritoneum using 5–0 absorbable vicryl suture and suture the skin using 5–0 nonabsorbable surgical Ethilon suture. Clean the skin with a cotton Q tip to remove any blood and apply tissue adhesive on the wound (*see* **Note 8**).
10. At this point, remove the anesthesia nose cone, place the mouse on a heat pad, and inject analgesic for recovery. When the mouse wakes up, it can be then return to its cage.

3.4 Visualization of the rAAV9-Luciferase Animals

1. As early as one week postinjection, animals injected with rAAV9-Luciferase can be visualized by bioluminescence using an *In Vivo* imaging system 10 min after intraperitoneal injection of luciferin (75 μL of a 30 mg/mL solution), (Fig. 2; *see* **Note 9**).

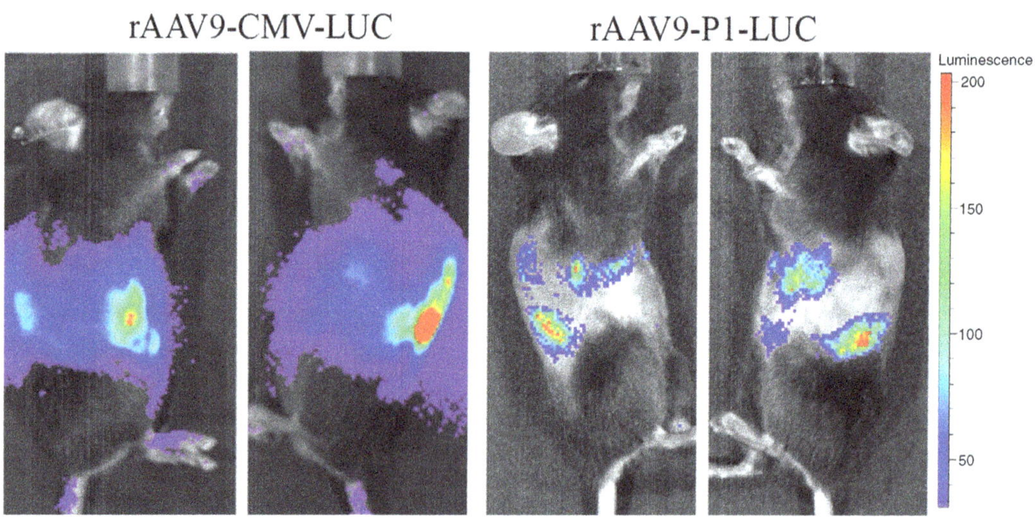

Fig. 2 Comparison of mice after retrograde renal vein injection of rAAV9-CMV-luciferase and rAAV9-P1-luciferase visualized by IVIS. We can observe that even if the retrograde renal vein injection was performed in the left kidney, both left and right kidneys express the transgene, as well as other organs like the liver and spleen. In contrast, using a kidney-specific promoter (P1 = parathyroid hormone receptor "kidney-specific" promoter), we can significantly restrict luciferase expression in kidneys (right panel) compared to a ubiquitous promoter (left panel)

4 Notes

1. Prepare the 100 μL virus preparation in a 1.5 mL Eppendorf tube because the needle does not fit in a 0.5 mL tube. Keep the tube on ice until the injection.
2. It is important to modulate the rate of anesthetic delivered during the surgery to make sure it is only enough to maintain the mouse in deep sleep. Monitor how the mouse is breathing. If it is too slow, remove the nose cone temporarily in order for the animal to get fresh air and return to a normal breathing. Then, replace the nose cone to keep the mouse anesthetized.
3. After the incision, separate the skin from the peritoneum to facilitate the sutures.
4. Make sure there is no bubble.
5. At this point, it is possible to see if the injection was successfully done. One indication is that the kidney turns pale because the blood inside the organ has been flushed out by the rAAV-containing solution. However, if the liquid incorrectly went into the capsule instead of the kidney, it will form a bubble around the kidney. Also, if there is too much bleeding at this point, it might affect the efficiency because of the loss of the rAAV solution into the blood.
6. During the 15 min, it is normal for the kidney to turn very dark because of the blood flow interruption. Again, if breathing is too slow, remove the nose cone in order for the animal to get fresh air and go back to normal breathing.
7. If the mouse is bleeding after removing the clamp, apply compression with cotton Q tip until the bleeding stops.
8. Monitor the sutured skin for a couple of days because the animals have a tendency to scratch and reopen the wound.
9. Luciferase expression gives a global view of where the transgene is expressed. For a cell-specific expression analysis, it is recommended to use a fluorescent transgene (i.e., eGFP), and analyze the tissues by microscopy post-processing.

References

1. Devuyst O, Knoers NV, Remuzzi G et al (2014) Rare inherited kidney diseases: challenges, opportunities, and perspectives. Lancet 383(9931):1844–1859
2. Knoll G (2008) Trends in kidney transplantation over the past decade. Drugs 68(Suppl 1):3–10
3. Davis CL, Delmonico FL (2005) Living-donor kidney transplantation: a review of the current practices for the live donor. J Am Soc Nephrol 16(7):2098–2110
4. Cecka JM (2008) Kidney transplantation in the United States. Clin Transpl 1–18
5. Rocca CJ, Ur SN, Harrison F et al (2014) rAAV9 combined with renal vein injection is optimal for kidney-targeted gene delivery: conclusion of a comparative study. Gene Ther 21 (6):618–628

6. Beckmann CF, Abrams HL (1980) Renal venography: anatomy, technique, applications, analysis of 132 venograms, and a review of the literature. Cardiovasc Intervent Radiol 3 (1):45–70
7. Maruyama H, Hiquchi N, Nishikawa Y et al (2002) Kidney-targeted naked DNA transfer by retrograde renal vein injection in rats. Hum Gene Ther 13(3):455–468
8. Chetboul V, Klonjkowski B, Lefebvre HP et al (2001) Short-term efficiency and safety of gene delivery into canine kidneys. Nephrol Dial Transplant 16(3):608–614
9. Lipkowitz MS, Hanss B, Tulchin N et al (1999) Transduction of renal cells in vitro and in vivo by adeno-associated virus gene therapy vectors. J Am Soc Nephrol 10(9):1908–1915
10. Chen S, Agarwal A, Glushakova OY et al (2003) Gene delivery in renal tubular epithelial cells using recombinant adeno-associated viral vectors. J Am Soc Nephrol 14(4):947–958
11. Takeda S, Takahashi M, Mizukami H et al (2004) Successful gene transfer using adeno-associated virus vectors into the kidney: comparison among adeno-associated virus serotype 1-5 vectors in vitro and in vivo. Nephron Exp Nephrol 96(4):e119–e126
12. Qi YF, Li QH, Shenoy V et al (2013) Comparison of the transduction efficiency of tyrosine-mutant adeno-associated virus serotype vectors in kidney. Clin Exp Pharmacol Physiol 40 (1):53–55

Chapter 15

Co-Delivery of a Short-Hairpin RNA and a shRNA-Resistant Replacement Gene with Adeno-Associated Virus: An Allele-Independent Strategy for Autosomal-Dominant Retinal Disorders

Michael T. Massengill, Brianna M. Young, Alfred S. Lewin, and Cristhian J. Ildefonso

Abstract

Recombinant adeno-associated virus (rAAV) has become an important gene delivery vector for the treatment of inherited retinal degenerative diseases. Many of the mutations leading to retinal degeneration are inherited in an autosomal-dominant pattern and can produce toxic gain-of-function and/or dominant-negative effects. Here we describe an allele-independent gene therapy strategy with rAAV to treat autosomal-dominant retinal degenerative diseases. In this methodology, we co-deliver a short-hairpin RNA (shRNA) to inhibit expression of both the toxic and (WT) copies of the gene as well as an shRNA-resistant cDNA for functional gene replacement with a rAAV.

Key words Gene therapy, Recombinant adeno-associated virus, Autosomal-dominant, Retinal degeneration, Allele-independent, Short-hairpin RNA, shRNA-resistant cDNA

1 Introduction

Mutations in over 250 genes can cause retinal degeneration [1]. In the case of autosomal-recessive or X-linked mutations, delivery of a wild-type (WT) copy of the gene with a rAAV should be sufficient to restore retinal physiology, which has been demonstrated in Leber Congenital Amaurosis 2 (LCA2) [2–4]. However, many of these mutations, such as those seen in rhodopsin (RHO) [5] and bestrophin (BEST1) [6], are inherited in an autosomal-dominant pattern and may demonstrate toxic gain-of-function and/or dominant-negative effects. Treatment of autosomal-dominant retinal degenerative disease entails removal of the expression of the mutated gene and replacing its lost expression to restore homeostasis.

Fredric P. Manfredsson and Matthew J. Benskey (eds.), *Viral Vectors for Gene Therapy: Methods and Protocols*, Methods in Molecular Biology, vol. 1937, https://doi.org/10.1007/978-1-4939-9065-8_15,

The RNA interference (RNAi) pathway in eukaryotic cells represents a powerful tool to inhibit the expression of specific genes [7]. The effector of the RNAi pathway is the RNA-induced silencing complex (RISC) composed of Argonaute proteins bound to an approximately 22 nucleotide single-stranded RNA (ssRNA): either a microRNA (miRNA) or a small-interfering RNA (siRNA). The RISC-bound ssRNA, also known as the guide strand of the siRNA or miRNA, can base-pair with a complimentary target site on an mRNA in order to inhibit expression of that mRNA (Fig. 1a).

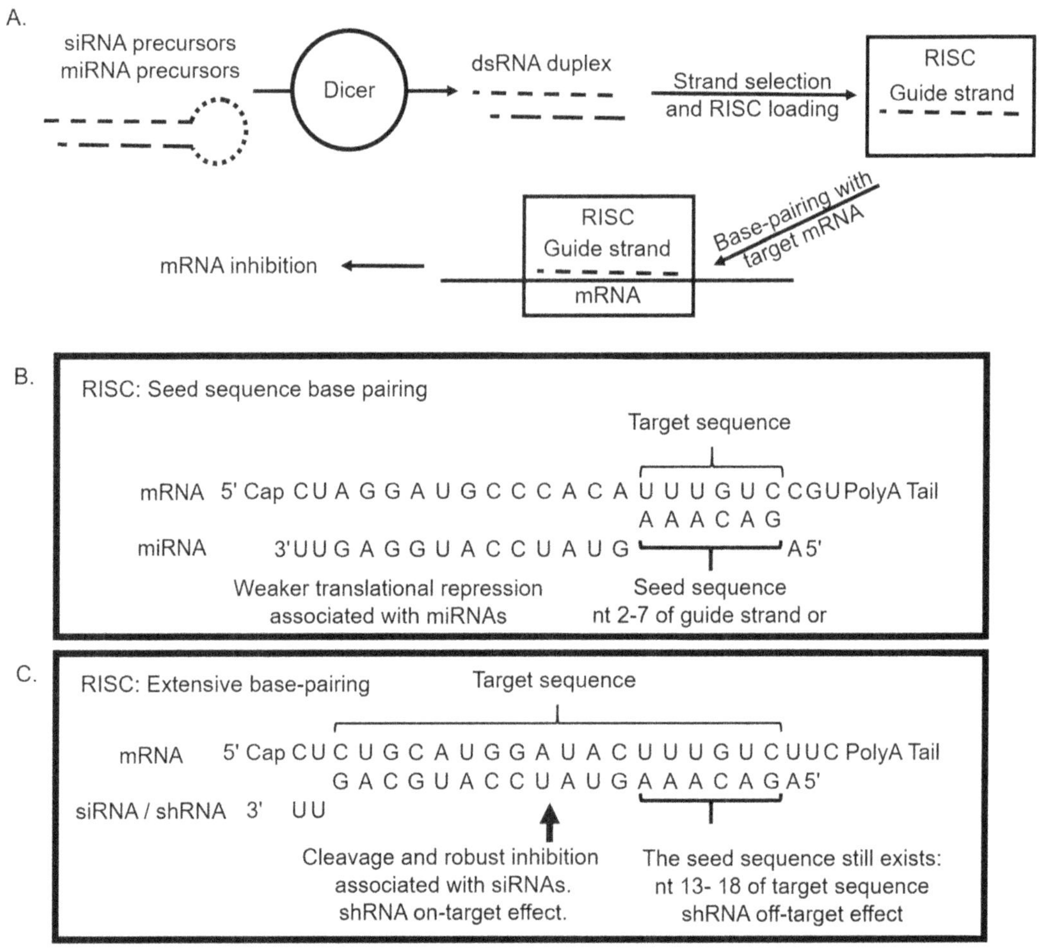

Fig. 1 The RISC complex and mRNA silencing. (**a**) siRNAs and miRNAs are generated from longer RNA precursors molecules that are processed by Dicer, an RNAseIII, into short ~20-nt dsRNA duplexes. shRNAs share a common structure with pre-miRNAs, allowing them to be processed by Dicer and enter the RNAi pathway. Only one strand of the RNA duplex is incorporated into RISC. The ssRNA in RISC can base-pair with complementary mRNAs to induce their inhibition. (**b**) The magnitude of mRNA inhibition is dependent on the extent of mRNA binding. If the shRNA guide strand binds like a miRNA only with a seed sequence, then inhibition is weaker. (**c**) However, if the shRNA binds to its mRNA extensively, like an siRNA, then cleavage of that mRNA and robust inhibition can occur. Even with extensive binding, the seed sequence can still bind to other mRNAs, producing off-target effects

The extent of base-pairing between the guide strand in RISC and its binding site on the mRNA (the target site) determines the efficiency of inhibition. Guide strands of miRNAs typically base-pair with their "seed sequence," which leads to modest inhibitory effects (Fig. 1b) while those associated with siRNAs can exhibit extensive base-pairing with their target mRNA, resulting in mRNA cleavage and robust inhibition (Fig. 1c). A detailed discussion of the RNAi pathway is out of the scope of this chapter; however, this pathway has been reviewed elsewhere [8, 9].

By delivering a carefully designed short-hairpin RNA that shares important features with miRNAs and siRNAs with a rAAV to a retinal cell, the expression of disease-associated proteins can be blocked to treat autosomal-dominant retinal disorders. These features include (reviewed Fakhr et al. [10]):

1. A dsRNA stem and loop structure of approximately 20 nucleotides forming a hairpin and a 3′ dinucleotide overhang. Since these structural features are similar to miRNA precursors (pre-miRNAs) (Fig. 2a), an shRNA can enter the RNAi pathway after its association with and processing by Dicer into a dsRNA duplex (Fig. 1a).

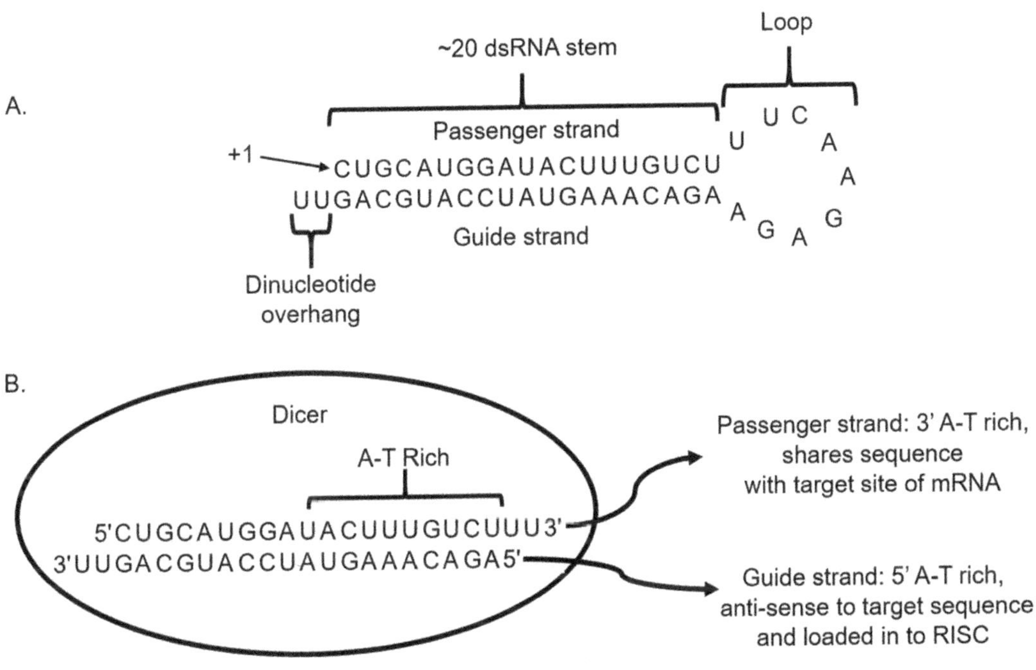

Fig. 2 Important features of shRNAs. (**a**) shRNA's have a hairpin structure and 3′ dinucleotide overhang that enables their processing by dicer. When expressed from a H1-shRNA cassette, transcription initiates at the first nucleotide of the 5′ arm of the shRNA and terminates with a string of five Ts (TTTTT). It is predicted that the UU overhang originates from the TTTTT termination signal. (**b**) The strand of the RNA duplex with the lesser stability at its 5′ end is typically loaded into RISC. When designing shRNAs, the guide strand is encoded on the 3′ arm

2. A guide strand with low 5′ thermodynamic stability. The strand of the dsRNA duplex that is typically selected for loading into RISC has lower thermodynamic stability at its 5′ end (due to a A-T rich sequence) (Fig. 2d).
3. A guide strand with complete complementarity to the target site of an mRNA. This will enable robust, siRNA-like inhibition of the target mRNA.
4. A guide strand whose "seed sequence" does not base-pair with mRNAs of important genes. Although the shRNA is like a siRNA in terms of base-pairing, a "seed sequence" will always be present (Fig. 1c).

In theory, the shRNA can be carefully designed to base-pair with the mutation site of the mRNA of the disease-causing gene, in effect preserving expression of the WT gene. However, many mutations within the same gene can cause retinal degeneration. For example, more than 100 mutations have been identified in RHO that can lead to autosomal-dominant Retinitis Pigmentosa (adRP) [1, 5]. Generating shRNAs to target each mutation individually would not be labor or cost efficient. Therefore, the ideal shRNA should be allele-independent, capable of targeting any mutant form of the gene. This allele-independent shRNA, however, would result in loss of expression of the associated endogenous WT gene, thus requiring the co-design of an shRNA-resistant cDNA.

Here, we will first describe how to design an shRNA against a mutant gene of interest (GOI). We will then explore how to generate a functional and shRNA-resistant cDNA of the GOI. Co-delivery of the shRNA and the shRNA-resistant cDNA to a diseased cell represents a methodology that could potentially treat any autosomal-dominant disorder of the retina [11–13]. This methodology has proved successful in the mutant RHO mouse model of autosomal-dominant Retinitis Pigmentosa (adRP) [14, 15].

2 Materials

2.1 Materials and Reagents

1. H1-shRNA Cassette Plasmids.
2. Mammalian Expression plasmid.
3. Pipettes and tips.
4. 1.5 mL tubes.
5. 12 well cell culture plates.
6. Polyethylenimine ~25,000 MW (PEI; 1 mg/mL).
7. Human Embryonic Kidney Cells 293 T (HEK293T) cells.
8. DMEM Cell Culture Media.
9. Fetal Bovine Serum.

10. Penicillin and Streptomycin.
11. NucBlue™ Live Cell Stain.
12. Mouse anti-turboGFP antibody (Origene).
13. Rabbit anti-β-Tubulin antibody.
14. Protease Inhibitor Cocktail (100×; Thermo-Halt).
15. Phosphate-buffered saline.
16. Cell scraper.
17. Sucrose.
18. 4× Loading dye buffer: 200 mM Tris-Cl (pH 6.8) + 400 mM DTT + 8% SDS + 40% glycerol + bromophenol blue.
19. 25-gauge insulin syringe.
20. 660 nm Protein Assay Reagent (Pierce™).
21. Ionic Detergent Compatibility Reagent.
22. Bovine Serum Albumin (BSA).
23. 10% Mini-PROTEAN® TGX™ Precast Protein Gels (Bio Rad).
24. Chameleon ladder (Li-COR).
25. iBlot PVDF Transfer Stacks.
26. Methanol.
27. Odyssey blocking buffer (LiCor).
28. IRDye 800CW Donkey-anti-rabbit (LiCor).
29. IRDye 680RD Goat-anti-mouse (LiCor).
30. Tween.
31. Q5® Site-Directed Mutagenesis Kit (NEB).
32. Primer Oligonucleotides.
33. rAAV Plasmid.
34. Restriction Enzymes: KpnI-HF (20 units/μL), XbaI (20 units/μL), NotI (10 units/μL), SalI (20 units/μL), and XmaI (10 units/μL).
35. Phusion polymerase (2 units/μL; NEB).
36. dNTPs (10 mM).
37. Agarose.
38. Scalpel.
39. Monarch® DNA Gel Extraction Kit (NEB).
40. Tris-Borate-EDTA (TBE) buffer.
41. Ethidium Bromide (10 mg/mL).
42. Calf-Intestinal Alkaline Phosphatase (CIAP, 1 unit/μL; Promega).

43. T4 Ligase (400 units/μL).
44. 10× Ligase Buffer.
45. Nuclease-free H_2O.
46. Monarch PCR Cleanup and Purification kit (NEB).
47. SURE2 Cells (Agilent).
48. Luria Broth.
49. Agar plates (50 μg/mL).
50. Ampicillin (100 mg/mL in 1:1 H_2O: isopropanol).
51. GeneJET Plasmid Miniprep kit (ThermoFisher).
52. Vetropolycin HC.
53. Proparacaine HCl Ophthalmic Solution (0.5%).
54. Antisedan (5 mg/mL).
55. Hypromellose 2.5%.
56. Phenylephrine HCl Ophthalmic Solution (2.5%).
57. Atropine sulfate Ophthalmic Solution (1%).
58. KetVed (100 mg/mL).
59. Xylazine (100 mg/mL).
60. Sterile Normal Saline.

2.2 Equipment

1. Computer with internet access.
2. ApE plasmid editor software (http://biologylabs.utah.edu/jorgensen/wayned/ape/).
3. ImageJ software.
4. Water bath.
5. Incubator at 37 °C with 5% CO_2.
6. Fluorescent Microscope.
7. Micro-centrifuge.
8. iBlot system (Invitrogen).
9. Electrophoresis apparatus (Bio Rad).
10. Odyssey CLx system (LiCor).
11. UV-transilluminator.
12. Heating block.
13. Electroporation Machine.
14. Hamilton syringe (85RN, 26S G).

3 Methods

3.1 Finding the shRNA Target Sites on Your GOI

1. Free online software is available for determination of potential shRNA target sequences within your GOI. We have constructed successful shRNAs based on the target sequences generated from siRNA design software as well, such as the Whitehead Institute siRNA Design Program (https://www.nature.com/cgt/journal/vaop/ncurrent/pdf/cgt20164a.pdf?origin=ppub): Select three versions of siRNA or shRNA design software such that the output can be compared.
2. If your GOI has been annotated, the sequence can be found by visiting www.pubmed.gov, selecting the "nucleotide" tab on the search bar dropdown menu, and searching for your GOI. On the results page, filter the results by selecting "mRNA" under "Molecule Types" and your target organism under "Top Organisms."
3. Input the coding sequence and 3′ untranslated region (UTR) into the three versions of siRNA or shRNA software. Do not input intronic sequences of genomic DNA.
4. Compare the output of the programs and select between four and ten 19-nt target sequences to be tested (*see* **Note 1**). Base the selection on the method parameters described by Fakhr et al. [10]:
   (a) Target sequences that appear in more than one program output.
   (b) Target sequences with an A-T rich 3′ end, such that the proper guide strand is loaded into the RISC complex.
   (c) Target sequences with limited potential off-target effects. Perform two BLAST searches (https://blast.ncbi.nlm.nih.gov/Blast.cgi?PAGE_TYPE=BlastSearch), for each target sequence. First, input the entire target sequence and exclude those target sequences that share extensive homology (>13 nucleotides (nt) in a row) with other genes. Second, input nt 13–18 of the target sequence (complimentary to the seed sequence of the guide strand, nt 2–7 (Fig. 1b)) and ensure that none of the hits are highly expressed in your cell type of interest.
   (d) The shRNAs to be tested should target different areas (i.e., 5′ vs. 3′) of the GOI's mRNA.

3.2 In Silico Design and Cloning of the H1-shRNA Cassette

1. shRNAs are efficiently expressed from the H1-promoter [16, 17]. The transcription start site for the H1 promoter is the 26th nucleotide downstream of the TATA box (TATAA). The H1 promoter has low preference for the +1 nucleotide, although the naturally occurring +1 nucleotide is adenosine

and transcription initiation may be variable [18, 19] (Table 1 Panel A) (*see* **Note 2**).

2. The elements of the H1-shRNA cassette are: (a) the modified H1-promoter containing a BamHI site at positions −6 to −1 (Table 1 Panel B), (b) the shRNA containing the Target Sequence, the Loop sequence [17], and the Guide Sequence (Table 1 Panel C), and (c) a modified transcription termination signal following the shRNA sequence.
3. The Final H1-shRNA Cassette (Table 1 Panel G) can be modified to contain the target and guide sequences for the GOI with the free ApE plasmid editor software (http://biologylabs.utah.edu/jorgensen/wayned/ape/). Paste the 19-nt target sequences from the shRNA design software output in place of the example target sequence (CTGCATGGATACTTTGTCT).

Table 1
Designing the H1-shRNA cassette

Panel A. Minimal H1-Promoter, Wild-type DNA Sequence
ATATTTGCATGTCGCTATGTGTTCTGGGAAATCACCATAAACGTGAAATGTCTTTGG ATTTGGGAATCT**TATAA**GTTCTGTATGAGACCACTCTTTCCC / A + 1
Panel B. Modified H1-Promoter DNA Sequence
TAAAACGACGGCCAGTGAATTCATATTTGCATGTCGCTATGTGTTCTGGGAAATCACCA TAAACGTGAAATGTCTTTGGATTTGGGAATCTTATAAGTTCTGTATGAGACCACT CGGATCC / +1 Modifications are underlined.
Panel C. shRNA DNA Sequence
19-nt Target Sequence – TTCAAGAGA (Loop sequence) – 19-nt Guide Sequence
Panel D. Example shRNA DNA Sequence with target sequence: CTGCATGGATACTTTGTCT
CTGCATGGATACTTTGTCTTTCAAGAGAAGACAAAGTATCCATGCAG
Panel E. Termination signal
TTTTT
Panel F. Modified Termination Signal
TTTTTAAGCTTTTTGGCGTAATCATG HindIII Restriction is underlined.
Panel G. Final H1-shRNA Cassette
GTCGACTAAAACGACGGCCAGTGAATTCATATTTGCATGTCGCTATGTGTTCTGGGAAAT CACCATAAACGTGAAATGTCTTTGGATTTGGGAATCTTATAAGTTCTGTATGAGACCAC TCGGATCCCTGCATGGATACTTTGTCTTTCAAGAGAAGACAAAGTATCCATGCAG TTTTTAAGCTTTTTGGCGTAATCATG***GTCGAC*** ***SalI site*** for cloning of rAAV plasmid. Target and guide sequences need to be replaced.

Copy the reverse complement of the 19-nt target sequence and paste it in the place of the example guide sequence (AGACAAAGTATCCATGCAG).

4. The completed gene for each of the potential target sites can be ordered directly from a manufacturer, making sure to include restriction sites at either side of the H1-shRNA cassette (here the SalI site: GTCGAC) for downstream cloning into the rAAV plasmid.
5. A scrambled negative control shRNA should also be designed. Online tools are available to scramble one of the target sequences, such as http://www.invivogen.com/sirnawizard/scrambled.php. For example, the example target sequence can be scrambled to produce GTCATTCGTTCTGGTCATA.

3.3 Finding the H1-shRNA Cassettes with the Highest Knockdown Efficiency of the WT GOI

1. Knockdown efficiency can be measured by performing co-transfections of the control and on-target shRNAs and a plasmid expressing the GOI.
2. Create an empty DNA control that does not express an shRNA.
3. The sequence of the WT GOI can be found as described in Subheading 3.1, **step 2**. Order or clone the GOI, into an expression plasmid with a highly expressed, ubiquitous promoter, such as the cytomegalovirus (CMV) promoter. Ensure that the GOI contains a segment of the 5′ untranslated region that contains the gene's endogenous Kozak sequence for efficient translation. Alternatively, a consensus Kozak sequence can be included—GCCACC(ATG) before the gene's open reading frame. If possible, add a green-fluorescent protein (GFP) tag, such as turboGFP.
4. The following protocol was used for measuring shRNA-mediated knockdown of turboGFP-tagged RHO by Western Blot. Optional companion fluorescent microscopy images can be taken. The protocol can be modified as needed for the GOI.
5. Transfect 500 ng of the CMV-GOI-GFP plasmid and 1 μg of each shRNA plasmid, including the scrambled shRNA and empty DNA controls, into separate wells of a 12 well plate containing HEK293T cells, or another cell line with a high transfection efficiency. Include additional wells for a non-transfected control. Perform the transfection in triplicate at room temperature. Briefly:
   (a) Plate 1×10^5 HEK293T cells in 1 mL of complete DMEM containing 10% FBS, 1% Penicillin/Streptomycin, and 0.1% Amphotericin B in each well of a 12 well plate. Incubate overnight at 37 °C with 5% CO_2.
   (b) The next day, the cells should be >70% confluent. Two hours prior to transfection, change the media of the HEK293T cells to pre-warmed DMEM containing only 10% FBS.

(c) For each shRNA, including the scrambled shRNA and empty DNA controls, mix 1 μg of the shRNA plasmid, 500 ng of the CMV-GOI-GFP plasmid, and 50 μL Opti-MEM I reduced serum medium media. For each shRNA and control plasmids, to a second tube, add 4.5 μL PEI to 50 μL Opti-MEM I reduced serum medium media (1 μg DNA: 3 μL PEI). Incubate for 5 min (*see* **Note 3**).

(d) Transfer the 50 μL of Opti-MEM I reduced serum medium containing the PEI to the corresponding 1.5 mL tube containing the shRNA or control plasmid, and incubate at room temperature for 15 min.

(e) Add the 100 μL PEI: plasmid mixture dropwise to the separate wells of the 12 well plate.

(f) Incubate at 37 °C with 5% CO_2 for 48 h.

6. *Optional.* Perform fluorescent microscopy to visualize the extent of GOI-turboGFP expression.

(a) Place a drop of NucBlue™ Live Cell Stain into each well of the 12 well plate, cover in tin foil, and incubate at 37 °C with 5% CO_2 for 15 min.

(b) Image each well at 4× and 10× magnification using the DAPI and Green fluorescent filters of a fluorescent microscope.

(c) Ensure that the exposure time and other settings are the same for each well.

7. Isolate protein and perform a Western Blot to measure the abundance of the GOI-GFP using the mouse anti-turboGFP antibody relative to the rabbit anti-β-Tubulin antibody loading control.

(a) Aspirate the media from each well and pipette 1 mL of ice-cold PBS 1× into each well.

(b) HEK293T cells are weakly adherent and can be removed from the plate by pipetting up and down. Alternatively, a cell scraper can be used. Transfer the 1 mL PBS containing the cells to fresh 1.5 mL tubes.

(c) Pellet the cells by centrifuging at 3000 × *g* for 10 min at 4 °C.

(d) Aspirate the PBS, leaving the cell pellet in the 1.5 mL tube.

(e) Resuspend the pellet in 0.23 M Sucrose with 1× Thermo Halt Protease Inhibitor Cocktail (100×) in PBS (typically 150–200 μL) and sonicate for 10 s.

(f) Add 4× loading dye buffer to each sample so that its final concentration is 1×.

(g) Since RHO-GFP is prone to aggregation, the samples were incubated at room temperature for >10 min and returned to ice. Other proteins can be boiled for 10 min to denature then returned to ice.

(h) Pass each sample through a 25-gauge insulin syringe five times to shear the DNA to decrease clogging of pipette tips when loading the protein gel.

(i) Centrifuge the sample at max speed for 10 min at 4 °C and transfer 75% of the upper sample to a new 1.5 mL tube, leaving the cell debris behind.

(j) Quantify the concentration of each sample using the Pierce™ 660 nm Protein Assay Reagent and the Ionic Detergent Compatibility Reagent relative to a twofold standard curve of BSA over a range of 1.5 to 0.094 μg per sample.

(k) Load 20 μg of sample into each well of a 10% Mini-PROTEAN® TGX™ Precast Protein Gels. Load one well with 5 μL the Li-COR Chameleon ladder.

(l) Run the Mini-PROTEAN gel at 100 V until the bromophenol blue of the loading dye has reached the bottom of the gel.

(m) Transfer the protein from the gel to a PVDF membrane using Invitrogen's iBlot system.

(n) Incubate the PVDF in methanol while shaking at room temperature for 5 min.

(o) Wash the membrane with diH_2O five times.

(p) Block the PVDF membrane in Odyssey blocking buffer for 1 h while shaking at room temperature.

(q) Wash three times with 0.1% Tween in PBS 1× while shaking for 5 min each.

(r) Dilute the mouse anti-turboGFP (1:2000) and rabbit anti-β-Tubulin (1:5000) in Odyssey blocking buffer. Apply to the PVDF membrane and incubate while shaking at room temperature for 2 h or 4 °C overnight.

(s) Wash three times with 0.1% Tween in PBS while shaking for 5 min each.

(t) Dilute IRDye 800CW Donkey-anti-rabbit (1:5000) and IRDye 680RD Goat-anti-mouse (1:5000) in Odyssey blocking buffer. Apply to the PVDF membrane and incubate while shaking at room temperature for 45 min.

(u) Wash three times with 0.1% Tween in PBS while shaking for 5 min each.

(v) Image gel with an Odyssey CLx Imaging system.

(w) Quantify the intensity of the band corresponding to GOI-GFP relative to the band corresponding to β-Tubulin for each sample using ImageJ software [20].

3.4 Validate that the H1-shRNA Cassettes Are Allele-Independent

1. Find disease-causing mutations associated with the GOI. Select at least one mutation to validate that the shRNAs are allele-independent (*see* **Note 4**). The mutation can be selected based on its prevalence or severity. For example, we chose to study the P23H mutation in RHO that causes adRP due to its high prevalence in the United States.
2. The mutation can be induced with the Q5® Site-Directed Mutagenesis Kit using the CMV-GOI-GFP plasmid.
3. Primers for this kit can be designed using the NEBasechanger tool (http://nebasechanger.neb.com/), which will also provide the Tm for the PCR reaction.
4. Generate a CMV-mutant GOI-GFP plasmid as per the manufacturer's instructions. For example, the CMV-P23H RHO-GFP plasmid was created by altering codon 23 of the open reading frame from CCC to CAC.
5. Repeat the experiments in Subheading 3.3, except using the CMV-mutant GOI-GFP plasmid in the place of the CMV-GOI-GFP plasmid.
6. Select the two shRNAs, herein named shRNA A and shRNA B, with the highest knockdown efficiency of both the WT and mutant GOI-GFP for further analysis. This is important in case shRNA A or B has unforeseen off-target effects in future experiments.

3.5 Generating shRNA-Resistant cDNAs of the GOI for Functional Gene Replacement

1. Create two shRNA-resistant cDNAs of the GOI (shr-GOI) corresponding to the shRNAs A and B.
2. shRNA resistance can be conferred to the replacement GOI by inducing silent mutations in the shRNA's target sequence by altering the codon wobble base positions.
3. If possible, four silent mutations should be induced, particularly in the wobble bases around nucleotides 9–11 (RISC-cleavage site) and 13–18 (seed sequence) in the target sequence of the WT GOI's mRNA (Fig. 2a).
4. Each shRNA's target sequence can be input into a codon optimization software, such as JCat (http://www.jcat.de/) [21]. Ensure that the target sequence is pasted "in-frame," otherwise the mutations will not be silent, and that the correct target organism is selected. For the example target sequence, CTGCATGGATACTTTGTCT, which is in-frame (i.e., the 5′ CTG is the first codon), the jcat software produced five silent mutations: two appear in the "seed sequence," one in the "cleavage site," and two in the 3′ end of the shRNA

(Fig. 2b). These alterations should be sufficient to render the codon-optimized version resistant to the shRNA (*see* **Note 5**).

5. If method 1 fails, then codons can be selected by hand as a secondary method. To select appropriate codons, reference a codon usage table for your organism of interest. Choose the silent mutation associated with the highest possible codon usage proportion of the same amino acid.
6. When using either method 1 or 2, A to G and C to T mutations should be avoided because guanidine in the mRNA or shRNA guide strand can wobble base-pair with uracil, resulting in less than maximal resistance (Fig. 3c). An example can be seen for valine in Fig. 3d. Consider using a lesser used codon in this circumstance.
7. The silent mutations can be introduced with the Q5® Site-Directed Mutagenesis Kit.

3.6 Validating that the shRNA-Resistant cDNA Will Co-express with the shRNA

1. Repeat the experiments in Subheading 3.3, except using the CMV-shr-GOI-GFP plasmids in the place of the CMV-GOI-GFP plasmid. Perform the experiment with both shRNA A and B versions of the shr-GOI-GFP.
2. At least five conditions should be studied for each CMV-shr-GOI-GFP plasmid. Into a 12 well plate, make triplicate co-transfections of CMV-shr-GOI-GFP plasmid and the following:
   (a) Negative scrambled control shRNA.
   (b) shRNA A.
   (c) shRNA B.
   (d) Empty DNA.
   (e) Include a no transfection control as well.
3. Measure the expression of the shr-GOI-GFP via Western blot and ImageJ analysis as described in Subheading 3.3. A shRNA A-resistant GOI-GFP should show minimal to no loss of expression when co-expressed with shRNA A, but should be reduced in expression when co-expressed with a separate validated shRNA B. The converse should be true for the shRNA B-resistant GOI.

3.7 Cloning the rAAV Plasmid to Contain the shRNA and the Corresponding shRNA-Resistant GOI

1. Using the ApE plasmid editor software, construct a map of the rAAV plasmids, which will contain standard plasmid elements (i.e., origin of replication, antibiotic resistance) as well as AAV2 inverted-terminal repeats (ITRs) flanking [the plasmid in Fig. 4 will be used as an example for cloning the plasmid]:
   (a) A cell-type-specific promoter with 5′ and 3′ restriction sites, here the human-opsin promoter (HOP) for rod

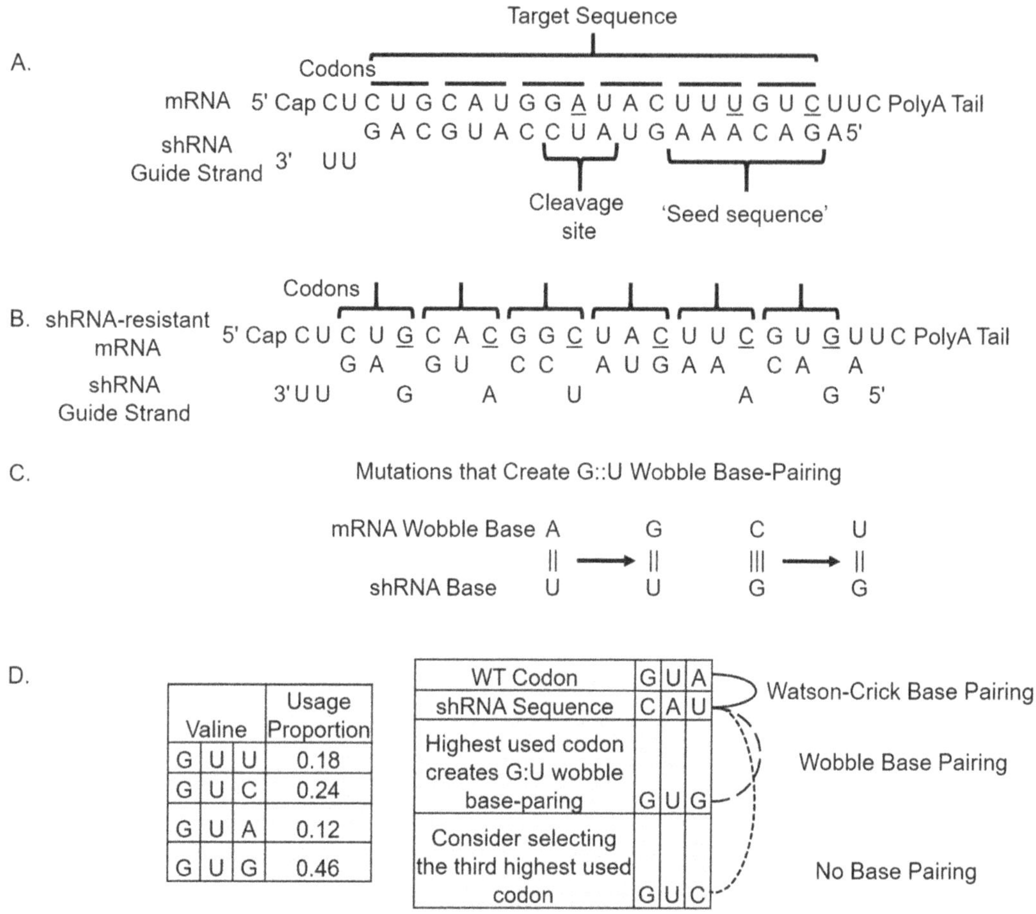

Fig. 3 Generating the shRNA-resistant GOI. (**a**) The example target sequence is bound extensively by the guide strand of the example shRNA. Alterations to the underlined wobble bases in the "seed sequence" and "cleavage site" are particularly important to mutate to prevent base-pairing between the guide shRNA and the mutagenized mRNA. (**b**) Codon optimization of the example target sequence yielded five silent mutations, situated both in the "seed sequence" and "cleavage site," which should provide shRNA resistance. (**c**) Mutations that create G::U wobble base-pairing should be avoided. (**d**) Algorithm for selecting the wobble base-pair by hand. Here, the WT codon is GUA and needs to be altered. Look for a codon of the same amino acid (valine) with the highest usage proportion. If the silent mutation leads to a change from A to G or C to T, then select the next available codon with the highest usage to avoid G:U wobble base-pairing between the guide strand and the shRNA-resistant target sequence

photoreceptors with a 5′ KpnI site and 3′ XbaI restriction sites (RSs).

(b) The shr-GOI open reading frame, containing the endogenous or consensus Kozak sequence, with 5′ and 3′ restriction sites, here NotI RSs.

(c) A suitable polyA signal, here the SV40 PolyA signal.

(d) The corresponding H1-shRNA cassette with surrounding restriction sites, here SalI RSs.

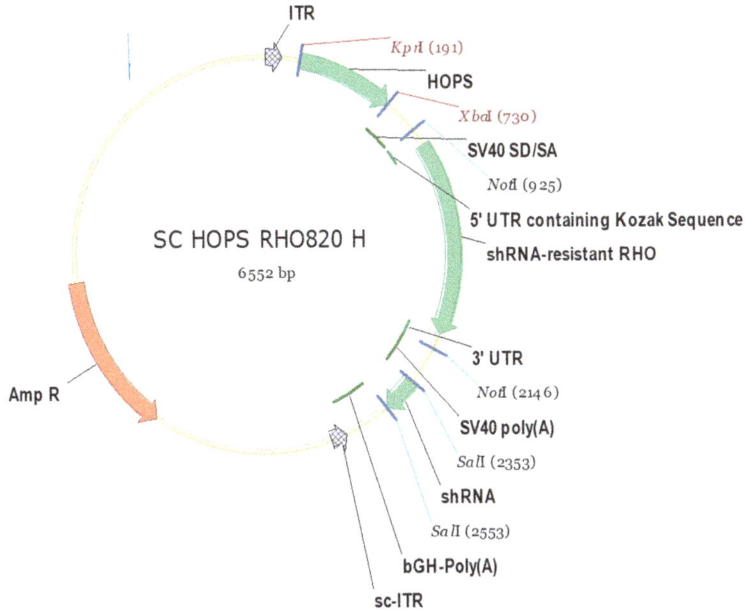

Fig. 4 Example of rAAV genome designed to treat adRP caused by mutant RHO [36]. The rAAV genome is situated between ITRs, with the self-complimentary ITR oriented at the 3′ terminus. The HOPs promoter (flanked by KpnI and XbaI sites) drives the expression of shRNA-resistant RHO mRNA (flanked by NotI sites), which includes the 5′ UTR with the endogenous Kozak sequence, with a SV40 PolyA signal. The H1-shRNA (flanked by SalI sites) is located near and transcribes toward the mutant ITR

2. If the proposed rAAV genome is <2.4 kb, then a self-complimentary (sc) rAAV can be used. The use of sc rAAV would result in faster transgene expression [22, 23]. Otherwise, use a single-stranded (ss) rAAV with WT ITRs.
3. Clone the two shr-GOIs into the NotI RS of the rAAV plasmid.
   (a) Perform PCR with NEB Phusion polymerase using the CMV-shRNA-resistant GOI-GFP plasmids as templates.
   (b) Design primers that anneal to the 5′ and 3′ end of the GOI with a Tm between 60 and 72 °C calculated with the NEB Tm Calculator (https://tmcalculator.neb.com/#!/). Add TAT-GCGGCCGC to the 5′ terminus of the forward primer. The additional TAT nucleotides allow efficient cutting at the NotI RS (GCGGCCGC). Again, ensure that either the endogenous or consensus Kozak sequence is included in the shr-GOI. The primer should begin to anneal either at the 5′ end of the 5′ UTR or the consensus Kozak sequence of the GOI. Add TAT-GCGGCCGC to the annealing sequence of the reverse primer, ensuring that the annealing sequence is antisense and begins with the antisense stop codon

Table 2
Primer design for cloning of the shRNA-resistant GOI

Forward Primer	TAT-GCGGCCGC-GCCACC-ATGNNNNNNNNNNNNN
Reverse Primer	TAT-GCGGCCGC-TTANNNNNNNNNNNNN

Ns in the forward primer should begin with the start codon and be sense, whereas those in the reverse primer should begin with the stop codon and be antisense

Table 3
PCR mixture for cloning of the shRNA-resistant GOI

Nuclease-free H_2O	Up to 50 μL
HF-Phusion Buffer	10 μL
10 mM dNTPs	1 μL
10 μM Forward Primer	2.5 μL
10 μM Reverse Primer	2.5 μL
shRNA-resistant GOI Plasmid	1 ng
Phusion Polymerase	0.5 μL

Table 4
PCR conditions for amplification of the shRNA-resistant GOI

	Initial denaturation	98 °C for 30 s
30–35 cycles	Denaturation Annealing Extension	98 °C for 10 s Tm for 30 s 72 °C for 30 s per kB of GOI
	Final Extension	98 °C for 7 min

(ACT, ATT, or ATC) (Table 2). The Tm should be calculated for only the portion of the primer that anneals to the GOI.

(c) Create a PCR mixture according to Table 3.

(d) Perform a PCR reaction with the parameters in Table 4.

(e) Run the PCR reaction with loading dye next to NEB's TriDye-2-log ladder on a 1% agarose gel in TBE at 120 V for 30 min. Excise the band corresponding to the anticipated size with a scalpel under a UV-transilluminator and extract the band with the Monarch® DNA Gel Extraction Kit according to the manufacturer's instructions, eluting in 50 μL of nuclease-free H_2O.

Table 5
Restriction digestions for shRNA-resistant GOI cloning

	Gel-purified PCR fragment of shRNA-resistant GOI	rAAV Plasmid
H_2O	0 μL	Up to 50 μL
Buffer 3.1	5 μL	5 μL
Template	44 μL of the Gel-Purified Product	2 μg of plasmid
NotI	1 μL	1 μL

Table 6
Ligation reactions of the shRNA-resistant GOI into the rAAV plasmid

	1:1 Ratio (μL)	1:2 Ratio (μL)	1:3 Ratio (μL)	Negative control (μL)
rAAV Backbone	3	3	3	3
GOI PCR Fragment	3	6	9	0
Ligase Buffer	2	2	2	2
T4 Ligase	1	1	1	1
H_2O	11	8	5	14

(f) To restriction digest the gel extracted PCR product and the rAAV plasmid for cloning, set up the reactions in Table 5. Incubate at 37 °C for >2 h. To the rAAV plasmid only, after 2 h, add 10 μL of Promega's CIAP and incubate at 37 °C for an additional >1 h to prevent self-ligation. Heat-inactivate both reactions at 65 °C for 20 min.

(g) Purify the NotI-digested PCR fragment of the shr-GOI with NEB's Monarch PCR Cleanup and Purification kit as per the manufacturer's instructions, eluting in 30 μL H_2O.

(h) Isolate the NotI-digested rAAV plasmid by gel extraction according to Subheading 3.6, **step 2(e)**, and elute in 30 μL H_2O.

(i) Make ligation reactions using NEB's T4 Ligase at different molar ratios of digested rAAV plasmid to shr-GOI PCR fragment according to Table 6.

(j) Incubate at 16 °C overnight.

(k) Transform ligation reaction into Agilent SURE2 cells and grow at 30 °C for approximately 16 h on 50 μg/mL Ampicillin Agar Plates.

(l) The next day, pick colonies and inoculate a 5 mL culture containing Luria Broth and 5 μL of 100 mg/mL Ampicillin stock solution and incubate for approximately 16 h at 30 °C in a water bath while shaking.

(m) Perform minipreps using Thermo's GeneJET Plasmid Miniprep kit on the resulting cultures. Perform two separate restriction digestions for each Miniprep using:

- NotI: to check that an insert of the appropriate size was ligated.
- XmaI: to check that at least 50% of the ITRs were retained. Retained ITRs will appear as two intense bands situated above and below 3 kB. Lost ITRs will appear as a single band around 6 kB, the size of the linearized rAAV plasmid. >90% Retained ITRs can be seen in Fig. 5.

Restriction digestions of the Miniprep DNA should be performed for >2 h. Run the samples in loading dye next to NEB's TriDye-2-log ladder on a 0.7% agarose gel in TBE at 120 V until the dye has reached the bottom of the gel. Image with a UV camera.

(n) For colonies containing an insert and retained ITRs, Sanger sequence with the appropriate primers to verify that the GOI has been inserted properly. For the example plasmid, primers that anneal within the SD/SA element in the forward direction (AAAGCTGCGGAATTG-TACCC) and SV40 PolyA element in the reverse direction (GCATTCTAGTTGTGGTTTGTCC) were used.

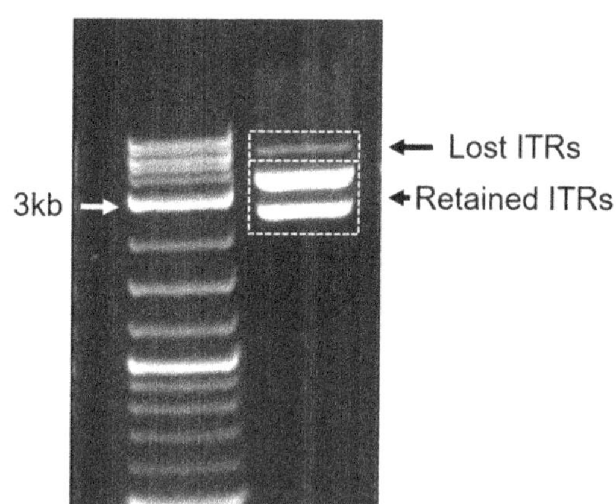

Fig. 5 XmaI digestion of a rAAV plasmid that retained >90% of its ITRs, suitable for capsid packaging. A plasmid that has lost ~90% of its ITRs can be recovered by retransforming and rescreening for ITRs with XmaI

(o) For correct colonies containing intact ITRs, perform an XmaI.

(p) In the next step, the shRNAs will be cloned into the plasmids isolated from the correct bacterial colonies. Double verify that these rAAV plasmids have >50% ITRs before continuing.

4. Clone the shRNAs into the SalI RS of the corresponding rAAV plasmids containing the shr-GOI obtained from Subheading 3.6, **step 3**.

(a) Digest the shRNA and the rAAV shr-GOI plasmids using the restriction enzymes and the reactions conditions outlined in Table 7. Incubate at 37 °C for >2 h. To the rAAV shr-GOI plasmid backbones only, after 2 h, add 10 μL of Promega's CIAP and incubate at 37 °C for an additional >1 h to prevent self-ligation. Heat-inactivate both reactions by incubating at 65 °C for 15 min.

(b) Isolate the shRNAs and rAAV shr-GOI plasmid backbones as described in Subheading 3.6, **step 3(e)**, except elute in 40 μL H_2O.

(c) Make ligation reactions at different molar ratios of digested rAAV shr-GOI plasmid backbones and the corresponding shRNAs according to Table 8.

Table 7
Restriction digestions for shRNA cloning into the rAAV shRNA-resistant GOI plasmid

	shRNA	rAAV containing resistant GOI
H_2O	Up to 50 μL	Up to 50 μL
Buffer 3.1	5 μL	5 μL
Template	10 μg of plasmid	2 μg of plasmid
SalI	1 μL	1 μL

Table 8
Ligation reactions of the shRNA and rAAV shRNA-resistant GOI plasmid

	1:2.5 Ratio (μL)	1:5 Ratio (μL)	1:7.5 Ratio (μL)	Negative control (μL)
rAAV Backbone	2	2	2	3
GOI PCR fragment	5	10	15	0
Ligase buffer	2	2	2	2
T4 ligase	1	1	1	1
H_2O	10	5	0	14

(d) Transform, inoculate, miniprep, and sequence as described in Subheading 3.6, **step 3(l, m)**, except using a primer that anneals within the SV40 PolyA sequence in the forward direction (GGACAAACCACAACTAGAATGC). Notify the sequencing service that the template is a shRNA construct with a strong hairpin structure so that the correct reaction conditions can be used.

(e) For correct colonies, perform an XmaI restriction digestion to ensure that the ITRs of the rAAV plasmid have not recombined. If <80% of the ITRs are intact, simply retransform in SURE2 cells and screen for >80% ITR retention by XmaI digestion. The rAAV is ready for packaging and testing (*see* **Note 6**).

3.8 Selection of rAAV Capsid for Ocular Delivery

1. Successful gene delivery to the retina requires the careful selection of a rAAV capsid together with the appropriate selection of delivery route. The most commonly used rAAV capsid serotypes used in ocular gene therapy are AAV1, AAV2, AAV5, AAV6, AAV8, and AAV9. Versions of these capsid serotypes have been engineered to improve the transduction efficiency or to modify the cell tropism within the retina. Below there is a list of rAAV capsid serotypes and their target cells within the retina:
   (a) AAV1—RPE transduction after subretinal delivery [24].
   (b) AAV2—RPE (subretinal), and RGC (Intravitreal) transduction [25].
   (c) AAV5—Photoreceptors and RPE transduction after subretinal delivery [26].
   (d) AAV6—RPE transduction after subretinal delivery [27].
   (e) AAV8—Photoreceptors after subretinal delivery [28].
   (f) AAV9—Photoreceptors after subretinal delivery [29].
   (g) AAV2(quadY-F) + T491 V—RGC, Bipolar cells, and VEC after intravitreal delivery [30, 31].

   Direct evolution has allowed the production of engineered capsid types leading to improved transduction of retinal cells [32]. Further capsid variants of the abovementioned are currently available and have been demonstrated to have improved transduction profiles. Further information can be found at the UF Ocular Gene Therapy Core [33]. It should be noted that the choice of capsid type depends on the species of experimental animal.

3.9 Ocular Injections in Mice

1. Mouse injections.
   (a) Mice should receive 2 drops of phenylephrine HCl ophthalmic solution (2.5%), atropine sulfate ophthalmic solution (1%), and proparacaine HCl ophthalmic solution

(0.5%). This will ensure proper dilation and will provide topical anesthesia before the procedure.

(b) Mice are anesthetized with an intraperitoneal injection (5 mL/kg) of a ketamine/xylazine mixture (20 mg/mL ketamine, 0.79 mg/mL xylazine, in normal saline).

(c) The plane of anesthesia is assessed by tail-pinching reflex until the mouse is completely anesthetized.

(d) Use a Hamilton syringe with a 26 G point/blunt needle and load it with 1 μL of diluted vector containing Akfluor® (Fluorescein). This reagent will allow visual assessment of the injection quality.

(e) Use an alcohol pad to clean the needle before proceeding with the injection.

(f) Place the anesthetized mouse under stereoscope with the desired eye facing the lens.

(g) Overlay a drop of Gonak (2.5% hypromellose ophthalmic solution) on the eye to protect it from dehydration.

(h) With the bevel of a 25G needle facing upward, an incision is made within the limbus zone carefully avoiding blood vessels surrounding the area and retracting the needle as soon as the bevel is seen through the pupil.

(i) Retrieve the Hamilton syringe and pass the needle through the incision carefully avoiding the lens.

(j) *For an intravitreal injection*: As soon as the needle is seen through the pupil, ask an assistant to push the needle plunger with a slow and steady rate. The injection should take between 15 and 20 s to complete. A bright green color, due to the fluorescein, should be seen inside the eye. Once completed hold the needle in place for 15 seconds to avoid immediate outflow of the vector injected.

(k) *For a subretinal injection*: As soon as the needle is seen through the pupil, incline the syringe 10–15% toward the posterior pole of the retina and introduce it further into until the globe seems to retract into the orbit. At this point, retract the needle until the globe seem to move forward to its original position. Then ask an assistant to push the plunger slow and steady. This injection should take between 1 and 2 min to avoid causing a retinal tear. The retina should acquire a pale green color as the injection proceeds. Once done carefully retract the needle.

(l) After the injection has been completed, apply vetropolycin HCl on the incision to avoid infections.

(m) Administer Antisedan (0.25 mg/kg) intraperitoneally followed by 0.5 mL of warm normal saline to help hydrate the mouse.

(n) Place the mouse in a warm pad and wait until the mouse is ambulatory before returning it to its housing location.

4 Notes

1. Slightly longer target sites of 20 or 21-nt target sequences can be used as well.
2. The U6-promoter can also be used to express small RNA molecules, such as shRNAs [18, 19]. The BamH1 and HindIII sites allow cloning of oligonucleotides into the target-loop-guide shRNA-TTTTT site of the H1-shRNA cassette. We chose to order the completed H1-shRNA genes as cloning can be expensive in terms of labor and cost. Finally, other loop sequences are available, such as AAGTTCTCT (Promega) and TTTGTGTAG [34, 35].
3. We recommend including the endogenous Kozak sequence in the 5′ UTR of the GOI over the consensus Kozak sequence.
4. The ratio of plasmid DNA to PEI can be altered to optimize the experiment; ratios up to 1 μg:1 μL are acceptable. The ratio of the shRNA plasmid to the GOI-GFP plasmid can be increased such that better knockdown can be achieved. If the GOI-GFP is not prone to aggregation, then the samples can be boiled for 15 min instead of incubating at room temperature. As an alternative approach, cells can be submitted for flow cytometry to measure mean fluorescence across a sample of cells for each condition.
5. Less than four silent mutations may be sufficient to render the GOI shRNA-resistant, particularly if they lie within the cleavage site and "seed sequence." On the other hand, too many silent mutations may destabilize the mRNA and lead to less efficient translation of the shr-GOI mRNA.
6. Since the XmaI digestions assess a population of bacterial colonies, the ITRs can be salvaged even when less than 10% of the colonies retain their ITRs by retransforming in SURE2 cells and rescreening colonies. In our hands, ITRs are typically lost on the Agar plate, not during growth in larger liquid cultures. Picking small colonies on the Agar plate may facilitate with maintaining ITRs. Using Terrific broth rather than Luria broth may also be helpful to maintain ITRs.

Acknowledgments

This work was funded by an F30 from the NEI (MM), an R01 from the NEI (ASL), a grant from the Bright Focus Foundation (CJI) and an unrestricted grant from the Research to Prevent Blindness (CJI).

References

1. RetNet: Summaries. https://sph.uth.edu/Retnet/sum-dis.htm. Accessed 26 Oct 2017
2. Maguire AM, Simonelli F, Pierce EA et al (2008) Safety and efficacy of gene transfer for Leber's congenital amaurosis. N Engl J Med 358:2240–2248. https://doi.org/10.1056/NEJMoa0802315
3. Hauswirth WW, Aleman TS, Kaushal S et al (2008) Treatment of Leber congenital amaurosis due to RPE65 mutations by ocular subretinal injection of adeno-associated virus gene vector: short-term results of a phase I trial. Hum Gene Ther 19:979–990. https://doi.org/10.1089/hum.2008.107
4. Jacobson SG, Cideciyan AV, Ratnakaram R et al (2012) Gene therapy for Leber congenital amaurosis caused by RPE65 mutations: safety and efficacy in 15 children and adults followed up to 3 years. Arch Ophthalmol 130:9–24. https://doi.org/10.1001/archophthalmol.2011.298
5. OMIM Entry - * 180380 - RHODOPSIN; RHO. http://www.omim.org/entry/180380. Accessed 26 Oct 2017
6. OMIM Entry - # 153700 - MACULAR DYSTROPHY, VITELLIFORM, 2; VMD2. http://www.omim.org/entry/153700#genotypePhenotypeCorrelations. Accessed 26 Oct 2017
7. Zeng Y, Wagner EJ, Cullen BR (2002) Both natural and designed micro RNAs can inhibit the expression of cognate mRNAs when expressed in human cells. Mol Cell 9:1327–1333. https://doi.org/10.1016/S1097-2765(02)00541-5
8. Carthew RW, Sontheimer EJ (2009) Origins and mechanisms of miRNAs and siRNAs. Cell 136:642–655. https://doi.org/10.1016/j.cell.2009.01.035
9. Pratt AJ, MacRae IJ (2009) The RNA-induced silencing complex: a versatile gene-silencing machine. J Biol Chem 284:17897–17901. https://doi.org/10.1074/jbc.R900012200
10. Fakhr E, Zare F, Teimoori-Toolabi L (2016) Precise and efficient siRNA design: a key point in competent gene silencing. Cancer Gene Ther 23:73–82. https://doi.org/10.1038/cgt.2016.4
11. Li C, Xiao P, Gray SJ et al (2011) Combination therapy utilizing shRNA knockdown and an optimized resistant transgene for rescue of diseases caused by misfolded proteins. Proc Natl Acad Sci U S A 108:14258–14263. https://doi.org/10.1073/pnas.1109522108
12. Millington-Ward S, Chadderton N, O'Reilly M et al (2011) Suppression and replacement gene therapy for autosomal dominant disease in a murine model of dominant retinitis pigmentosa. Mol Ther 19:642–649. https://doi.org/10.1038/mt.2010.293
13. Mueller C, Tang Q, Gruntman A et al (2012) Sustained miRNA-mediated knockdown of mutant AAT with simultaneous augmentation of wild-type AAT has minimal effect on global liver miRNA profiles. Mol Ther 20:590–600. https://doi.org/10.1038/mt.2011.292
14. Mao H, Gorbatyuk MS, Rossmiller B et al (2012) Long-term rescue of retinal structure and function by rhodopsin RNA replacement with a single adeno-associated viral vector in P23H RHO transgenic mice. Hum Gene Ther 23:356–366. https://doi.org/10.1089/hum.2011.213
15. O'Reilly M, Palfi A, Chadderton N et al (2007) RNA interference–mediated suppression and replacement of human rhodopsin in vivo. Am J Hum Genet 81:127–135. https://doi.org/10.1086/519025
16. Baer M, Nilsen TW, Costigan C, Altman S (1990) Structure and transcription of a human gene for H1 RNA, the RNA component of human RNase P. Nucleic Acids Res 18(1):97
17. Brummelkamp TR, Bernards R, Agami R (2002) A system for stable expression of short interfering RNAs in mammalian cells. Science 296:550–553. https://doi.org/10.1126/science.1068999
18. Ma H, Wu Y, Dang Y et al (2014) Pol III promoters to express small RNAs: delineation

of transcription initiation. Mol Ther Nucleic Acids 3:e161. https://doi.org/10.1038/mtna.2014.12
19. Gao Z, Harwig A, Berkhout B et al (2017) Mutation of nucleotides around the +1 position of type 3 polymerase III promoters: The effect on transcriptional activity and start site usage. Transcription 8(5):275–287. https://doi.org/10.1080/21541264.2017.1322170
20. Schneider CA, Rasband WS, Eliceiri KW (2012) NIH Image to ImageJ: 25 years of image analysis. Nat Methods 9:671–675. https://doi.org/10.1038/nmeth.2089
21. Grote A, Hiller K, Scheer M et al (2005) JCat: a novel tool to adapt codon usage of a target gene to its potential expression host. Nucleic Acids Res 33:W526–W531. https://doi.org/10.1093/nar/gki376
22. Wang XS, Ponnazhagan S, Srivastava A (1996) Rescue and replication of adeno-associated virus type 2 as well as vector DNA sequences from recombinant plasmids containing deletions in the viral inverted terminal repeats: selective encapsidation of viral genomes in progeny virions. J Virol 70:1668–1677
23. Zhou Q, Tian W, Liu C et al (2017) Deletion of the B-B' and C-C' regions of inverted terminal repeats reduces rAAV productivity but increases transgene expression. Sci Rep 7 (1):5432. https://doi.org/10.1038/s41598-017-04054-4
24. Biswal MR, Han P, Zhu P et al (2017) Timing of antioxidant gene therapy: implications for treating dry AMD. Invest Ophthalmol Vis Sci 58:1237–1245. https://doi.org/10.1167/iovs.16-21272
25. Guy J, Qi X, Koilkonda RD et al (2009) Efficiency and safety of AAV-mediated gene delivery of the human ND4 complex I subunit in the mouse visual system. Invest Ophthalmol Vis Sci 50:4205–4214. https://doi.org/10.1167/iovs.08-3214
26. Boye SE, Alexander JJ, Boye SL et al (2012) The human rhodopsin kinase promoter in an AAV5 vector confers rod- and cone-specific expression in the primate retina. Hum Gene Ther 23:1101–1115. https://doi.org/10.1089/hum.2012.125
27. Yang GS, Schmidt M, Yan Z et al (2002) Virus-mediated transduction of murine retina with adeno-associated virus: effects of viral capsid and genome size. J Virol 76:7651–7660. https://doi.org/10.1128/JVI.76.15.7651-7660.2002
28. Allocca M, Mussolino C, Garcia-Hoyos M et al (2007) Novel adeno-associated virus serotypes efficiently transduce murine photoreceptors. J Virol 81:11372–11380. https://doi.org/10.1128/JVI.01327-07
29. Lebherz C, Maguire A, Tang W et al (2008) Novel AAV serotypes for improved ocular gene transfer. J Gene Med 10:375–382. https://doi.org/10.1002/jgm.1126
30. Scalabrino ML, Boye SL, Fransen KMH et al (2015) Intravitreal delivery of a novel AAV vector targets ON bipolar cells and restores visual function in a mouse model of complete congenital stationary night blindness. Hum Mol Genet 24:6229–6239. https://doi.org/10.1093/hmg/ddv341
31. Kay CN, Ryals RC, Aslanidi GV et al (2013) Targeting photoreceptors via intravitreal delivery using novel, capsid-mutated AAV vectors. PLoS One 8(4):e62097. https://doi.org/10.1371/journal.pone.0062097
32. Dalkara D, Byrne LC, Klimczak RR et al (2013) In vivo–directed evolution of a new adeno-associated virus for therapeutic outer retinal gene delivery from the vitreous. Sci Transl Med 5:189ra76–189ra76. https://doi.org/10.1126/scitranslmed.3005708
33. Vector production options. http://ogtc.eye.ufl.edu/2016/04/20/vector-production-options/. Accessed 16 Nov 2017
34. Song H, Yang P-C (2010) Construction of shRNA lentiviral vector. North Am J Med Sci 2:598–601. https://doi.org/10.4297/najms.2010.2598
35. Mello CC, Conte D (2004) Revealing the world of RNA interference. Nature 431:338–342. https://doi.org/10.1038/nature02872
36. Cideciyan AV, Sudharsan R, Dufour VL, Massengill MT, Iwabe S, Swider M, Lisi B et al (2018) Mutation-independent rhodopsin gene therapy by knockdown and replacement with a single AAV vector. Proc Natl Acad Sci 115(36):E8547. https://doi.org/10.1073/pnas.1805055115

Chapter 16

Localized Intra-Arterial Gene Delivery Using AAV

Koji Hosaka, Fredric P. Manfredsson, and Brian L. Hoh

Abstract

In vivo gene therapy is a tremendous tool for a wide variety of genetic modifications. However, often a specific and precise local administration of the viral vector is necessary to deliver the genetic payload in vivo. For many animal studies using viral vectors, such as those investigating neurological disorders, the vector is targeted directly into the tissue/organ of interest. On the other hand, in vascular disease research, viral vectors are administered systemically, either via a tail vein injection or through catheter-mediated infusion, which results in off-target transduction of cells and tissues. Targeting cells in the vascular wall without off-target activity, however, requires localized delivery in order to efficiently target cells of the internal vasculature. Here we describe a novel murine in vivo targeted intra-arterial viral vector delivery method, which has been developed in order to be able to perform more intricate studies in cardiovascular disease.

Key words In vivo transduction, Intra-arterial transduction, Viral vector, AAV, Cardiovascular diseases

1 Introduction

Viral vectors are an effective means of gene transfer in animal studies targeting a wide variety of cells and organs. Moreover, viral gene therapy has also been used in human clinical trials [1]. Viral vectors such as adeno-associated virus (AAV) efficiently transduce both dividing and quiescent cells. In terms of the genetic payload, genetic manipulation such as overexpression, silencing, or CRISPR-mediated knockout is commonplace. AAV has some advantages compared to other viral vectors, which include (1) low or no toxicity, (2) minimal immune response after administration, (3) high efficacy and longevity of transduction, and (4) the maintenance of an episomal genomic state negates any concerns with chromosomal integration [2, 3]. In vascular disease research, knockout animals, transgenic animals, liposome-mediated gene modification, antibody or antagonist-mediated modulations of protein of interest, or systemic injection of viral vectors have been the primary means to investigate disease mechanisms and for

Fredric P. Manfredsson and Matthew J. Benskey (eds.), *Viral Vectors for Gene Therapy: Methods and Protocols*, Methods in Molecular Biology, vol. 1937, https://doi.org/10.1007/978-1-4939-9065-8_16,

preclinical applications [4–7]. These methods, however, are not suitable for studying genes and proteins which have an anatomically restricted function. For instance, if one aims to study local vascular injury, systemic gene expression may confound any interpretations of results. To that end, here we developed a novel murine in vivo site-specific intra-arterial transduction method using AAV in order to facilitate future cardiovascular studies requiring targeted and local gene manipulations.

2 Materials

2.1 Animals

1. Adult mouse or rat (*see* **Note 1**).

2.2 Pre-surgery

1. Isoflurane or injectable anesthetic.
2. Betadine or comparable agent.
3. Saline solution (0.9% NaCl).
4. Cotton-tipped applicators.
5. Sterile eye lubricant.
6. Shaver.
7. Heating pad.

2.3 Artery Dissection Surgery

1. General surgical tools (scissors, tweezers, curved tweezers, and needle holder) (*see* **Note 2**).
2. Dissecting microscope (Leica MZ 125 (Leica Microsystems GmbH, Wetzlar, Germany) or equivalent).
3. Micro clip (Roboz RS-5426 (Roboz, Gaithersburg, MD, USA) or equivalent).
4. Flash beads sterilizer.
5. Sterile silk 3–0 suture.
6. Sterile nylon 7–0 suture.
7. Tape.
8. Saline (0.9% NaCl).
9. Latex cuff (*see* **Note 3**).

2.4 AAV Injection

1. AAV.
2. Parafilm or comparable item.
3. 30 gauge or smaller needle, or micro-surgical knife.
4. Micro glass needles.
5. Micro-injector with needle holder.
6. Bipolar cauterizer (AESCULAP GN60 (Aesculap, Center Valley, PA, USA) or equivalent).

7. Sterile monofilament nonabsorbable 5-0 suture or surgical staples.
8. Sterile saline (0.9% NaCl) or Lactated Ringer's solution.
9. Carprofen (5 mg/kg) or comparable analgesic.

2.5 Euthanasia

1. General surgical tools (scissors and tweezers).
2. Injectable anesthetic.
3. Perfusion platform.
4. 4% paraformaldehyde solution in PBS (pH 7.4).

3 Methods

3.1 Preoperational Preparation

1. All instruments are steam sterilized in packs prior to each initial surgery and subsequently flash sterilized between individual surgeries using a dry bead sterilizer.
2. Surgeons wear sterile gloves (for each animal) and surgical scrubs when performing surgeries.
3. Mice are anesthetized with isoflurane with a scavenging system. Injectable anesthesia can also be used.
4. When the mouse is sufficiently anesthetized (*see* **Note 4**), the fur is removed around the surgical site.
5. The site is scrubbed with Betadine™ or comparable agent followed by a 2× saline scrub.
6. Eye dehydration should be prevented by placing sterile eye lubricant in the eyes.
7. Level of anesthesia is assessed by continuous monitoring of respiratory rate, pulse oximetry, toe pinch, corneal reflex, and muscular relaxation.
8. The mouse is positioned for surgery under a dissecting microscope with isoflurane tubing and face mask (if isoflurane is used).
9. A heating pad is placed under the animal and a thermometer is placed between the pad and the animal to ensure that the temperature does not exceed 39 °C (*see* **Note 5**).

3.2 Right Common Carotid Artery Exposure and Preparation

1. After an appropriate level of anesthesia is induced, the mouse is placed in a supine position on the operating platform under dissecting microscope.
2. An incision is made near the clavicles and advanced cranially, for approximately 1.5 cm (*see* Fig. 1 and **Note 6**).
3. After locating the submandibular gland (*see* Fig. 1), a cut is made just below, taking care to avoid brachiocephalic veins.

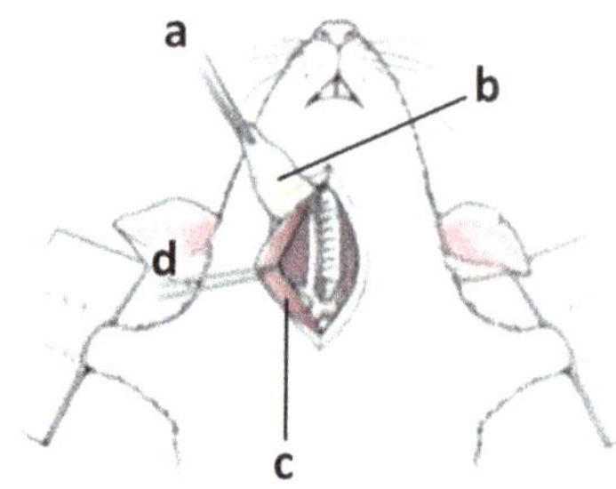

Fig. 1 How to expose right common carotid artery. A micro clip (**a**) is attached to the submandibular gland (**b**) and reflect cranially and slightly to the right. Next, the right sternocleidomastoid muscle (SCM) (**c**) is dissected and a silk suture (**d**) is used to reflect the SCM laterally

4. Once the area near the carotid is exposed, attach micro (aneurysm) clip to the submandibular gland and reflect cranially and slightly to the right.
5. The right sternocleidomastoid muscle (SCM) is dissected—dissection will be performed medially first, then tweezers will hold brachiocephalic veins laterally as the lateral aspect of SCM is exposed.
6. Using the curved part of tweezers facing downward, a scooping motion will isolate the muscle from the medial to the lateral side and silk suture is used to reflect the SCM laterally (*see* Fig. 1).
7. The right common carotid artery (RCCA) is exposed to the level of the bifurcation, taking care to stay parallel to the vessel.
8. A 3-0 silk suture is placed under the RCCA, then a latex cuff is placed below the RCCA to protect the trachea, etc.
9. The suture is positioned between the latex cuff and RCCA, in order to keep the RCCA elevated above the latex cuff.
10. Once the RCCA is exposed, two pieces of 7-0 nylon sutures are placed in the gap between the RCCA and a latex cuff distally and proximally, respectively.
11. RCCA is tied distally by the 7-0 nylon suture first by a bow knot method (*see* **Note 7**).
12. RCCA is tied proximally by the 7-0 nylon suture to halt the blood flow temporarily, and the "pocket" of RCCA (approximately 0.5 cm) will be created.
13. Keep RCCA moist by saline solution while you are preparing AAV in the injector.

3.3 AAV Injection into Right Common Carotid Artery

1. Once the pocket of RCCA is exposed, the viral vector is prepared. Place a droplet (5–10 μL) of solution containing AAV (3.5–5.0 × 10^{12} vector genomes(vg)/mL) with the gene of interest on sanitized parafilm.
2. Aspirate and hold the entire amount of solution in a micro glass needle attached with micro-injector.
3. A very small incision on the pocket of RCCA wall is made using a 30 gauge needle or a micro-surgical knife close to the suture that was tied proximally.
4. Blood is squeezed out from the pocket by using tweezers.
5. The glass needle with viral vector is inserted into the pocket of RCCA via incision (*see* **Note 8**).
6. The solution is injected into the pocket of RCCA.
7. Once the injection is completed, hold a bipolar cauterizer in the hand not holding the glass needle (*see* **Note 9**).
8. The bipolar cauterizer is placed over the incision with the glass needle.
9. Gently remove the needle from the incision, and cauterize and close the incision quickly (*see* **Note 10**).
10. Once the incision is closed, incubate the pocket of RCCA with viral vector for 30 min (*see* **Note 11**).
11. After the incubation, the distal suture is removed first.
12. Remove the remaining suture.
13. The skin incision is closed with appropriately sized (5-0 or 4-0; following NIH guidelines) monofilament nonabsorbable suture or surgical staples.
14. Administer warm postoperative fluids (sterile 0.9% saline or Lactated Ringer's Solution) at 0.01–0.02 mL/g of body weight intraperitoneally.
15. The animals is allowed to regain consciousness and become ambulatory on a slide warmer/heat pad to assist in appropriate thermoregulation.
16. Administer a single dose of Carprofen (5 mg/kg IP) to provide analgesia twice in 48 h (*see* **Note 12**).
17. The animals are examined on a daily basis to assure that there is no post-surgical complications or undue distress to the animal for a period of no less than 5 days.
18. The suture or staple is removed 14 days after surgery.

3.4 Euthanasia Using 4% Paraformaldehyde

1. After an appropriate level of anesthesia is induced by injectable anesthetic, the mouse is placed in a supine position on the perfusion platform in a perfusion station.

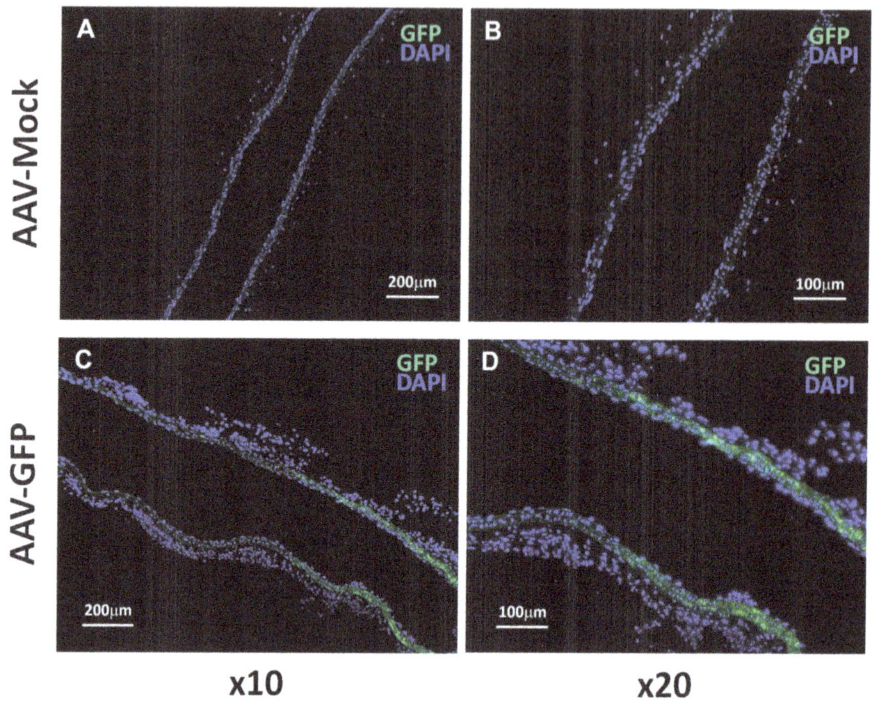

Fig. 2 Localized intra-artery viral transduction using AAV5-GFP in mouse common carotid artery. Longitudinal section of right common carotid arteries using AAV-Mock (**a**, **b**) or AAV-GFP (**c**, **d**). Cells in the smooth muscle layer of the artery are successfully transduced (GFP positive)

2. A midline ventral abdominal incision is made and advanced cranially and bilaterally to expose the heart.
3. 3 mL of 4% paraformaldehyde solution in PBS is injected into the left ventricle after puncturing the right atrium for euthanasia.
4. The RCCA is collected and prepared for embedding (either paraffin or freeze) in a block for sectioning (Fig. 2).

4 Notes

1. All animal procedures must be performed under the approval of your institution's Animal Care and Use Committee.
2. Use the micro-dissecting tweezers for RCCA separation.
3. You can make latex cuffs from a sterile latex glove. Cut out a rectangle-shaped cuff (approximately 0.5 cm × 2.0 cm).
4. Sufficiency of anesthesia can be determined by tail pinch, toe pinch, corneal reflex, jaw laxity, and muscular relaxation.
5. Level of anesthesia has to be monitored throughout surgery. This includes observation of skin tone at the periphery,

assessing overall temperature, regular breathing pattern, and failure to respond to occasional reflex tests.

6. Subcutaneous tissues along the incision should be reflected. This will make closing easier.
7. The bow knot method allows you to untie the sutures much easier.
8. If you are right handed, it is recommended that you hold the glass needle holder with your left hand and control the injector with your right hand.
9. Keep the needle in the RCCA pocket while doing this.
10. If the incision is not closed properly, you will see a "deflation" of the pocket.
11. The artery and surrounding tissue has to be kept moist.
12. Additional dose of analgesic should be given as needed.

Acknowledgment

This work was supported by NIH R01 grant (5R01NS083673, PI: Hoh).

References

1. Kotterman MA, Schaffer DV (2014) Engineering adeno-associated viruses for clinical gene therapy. Nat Rev Genet 15:445–451
2. Dyka FM, Boye SL, Chiodo VA et al (2014) Dual adeno-associated virus vectors result in efficient in vitro and in vivo expression of an oversized gene, MYO7A. Hum Gene Ther Methods 25:166–177
3. Bennett A, Patel S, Mietzsch M et al (2017) Thermal stability as a determinant of AAV serotype identity. Mol Ther Methods Clin Dev 6:171–182
4. Bader M, Bohnemeier H, Zollmann FS et al (2000) Transgenic animals in cardiovascular disease research. Exp Physiol 85:713–731
5. Derrett-Smith EC, Denton CP, Sonnylal S (2009) Animal models of scleroderma: lessons from transgenic and knockout mice. Curr Opin Rheumatol 21:630–635
6. Armeanu S, Pelisek J, Krausz E et al (2000) Optimization of nonviral gene transfer of vascular smooth muscle cells in vitro and in vivo. Mol Ther 1:366–375
7. Hosaka K, Rojas K, Fazal HZ et al (2017) Monocyte chemotactic protein-1-interleukin-6-osteopontin pathway of intra-aneurysmal tissue healing. Stroke 48:1052–1060

Chapter 17

Stable Genetic Modification of Mesenchymal Stromal Cells Using Lentiviral Vectors

Francisco Martín, María Tristán-Manzano, Noelia Maldonado-Pérez, Sabina Sánchez-Hernández, Karim Benabdellah, and Marién Cobo

Abstract

Mesenchymal stromal cell (MSC) therapy has produced very promising results for multiple diseases in animal models, with over 780 clinical trials on going or completed. However, most of the human clinical trials have not been as successful as trials using preclinical models. To improve the therapeutic potential of MSCs, different research groups have used gene transfer vectors to express factors involved in migration, survival, differentiation, and immunomodulation. The ideal gene transfer vector for most applications should achieve long-term, stable (constitutive or inducible) transgene expression in MSCs and their progeny. Given their efficiency and low impact on transduced cells, lentiviral vectors (LVs) are the vectors of choice. In this chapter we will describe a detailed protocol for the generation of genetically modified MSCs using lentiviral vectors (LVs). Although this protocol has been optimized for MSC lentiviral transduction, it can be easily adapted to other stem cells by changing culture conditions while maintaining volumes and incubation times.

Key words Mesenchymal stromal cells, MSCs, ASCs, Lentiviral vectors, Constitutive expression, Inducible expression, Transduction, Genetic modification

1 Introduction

Multipotent mesenchymal stromal cells (MSCs) are non-hematopoietic cells found in virtually all organs and tissues. MSCs have been defined as plastic adherent, $CD73^{+}CD90^{+}CD105^{+}CD11b^{-}$ $CD31^{-}CD45^{-}$ cells that can differentiate into adipocytes, chondroblasts, and osteoblasts in vitro [1]. MSCs have shown very promising results in animal models for numerous conditions, including stroke, graft-versus-host disease (GVHD), Crohn's disease, liver transplantation, spinal cord injury, osteoarthritis, multiple sclerosis (MS), diabetes, and ulcerative colitis. Based on these results, there are over 780 clinical trials ongoing or completed, 45 of which have reached phase III or IV (data obtained from

Fredric P. Manfredsson and Matthew J. Benskey (eds.), *Viral Vectors for Gene Therapy: Methods and Protocols*, Methods in Molecular Biology, vol. 1937, https://doi.org/10.1007/978-1-4939-9065-8_17,

www.clinicaltrials.gov). However, most of the human clinical trials have not been as successful as trials involving preclinical models. In order to improve the therapeutic potential of MSCs, genetically modified MSCs (GM-MSCs) expressing different factors have been developed (see Hwang, B et al., [2]). Various research groups have generated GM-MSCs with improved migration potential [3], anti-tumor effects [4–11], survival [12, 13], differentiation potential in relation to certain tissues [14–20], immunomodulatory properties [20–22], and/or secretion of trophic factors [13, 21, 23–25].

The ideal delivery methods for expressing the desired transgene in MSCs depend on the targeted disease and the therapeutic strategy. For some applications, a short-lived expression of the transgene can be the more appropriate strategy (i.e., to enhance differentiation potential to certain tissues).

However, for most applications, long-term and stable (constitutive or inducible) transgene expression in MSCs and their progeny is the best option. Most of the viral vectors described in this book, including adenovirus vectors (AdVs), adeno-associated vectors (AAVs), baculovirus and retrovirus vectors (ɣ-retrovirus (ɣ-RV) and lentiviral vectors (LVs)), have been used to generate GM-MSCs. A wide range of nonviral methods, involving cationic polymers, nanoparticles, electroporation, and sonotransfection, have also been used to improve the osteogenic and chondrogenic potential of MSCs (for review, *see* ref. 26).

LVs are probably the best tool, in term of safety and efficacy, for stable genetic modification of stem cells. They are very efficient, have a low impact in the biology of the transduced cells, and are also very versatile, allowing multiple modifications in their backbone and the use of multiple envelopes (pseudotyping). Most LVs include a deletion in the 3′LTR that make the LVs safer and allows the transgene to be expressed by the desired promoter. This is very useful when tissue-specific or inducible expression is required. Indeed, there are a wide variety of LVs that have been shown to be efficient tools for constitutive transgene expression or inducer-regulated transgene expression in MSCs. Choosing the appropriate LV for your application constitutes a critical step for generating genetically modified MSCs regarding protein stability, toxicity, or controlling expression levels. In this chapter we will describe a detailed protocol to generate genetically modified MSCs using LVs. The objective will be to maximize the percent of cells expressing the transgene while minimizing the effects on the MSCs intrinsic properties. We will also provide information on LVs available on the market and in research groups in order to achieve constitutive or inducible transgene expression in MSCs (Tables 1 and 2).

Table 1
Constitutive promoters used for transgene expression in MSCs derived from different species

	Source	Promoter	Transgene	Vector name	Ref.
Mouse	Ad	**CMV**	IL-35	*pCCS-IL35-Lv201*	[27]
	Ad	**CMV**	eGFP, VIP	*pHRSINncPPT_CEWP. pHRSINncPPT_LentiVIP*	[21]
	Ad	**CMV**	Interferon b1	*pHRSINncPPT_CIFNbWP*	[28]
	BM	**EF-1α**	β-Catenin, ROR2	*EF1a-Ctnnb1-IRES-eGFP, EF1a-ROR2-IRES-eGFP*	[29]
	BM	**EF-1α**	CXCR4, CXCR7	*pLV-CXCR4-IRES-GFP* *pLV-CXCR7-IRES-GFP*	[30]
	BM	**MSCV**	CCR1, CXCR2	*pMSCV-CCR1/CXCR2-IRES-GFP*	[31]
Rat	BM	**EF-1α**	VEGF, Bcl2	*pLV-CMV-EF1-fLuc-T2A-puro*	[32]
Rabbit	BM	**CMV**	Nerve growth factor β	*pDC316-hNGFb-mCMV-EGFP*	[33]
Human	Synovial-derived	**EF-1α**	Insulin-like growth factor	*pLVSIN-IRES-IGF-1 ZsGreen*	[34]
	BM	**SFFV, CMV**	Ds-Red, Ds-eGFP	*pHR_SINcPPT_SIEW, pHR_CMV_DRep*	[35]
	BM	**PGK**	BMP2	*Lenti PGK-GFP-mCM-BMP2-IRES puro*	[36]
	UCB	**CMV**	Coagulation factor I (FIX)	*PLVX-FIXi*	[37]
	iPSCs-derived	**EF-1α**	NEL-like protein 1, BMP2	*Lv-EF-1α-NELLP1, LV- EF-1α-* BMP2	[38]
	MSCs-BM	**CMV, EF, CAG**	eGFP	*pNL-EGFP/CMV, pNL-EGFP/EF, pNL-EGFP/CAG*	[39]

Ad adipose tissue, *BM* bone marrow, *UCB* umbilical cord blood, *iPSCs* induced pluripotent stem cells, *MSCs* mesenchymal stromal cells

Table 2
Lentiviral vectors for inducible expression on human and nonhuman MSCs

	Source[a]	System	TransAct[b]	Vector structure[c]	DOX[d]	Leaking[e]	Ref.
Rat	Ad	**Tet-On** *Binary*	Yes	*#LV1: **EF-1α**-rtTA* *#LV2: **TRE**-eGFP*	45 ng/mL	Low	[40]
Human	BM	**Tet-On** *Binary*	Yes	*#LV1: **CMV-IE**-rtTA2S-M2* *#LV2: **TRE/Pitt-CMV**-hGDNF/hrGFP*	100 ng/mL	Low	[41]
	hESC-derived	**Tet-On** *Binary*	Yes	*#LV1: Le-rtTA* *#LV2: Le-**TRETight**-VEGF*	1 mg/mL	Moderate	[42]
	UCB	**Tet-On** *All in one*	Yes	***CMVminTetO**-HGF-pA-**EF-1α** –eGFP-rtTA*	5 μg/mL	Moderate	[43]
	BM	**Tet-On** *All in one*	No	***CMVTetO**-eGFP* ***SFFV**-TetR*	100 ng/mL	Moderate	[44]
	BM, hESCs-derived	**Tet-On** *All in one*	No	***CMVTetO**-eGFP* ***EF-1α**–TetR* *+/–Insulator IS2*	100 ng/mL	Low	[45, 48]
	Ad	**Tet-Off** *All in one*	Yes	***TRE**-eGFP-**EF-1α**-tTA-2A-Bsd,* ***TRE**-eGFP-**CMV**-tTA-2A-Bsd*	1 μg/mL	Low	[46]

[a]MSCs were obtained from: *Ad* adipose tissue, *BM* bone marrow, *UCB* umbilical cord blood, *hESCs* human embryonic stem cells
[b]Trans-activator presence (tTA/rtTA) (*see* **Note 1**)
[c]Promoters are indicated in bold. Transactivators (tTA/rtTA) are highlighted in red; Tet repressor (TetR) in blue; transgene of interest in green; insulator presence in violet
[d]Optimal doxycycline dose established by the authors
[e]Residual expression in the absence of DOX

2 Materials

2.1 Cell Culture

1. Fully equipped Tissue culture room Category II.
2. Refrigerated Centrifuge.
3. Microscope.
4. Tissue culture flasks, pipettes, gloves, etc.
5. Human Embryonic Kidney 293T cell line (ATCC® 3216 ™).
6. Complete HEK293T media: High glucose Dulbecco's Modified Eagle's Medium (DMEM) supplemented with 10% fetal bovine serum (FBS) and 100 U/mL Penicillin/Streptomycin.
7. Trypsin-EDTA solution (0.12% trypsin and 0.02% EDTA) or commercial TrypLE™ (Thermo Fisher Scientific).
8. Phosphate buffered saline (PBS).
9. Complete Human MSCs media: Advanced DMEM (Gibco), supplemented with 10% FBS, Glutamax (Gibco) and 100 U/mL penicillin/streptomycin (Gibco).
10. Complete Mouse MSCs media: Specialized MesenCult medium (Stem Cell) containing 100 U/mL penicillin/streptomycin and 20% mouse mesenchymal supplements (Stem Cell).

2.2 Lentiviral Production and Harvest

1. LipoD293™ in vitro Transfection Reagent (Signagen Lab).
2. The human immunodeficiency virus (HIV) packaging (pCMV_R8.9) and VSV-G (pMD2.G) plasmids (Addgene).
3. Serum-Free high glucose DMEM.
4. 100 mm Petri dishes.
5. Syringes without needle (10, 20 mL, depending on the volume).
6. 0.45 μm sterile filter (Millex-HV Syringe Filter Unit, Merck, SLHV033RS).
7. DMEM supplemented with 10 or 2% FBS.
8. Alternatively, Opti-MEM™ Reduced Serum Medium (Gibco).

2.3 Ultrafiltration

1. 70% ethanol in sterile cell culture water.
2. 1× sterile PBS.
3. Amicon Ultra-15, PLHK Ultracel-PL membrane, 100 kDa (Merck, UFC910024).

3 Methods

3.1 LVs Preparation

LVs particles are generally produced by transient transfection of HEK293T cells. Bellow, we briefly describe our protocols to obtain high titer and concentrated LVs required for achieving good transduction efficiency in MSCs without affecting their phenotype.

3.1.1 Lentiviral Vectors Production

There are different ways to obtain LVs particles. We use transient transfection of HEK293T cells using LipoD293, because of its reproducibility and high yield. We routinely achieve 3–5 × 10^7 TU/mL using standard LVs.

1. *Preparation of 293T packaging cells*: Make sure cells are mycoplasma negative. After 2–3 consecutive passages, the 293T cells are plated for transfection.
2. 12–16 h before transfection, seed 4–7 × 10^6 cells in a 10 cm^2 Petri dish to obtain 80–90% confluency on the day of the transfection (*see* **Note 2**).
3. 30–60 min before transfection, remove old media and add 4 mL of fresh complete HEK293T media at 37 °C. Then proceed to **step 4** as quickly as possible.
4. *Transfection using LipoD293*: In sterile conditions prepare the DNA mixture using the different plasmids required for your LVs production. For second generation LV production, you should use three different plasmids: packaging (pCMV_R8.9), envelope (VSV-G, pMD2.G), and plasmid of interest (*see* the Chap. 7 for alternative production techniques). For a 10 cm^2 dish, add 18 μg of total plasmid DNA (3:2:1 desired plasmid: packaging:envelope ratio, therefore 9 μg your plasmid, 6 μg pCMV_R8.9, and 3 μg VSV-G envelope) to 500 μL serum-free DMEM with high glucose. It is critical to use serum-free media for in this step.
5. Prepare the LipoD293 solution: For each 10 cm dish, dilute 45 μL LipoD293 reagent in 500 μL serum-free DMEM with high glucose and vortex gently mix. Do not use Opti-MEM or similar to dilute plasmid DNA as it will disrupt the transfection complex.
6. Immediately add the LipoD293 solution to the DNA mixture (and not in reverse order), vortex briefly (or pipette up and down 4 times), and incubate for 10 min at RT.
7. Add the DNA/LipoD293 mixture dropwise to the culture medium in the 10 cm dish and homogenize the mixture by gently swirling the plates. Incubate for 5–18 h at 37 °C.
8. Remove the medium containing DNA/LipoD and add 7–9 mL of any of the following fresh media: (1) DMEM

+10% FBS, (2) DMEM +2% FBS, (3) Opti-MEM or other serum-free media suitable for 293T.

9. *Recovery of viral particles*: Harvest the supernatant containing LV particles 36 h post transfection and clean it using a 0.45 μm sterile filter to remove cell debris.
10. Add 7–8 mL of DMEM (2–10%FBS) or Opti-MEM to the producer cells to facilitate a second harvest.
11. Collect the supernatant again 10–20 h after the first harvest and clean it using a 0.45 μm sterile filter (*see* **Note 3**).
12. After each harvest, LVs can be either used immediately for transduction/titration, kept at 4 °C for 24 h (without losing substantial titter), aliquoted and conserved at −80 °C, or concentrated (*see* Subheading 3.1.2).

3.1.2 Lentiviral Vector Concentration by Ultrafiltration

We routinely concentrate the first and second harvests by ultrafiltration which allows the viral particles to be concentrated and minimizes concentration of impurities:

1. Add 6 mL of 70% ethanol to the 100 kDa filtration device. Mix slowly to ensure that the ethanol spreads through the entire device.
2. Centrifuge at 1800–4000 × *g* for 5 min so that the ethanol passes through the filter.
3. Add 6 mL of sterile PBS to the device and centrifuge at the same previous speed until all PBS has passed through the filter.
4. Repeat **step 3** two more times.
5. Fill the filtration device with the 15 mL of supernatant previously collected and filtered.
6. Centrifuge at 4000 × *g* for 10 min at 4 °C. Determine the concentration achieved (possible by observing the device). If sufficient concentration has been reached, follow the next step. If not, centrifuge at 4000 × *g* for a further 5–15 min depending on the concentration factor required.
7. Once the desired concentration factor has been achieved, discard the flow-through and collect the concentrated LVs retained in the device. Pipette up and down slowly 10–15 times for LV particle collection (*see* **Note 4**).
8. The concentrated viruses can be used directly or aliquoted and stored at −80 °C.

3.2 Preparation of MSCs to Optimize Transduction

The MSCs must be in optimal conditions to achieve high transduction efficiencies (*see* **Note 5**). Although we recommend transduction at the earliest possible passage, the cells can be genetically modified using LVs at any time. Below, we present a protocol for the transduction of 500,000 MSCs. Transduction of larger or smaller numbers of MSCs can be easily achieved by scaling volumes. To prepare cells for LV transduction, follow these steps:

1. Plate MSCs into a 10 cm^2 dish at a density of around 50,000 cells/cm^2 (500,000 MSCs/dish) and allow the MSCs to reach approximately 80% confluency (usually next day since the amount of cells plated is 10 times the amount used for standard MSC culture).
2. Remove the medium and wash cells twice with PBS.
3. Add a small volume (1/10 of initial culture media) of trypsin-EDTA solution or derivates such as TrypLE™ to detach the cells from the plastic. Incubate for 4–10 min at 37 °C with sporadic shaking until the cells are detached.
4. Recover the detached MSCs by adding 1–5 mL of PBS containing 2% FBS to inactivate the trypsin, collect in a centrifuge tube, and then count the cells.
5. Centrifuge for 5 min at 250 × *g*, discard the supernatant, resuspend the pellet in complete medium to obtain a final concentration of 1×10^6 MSCs/mL.
6. The MSCs are now ready for LV transduction (detailed in Subheading 3.3).

3.3 Genetic Modification of MSCs

Transduction of MSCs can be achieved with either fresh or frozen LVs. If using frozen virus, we recommend performing a further titration after thawing (*see* **Note 6**). To transduce the MSCs, follow these steps:

1. Start with in Subheading 3.2, **step 6** (1×10^6 MSCs in 1 mL of complete MSC medium).
2. Add 0.5 mL (500,000 MSCs) to a 15 mL tube.
3. Centrifuge for 5 min at 250 × *g* and then discard the supernatant.
4. *Important considerations*: Determine the MOI (number of LVs/number of MSCs) used in order to have a controlled experiment (*see* **Note 7**). Initial titers of approximately 10^7 TU/mL require low LV concentration (20×) to achieve high efficiencies. For example, 100 μL of a 10× concentrated virus at 2×10^8 TU/mL renders a MOI = 40, which is sufficient to achieve efficiencies of approximately 50% or higher. However, if initial LV titers are 10^6 TU/mL or lower, it is important to concentrate up to 100× in order to achieve good transduction efficiency (*see* **Note 8**).
5. Resuspend the pellet in 100 μL of concentrated LV supernatant (*see* Subheading 3.1) and incubate for 10 min at room temperature.

6. *Optional*: If viral titers are too low, consider using spinoculation to increase transduction efficacy. To do this, simply centrifuge the MSC/LV mixture at 800 × *g* for 30 min at 32 °C.
7. Following initial viral incubation, add 600 μL of fresh media to the MSC/LV mixture and seed in a 6-well plate well (final cellular density of 50,000 cells/cm^2). These high cellular densities and small volumes enable higher LV concentration per cell during transduction.
8. Incubate for 3–6 h at 37 °C. It is important to maintain short incubation times to minimize LV toxicity and the deleterious effects of high MSC density.
9. Remove the LV supernatant and detach the MSCs (as described in Subheading 3.2, **steps 2–5**).
10. Resuspend the transduced MSCs in 12 mL of fresh complete MSC media and seed in a 10 cm plate (final density of 8000 cells/cm^2).

3.4 Analysis of Transduction Efficacy and Transgene Expression Levels

After transduction of MSCs with LVs, it is necessary to analyze: (1) The percentage of cells that express the transgene and (2) the expression levels achieved in the cells transduced. We recommend carrying out the first analysis 4–7 days post-transduction and every 7–14 days thereafter in order to determine stability (*see* **Note 9**). There are several ways to analyze transduction efficacy and transgene expression levels:

3.4.1 Flow Cytometry

FACS analysis simultaneously provides information regarding both transduction efficiency and expression levels.

1. *LVs expressing fluorescent proteins such as eGFP and dsREd.* When LVs express a fluorescent protein, transduction efficiency and expression levels can be easily determined by flow cytometry [47]. *Transduction efficiency* is usually expressed as the percentage of cells expressing fluorescent proteins 4–7 days post-transduction in relation to untransduced control cells (*see* **Note 10**). *Expression levels* can be estimated in terms of the increment in the mean fluorescence intensity (MFI) of positive cells in relation to control cells (*see* **Note 11**). The higher the MFI increment, the higher the expression levels achieved in the MSCs. Constitutive LVs are expected to render stable transgene expression over time, while inducible promoters should express the transgene only when the inducer is added.
2. *LVs expressing nonfluorescent proteins.* If the LVs do not express a fluorescence marker and specific antibodies are available, flow cytometry can be used to measure transduction efficiency and expression levels. To do so, just follow an appropriate staining protocol for the protein of interest and use the procedures

described for fluorescent proteins. Additionally, with regard to secreted proteins, the vector copy number per cell (vcn/c) is generally used to determine transduction efficiency (see below), while ELISA and/or Western blot analyses of cell lysates and supernatants are used to determine expression levels [21].

3.4.2 Quantitative PCR (Q-PCR)

This is the gold standard for measuring transduction efficiency by determining the vector copy number per cell (vcn/c). This is the best option for measuring LV transduction efficiency and can also provide additional information. For example, if the cellular population has a vcn/c of 0.6 and 60% of cells express the protein; it can be assumed that the LVs are not silenced over time. However, if the percentage of cells expressing the protein drops to 10% and the vcn/c remain at 0.6, promoter silencing can be assumed. To measure the vcn/c ratio, follow the protocol described in [44], use genomic DNA extracted from MSCs at least 7 days post-transduction.

3.4.3 RT-qPCR, Western Blot, and ELISA to Measure Expression Levels

In addition to flow cytometry, the following analyses to further characterize GM-MSCs are recommended:

1. *RT-qPCR* is the best method to quantify mRNA expression levels. Standard procedures should be used to isolate MSCs mRNA, to generate the cDNA and to perform the RT-qPCR.
2. *Western blot* can provide additional information on protein expression levels and potential alterations in size/variants. Western blot can be performed on cell lysates or on supernatants when the protein is secreted.
3. *ELISA analysis* of supernatants of transduced MSCs can be highly informative and quantitatively interesting when the protein is secreted (*see* **Note 12**).

3.5 Analysis of Potential Side Effects of LVs Transduction on MSCs

Transduction of MSCs with highly concentrated LVs can lead to diverse effects on the GM-MSCs, such as cell death and differentiation. In addition, the expression of the transgene can also influence the biology of MSCs. It is therefore important to analyze the gene-modified MSCs for:

1. *Cell viability.* 7AAD/PS analysis should be carried out 5–7 days post-transduction.
2. *MSC markers pattern.* GM-MSCs must preserve their typical phenotype: positive for markers like CD90, CD73, CD44, CD105, and negative for CD45, CD34 or CD11 (1).
3. *MSCs differentiation potential.* GM-MSCs must keep their multipotency. Use commercially available kits to differentiate the MSCs to adipocyte, chondrocyte, and osteocytes and perform the analysis as indicated by the company's web site.

4 Notes

1. Tetracycline-regulated gene expression Tet-On/Tet-Off systems have been used successfully for conditional gene expression in MSCs (Table II). The majority of these Tet systems require the presence of a tetracycline-dependant transactivator (rtTA) to achieve robust transgene expression. However, rtTA contains the activating domain of herpes virus simplex viral protein 16 (VP-16) which is linked to the TetR repressor and can have unpredictable effects. As an alternative, several research groups have developed regulatable LVs using the original TetR repressor. In both cases, gene expression can be regulated by LV co-transduction, with one LV carrying the regulatable promoter and the transgene and the other expressing the repressor, or by transduction, with an all-in-one LVs containing all these elements in one vector.
2. The quality of the monolayer is critical in order to obtain high LV titers. Check that the 293T cells are in good conditions: free of mycoplasma and expressing the SV40 large T antigen. A good method for doing this is to culture 293T cells in the presence of neomycin.
3. A third harvest can be obtained using the same procedure. As LV particles are highly unstable at 37 °C, an increasingly large proportion of LVs in the culture are degraded over time. Therefore, the first harvests contain larger amounts of degraded LV particles compared to second and third harvest. Consequently, at equal titer, it is preferable to use second and third harvests which have lower cytotoxicity.
4. LV particles are very sensitive to heat and stress forces. Carry out the resuspension process very carefully and slowly and keep the LVs at 4 °C throughout.
5. MSCs from different species require different culture conditions: (1) *Murine mesenchymal stromal cells (mMSCs):* We recommend using commercially available specialized media to culture mMSCs under moderate hypoxic conditions (5% O_2) or under normoxic conditions. (2) *Human mesenchymal stromal cells (hMSCs):* These cells can be grown in commercially available specialized media or in DMEM +10%FCS. They can be cultured under normoxic or hypoxia conditions (*see* ref. 49 for details).
6. As LV particles are highly sensitive, small differences in freezing/thawing conditions can affect their viability. It is therefore important to examine the LV titer of each aliquot after thawing.

7. As MSCs from different origins differ in their resistance to genetic modification, the MOI used to achieve the desire efficiency must be determined for each MSC culture.
8. Most concentration strategies also concentrate VSVg pseudo particles which will increase the toxicity of the final LV preparation. Therefore, if using an LV concentration of 100× or higher, reduce LV incubation time to 2 h.
9. When using constitutive LVs, transduction efficiencies and expression levels should be kept stable over time in culture unless the promoter is silenced or the protein is toxic. If so, use different LVs containing different promoters and/or insulators or use inducible LVs.
10. If the percentage of transgene positive cells exceeds 60%, it will be necessary to determine the vcn/c to properly estimate transduction efficiency.
11. If you want to compare the expression levels of different LV backbones, it is important to ensure that the cell populations compared have similar vcn/c or similar percentages of positive cells (ideally under 40%).
12. Always use non-transduced cells as a negative control of gene expression and/or transduction.

References

1. Dominici M, Le Blanc K, Mueller I et al (2006) Minimal criteria for defining multipotent mesenchymal stromal cells. The international society for cellular therapy position statement. Cytotherapy 8(4):315–317
2. Hwang BW, Kim SJ, Park KM et al (2015) Genetically engineered mesenchymal stem cell therapy using self-assembling supramolecular hydrogels. J Control Release 220 (Pt A):119–129
3. Park SA, Ryu CH, Kim SM et al (2011) CXCR4-transfected human umbilical cord blood-derived mesenchymal stem cells exhibit enhanced migratory capacity toward gliomas. Int J Oncol 38(1):97–103
4. Tyciakova S, Matuskova M, Bohovic R et al (2015) Genetically engineered mesenchymal stromal cells producing TNFalpha have tumour suppressing effect on human melanoma xenograft. J Gene Med 17(1–2):54–67
5. Sasportas LS, Kasmieh R, Wakimoto H et al (2009) Assessment of therapeutic efficacy and fate of engineered human mesenchymal stem cells for cancer therapy. Proc Natl Acad Sci U S A 106(12):4822–4827
6. Shah K (2012) Mesenchymal stem cells engineered for cancer therapy. Adv Drug Deliv Rev 64(8):739–748
7. Knoop K, Kolokythas M, Klutz K et al (2011) Image-guided, tumor stroma-targeted 131I therapy of hepatocellular cancer after systemic mesenchymal stem cell-mediated NIS gene delivery. Mol Ther 19(9):1704–1713
8. Kucerova L, Altanerova V, Matuskova M et al (2007) Adipose tissue-derived human mesenchymal stem cells mediated prodrug cancer gene therapy. Cancer Res 67(13):6304–6313
9. Ren C, Kumar S, Chanda D et al (2008) Cancer gene therapy using mesenchymal stem cells expressing interferon-beta in a mouse prostate cancer lung metastasis model. Gene Ther 15 (21):1446–1453
10. Kim SM, Lim JY, Park SI et al (2008) Gene therapy using TRAIL-secreting human umbilical cord blood-derived mesenchymal stem cells against intracranial glioma. Cancer Res 68 (23):9614–9623
11. Kim SW, Kim SJ, Park SH et al (2013) Complete regression of metastatic renal cell carcinoma by multiple injections of engineered

mesenchymal stem cells expressing dodecameric TRAIL and HSV-TK. Clin Cancer Res 19(2):415–427

12. McGinley L, McMahon J, Strappe P et al (2011) Lentiviral vector mediated modification of mesenchymal stem cells & enhanced survival in an in vitro model of ischaemia. Stem Cell Res Ther 2(2):12
13. Xiang Q, Hong D, Liao Y et al (2017) Overexpression of Gremlin1 in mesenchymal stem cells improves hindlimb ischemia in mice by enhancing cell survival. J Cell Physiol 232 (5):996–1007
14. Nakashima M, Tachibana K, Iohara K et al (2003) Induction of reparative dentin formation by ultrasound-mediated gene delivery of growth/differentiation factor 11. Hum Gene Ther 14(6):591–597
15. Sheyn D, Pelled G, Zilberman Y et al (2008) Nonvirally engineered porcine adipose tissue-derived stem cells: use in posterior spinal fusion. Stem Cells 26(4):1056–1064
16. Kim HJ, Im GI (2011) Electroporation-mediated transfer of SOX trio genes (SOX-5, SOX-6, and SOX-9) to enhance the chondrogenesis of mesenchymal stem cells. Stem Cells Dev 20(12):2103–2114
17. Kim TH, Kim M, Eltohamy M et al (2013) Efficacy of mesoporous silica nanoparticles in delivering BMP-2 plasmid DNA for in vitro osteogenic stimulation of mesenchymal stem cells. J Biomed Mater Res A 101 (6):1651–1660
18. Frisch J, Venkatesan JK, Rey-Rico A et al (2014) Determination of the chondrogenic differentiation processes in human bone marrow-derived mesenchymal stem cells genetically modified to overexpress transforming growth factor-beta via recombinant adeno-associated viral vectors. Hum Gene Ther 25 (12):1050–1060
19. Joydeep D, Choi YJ, Yasuda H et al (2016) Efficient delivery of C/EBP beta gene into human mesenchymal stem cells via polyethylenimine-coated gold nanoparticles enhances adipogenic differentiation. Sci Rep 6:33784
20. Carrillo-Galvez AB, Galvez-Peisl S, Ayllón V et al (2018) Glycoprotein A repetitions predominant (GARP) protects MSCs from mtROS-mediated DNA damage and apoptosis via regulation of TGF-β Under review
21. Cobo M, Anderson P, Benabdellah K et al (2013) Mesenchymal stem cells expressing vasoactive intestinal peptide ameliorate symptoms in a model of chronic multiple sclerosis. Cell Transplant 22(5):839–854
22. Carrillo-Galvez AB, Cobo M, Cuevas-Ocana S et al (2015) Mesenchymal stromal cells express GARP/LRRC32 on their surface: effects on their biology and immunomodulatory capacity. Stem Cells 33(1):183–195
23. Aslan H, Sheyn D, Gazit D (2009) Genetically engineered mesenchymal stem cells: applications in spine therapy. Regen Med 4 (1):99–108
24. Dey ND, Bombard MC, Roland BP et al (2010) Genetically engineered mesenchymal stem cells reduce behavioral deficits in the YAC 128 mouse model of Huntington's disease. Behav Brain Res 214(2):193–200
25. Hu J, Lang Y, Zhang T et al (2016) Lentivirus-mediated PGC-1alpha overexpression protects against traumatic spinal cord injury in rats. Neuroscience 328:40–49
26. Oggu GS, Sasikumar S, Reddy N et al (2017) Gene delivery approaches for mesenchymal stem cell therapy: strategies to increase efficiency and specificity. Stem Cell Rev 13 (6):725–740
27. Guo H, Zhao N, Gao H et al (2017) Mesenchymal stem cells overexpressing interleukin-35 propagate immunosuppressive effects in mice. Scand J Immunol 86(5):389–395
28. Marin-Banasco C, Benabdellah K, Melero-Jerez C et al (2017) Gene therapy with mesenchymal stem cells expressing IFN-ss ameliorates neuroinflammation in experimental models of multiple sclerosis. Br J Pharmacol 174(3):238–253
29. Cai SX, Liu AR, He HL et al (2014) Stable genetic alterations of beta-catenin and ROR2 regulate the Wnt pathway, affect the fate of MSCs. J Cell Physiol 229(6):791–800
30. Gheisari Y, Azadmanesh K, Ahmadbeigi N et al (2012) Genetic modification of mesenchymal stem cells to overexpress CXCR4 and CXCR7 does not improve the homing and therapeutic potentials of these cells in experimental acute kidney injury. Stem Cells Dev 21 (16):2969–2980
31. Huang J, Zhang Z, Guo J et al (2010) Genetic modification of mesenchymal stem cells overexpressing CCR1 increases cell viability, migration, engraftment, and capillary density in the injured myocardium. Circ Res 106 (11):1753–1762
32. Ni X, Ou C, Guo J et al (2017) Lentiviral vector-mediated co-overexpression of VEGF and Bcl-2 improves mesenchymal stem cell survival and enhances paracrine effects in vitro. Int J Mol Med 40(2):418–426
33. Wang L, Zhao Y, Cao J et al (2015) Mesenchymal stem cells modified with nerve growth

factor improve recovery of the inferior alveolar nerve after mandibular distraction osteogenesis in rabbits. Br J Oral Maxillofac Surg 53 (3):279–284
34. Ikeda Y, Sakaue M, Chijimatsu R et al (2017) IGF-1 gene transfer to human synovial MSCs promotes their chondrogenic differentiation potential without induction of the hypertrophic phenotype. Stem Cells Int 2017:5804147
35. Grinev VV, Severin IN, Posrednik DV et al (2012) Highly efficient transfer and stable expression of two genes upon lentivirus transduction of mesenchymal stem cells from human bone marrow. Genetika 48(3):389–400
36. Choi KS, Ahn SY, Kim TS et al (2011) Characterization and biodistribution of human mesenchymal stem cells transduced with lentiviral-mediated BMP2. Arch Pharm Res 34 (4):599–606
37. Dodd M, Marquez-Curtis L, Janowska-Wieczorek A et al (2014) Sustained expression of coagulation factor IX by modified cord blood-derived mesenchymal stromal cells. J Gene Med 16(5–6):131–142
38. Liu J, Chen W, Zhao Z et al (2013) Reprogramming of mesenchymal stem cells derived from iPSCs seeded on biofunctionalized calcium phosphate scaffold for bone engineering. Biomaterials 34(32):7862–7872
39. Zhang XY, La Russa VF, Reiser J (2004) Transduction of bone-marrow-derived mesenchymal stem cells by using lentivirus vectors pseudotyped with modified RD114 envelope glycoproteins. J Virol 78(3):1219–1229
40. Qin JY, Zhang L, Clift KL et al (2010) Systematic comparison of constitutive promoters and the doxycycline-inducible promoter. PLoS One 5(5):e10611
41. Yang WH, Yang C, Xue YQ et al (2013) Regulated expression of lentivirus-mediated GDNF in human bone marrow-derived mesenchymal stem cells and its neuroprotection on dopaminergic cells in vitro. PLoS One 8(5):e64389
42. Hajizadeh-Saffar E, Tahamtani Y, Aghdami N et al (2015) Inducible VEGF expression by human embryonic stem cell-derived mesenchymal stromal cells reduces the minimal islet mass required to reverse diabetes. Sci Rep 5:9322
43. Chang HK, Kim PH, Cho HM et al (2016) Inducible HGF-secreting human umbilical cord blood-derived MSCs produced via TALEN-mediated genome editing promoted angiogenesis. Mol Ther 24(9):1644–1654
44. Benabdellah K, Cobo M, Munoz P et al (2011) Development of an all-in-one lentiviral vector system based on the original TetR for the easy generation of Tet-ON cell lines. PLoS One 6 (8):e23734
45. Benabdellah K, Munoz P, Cobo M et al (2016) Lent-On-Plus Lentiviral vectors for conditional expression in human stem cells. Sci Rep 6:37289
46. Moriyama H, Moriyama M, Sawaragi K et al (2013) Tightly regulated and homogeneous transgene expression in human adipose-derived mesenchymal stem cells by lentivirus with tet-off system. PLoS One 8(6):e66274
47. Toscano MG, Frecha C, Ortega C et al (2004) Efficient lentiviral transduction of Herpesvirus saimiri immortalized T cells as a model for gene therapy in primary immunodeficiencies. Gene Ther 11(12):956–961
48. Benabdellah K, Gutierrez-Guerrero A, Cobo M et al (2014) A chimeric HS4-SAR insulator (IS2) that prevents silencing and enhances expression of lentiviral vectors in pluripotent stem cells. PLoS One 9(1):e84268
49. Choi JR, Yong KW, Wan Safwani WKZ (2017) Effect of hypoxia on human adipose-derived mesenchymal stem cells and its potential clinical applications. Cell Mol Life Sci 74 (14):2587–2600

Chapter 18

Systemic Delivery of Adeno-Associated Viral Vectors in Mice and Dogs

Lakmini P. Wasala, Chady H. Hakim, Yongping Yue, N. Nora Yang, and Dongsheng Duan

Abstract

Many diseases affect multiple tissues and/or organ systems, or affect tissues that are broadly distributed. For these diseases, an effective gene therapy will require systemic delivery of the therapeutic vector to all affected locations. Adeno-associated virus (AAV) has been used as a gene therapy vector for decades in preclinical studies and human trials. These studies have shown outstanding safety and efficacy of the AAV vector for gene therapy. Recent studies have revealed yet another unique feature of the AAV vector. Specifically, AAV can lead to bodywide gene transfer following a single intravascular injection. Here we describe the protocols for effective systemic delivery of AAV in both neonatal and adult mice and dogs. We also share lessons we learned from systemic gene therapy in the murine and canine models of Duchenne muscular dystrophy.

Key words DMD, Systemic delivery, AAV, Neonatal mice, Neonatal dogs, Adult mice, Adult dogs

1 Introduction

Based on the delivery method, gene therapy can be divided into local therapy and systemic therapy. Local gene therapy refers to direct injection of a therapeutic vector to the diseased tissue in situ. Such an approach works extremely well for diseases that affect a confined region such as congenital blindness caused by mutations in the retinal gene. In this case, a simple one-time subretinal injection of an adeno-associated virus (AAV) vector can result in marvelous vision improvement [1]. As a matter of fact, a gene therapy drug (Luxturna) for treating an inherited form of blindness received Food and Drug Administration (FDA) approval on December 19, 2017, for commercial use [2]. However, many diseases affect multiple body systems or affect tissues that are widely distributed. An effective therapy for these diseases will require systemic delivery. In this case, a therapeutic vector is injected into the circulation to allow it to spread throughout the whole body.

Fredric P. Manfredsson and Matthew J. Benskey (eds.), *Viral Vectors for Gene Therapy: Methods and Protocols*, Methods in Molecular Biology, vol. 1937, https://doi.org/10.1007/978-1-4939-9065-8_18,

Systemic delivery is much more challenging because the vector has to escape from the vasculature and still maintain its infectivity to transduce the target cells. Systemic delivery is also more risky due to the administration of a huge quantity of the vector ($\geq 10^{14}$ particles/kg) over a short period of time and inevitable spreading of the vector to untoward tissues.

A variety of viral and nonviral vector systems have been explored for systemic delivery (reviewed in Ref. [3]). Unfortunately, most of them have failed to result in robust, persistent bodywide gene transfer following intravascular delivery. Currently, AAV is the only vector that can lead to whole body systemic transduction at a level that can meet the needs of clinical gene therapy in human patients [4]. The exact molecular mechanisms underlying AAV-mediated systemic gene transfer remain to be elucidated. However, a number of factors may contribute to systemic AAV delivery, such as the small size of the viral particle (AAV is one of the smallest viruses), the reported transcytosis of certain AAV serotypes, and the unique viral capsid structure of certain AAV serotypes. The first successful systemic delivery in rodent (normal and diseased mice) was reported in 2004 by the Chamberlain lab [5]. The first systemic delivery in large mammals was performed in neonatal dogs in 2008 by the Duan lab [6]. The first systemic delivery in an adult diseased large mammal was performed in the canine model of Duchenne muscular dystrophy (DMD) in 2015 by the Duan lab [7]. The first systemic delivery in human patients was performed by the Mendell lab in 2017 for treating type I spinal muscular atrophy [4].

Over the last 11 years, we have conducted extensive preclinical systemic AAV delivery in normal mice and dogs, and murine and canine DMD models [3, 6–18]. In these studies, we tested both reporter and therapeutic vectors and explored a variety of AAV serotypes. Here we describe detailed protocols we have developed for systemic AAV delivery in mice and dogs.

2 Materials

2.1 Systemic Delivery of AAV to Neonatal Mice (Fig. 1a)

1. Personal protective equipment (PPE) including disposable gowns and/or dedicated scrubs, head cover, shoe covers, surgical mask, and gloves.
2. 30G Hamilton syringe (Hamilton, Reno, NV, USA).
3. PE10 tubing (Hamilton, Reno, NV, USA).
4. Wee Sight transilluminator vein finder (Philips, Omaha, NE, USA).
5. Food dye (Wilton, Darien, IL, USA) sterilized with 0.2 μm filter before use.

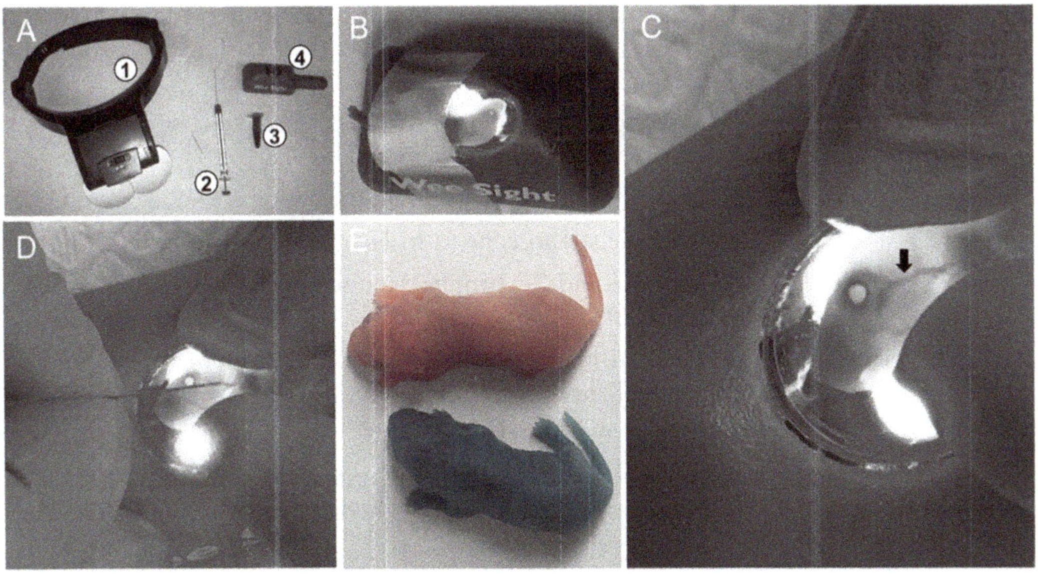

Fig. 1 Systemic injection in newborn mice via the facial vein. (**a**) Material required for the procedure: (1) loop headband magnifier, (2) Hamilton syringe with attached tubing and needle, (3) food dye, (4) Wee Sight transilluminator. (**b**) A newborn pup taped to the Wee Sight illuminator. (**c**) Position of the facial vein (arrow). (**d**) A photograph showing the placement of the injection needle in the facial vein. (**e**) Representative photographs showing an un-injected puppy (top) and an injected puppy (bottom)

6. Wet ice bucket.
7. Sterile gauge sponges (2 in. × 2 in.).
8. Medical tape (3 M, Maplewood, MN, USA).
9. Optivisor adjustable headband magnifier with a 2.5 × lens (Donegan Optical Company Inc., Lenexa, KS, USA).
10. Recombinant AAV vector (*see* **Note 1**).

2.2 Systemic Delivery of AAV to Adult Mice

1. Personal protective equipment (PPE) including disposable gowns and/or dedicated scrubs, head cover, shoe covers, surgical mask, and gloves.
2. Flat bottom rodent restrainer (Plas Labs Inc., Lansing, MI, USA).
3. BD Insulin syringe with a capacity of 100 μL and a 30G × 0.5 in. needle.
4. Thermophore heating pad (Medwing, Columbia, SC, USA). Webcol alcohol prep pad (Covidien, Dublin, Republic of Ireland).
5. Digital scale.
6. Biosafety hood.
7. Recombinant AAV vector (*see* **Note 1**).

2.3 Systemic Delivery of an AAV Vector to Neonatal Dogs (Fig. 2a)

1. Personal protective equipment (PPE) including disposable gowns and/or dedicated scrubs, head cover, shoe covers, surgical mask, and gloves.
2. MyWeigh SCM2600BLACK iBalance tabletop digital scale (MyWeigh, Phoenix, Arizona, USA).
3. Digital thermometer (ReliOn Inc., Spokane Valley, WA, USA).
4. Pediatric stethoscope (3 M Littmann, St. Paul, MN, USA).
5. Masimo Rad-57 handheld pulse oximeter (DRE, Louisville, KY, USA).
6. Thermal blanket (100% cotton) (Elite Home Products Inc., Saddle Brook, NJ, USA).
7. Animal fine hair clipper.
8. Infusion set (23G × 3/4 in. butterfly needle attached to a 7.5″ infusion line with Luer lock connector) (Greiner Bio-One, Kremsmünster, Austria).
9. Syringe with Luer-Lock Tip (3 mL) (Becton Dickinson company, Franklin Lakes, NJ, USA).
10. BD PosiFlush Pre-filed saline (0.9% sodium chloride) syringe for intravenous (IV) injection (Becton Dickinson company, Franklin Lakes, NJ, USA).
11. Sterile gauge sponges (2 in. × 2 in.).
12. ChloraPrep one step (2% Chlorhexidine gluconate and 70% Isopropyl alcohol) (CareFusion, Leawood, KS, USA).
13. Recombinant AAV vector (*see* **Note 1**).

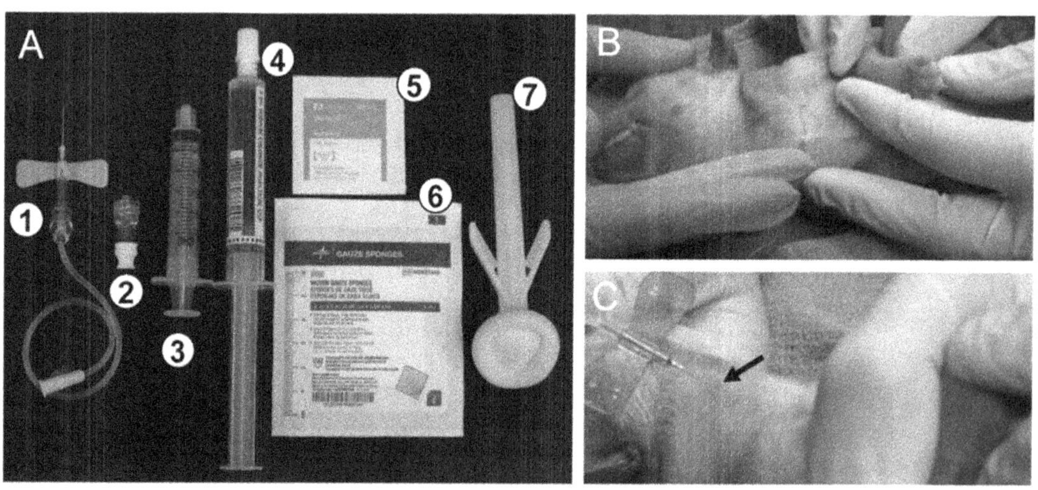

Fig. 2 Systemic injection of neonatal dog puppies via the jugular vein. (**a**) Materials required for the procedure (1). a butterfly needle attached to the infusion line, (2) a Luer lock connector, (3) the 3-mL syringe containing the AAV vector, (4) the prefilled saline syringe, (5), alcohol wipe, (6) gauze sponge, (7) a ChloraPrep one step. (**b**) A photograph showing the position of the puppy for injection. (**c**) A photograph showing the placement of the injection needle in the jugular vein

2.4 Systemic Delivery of AAV to Young/Adult Dogs (Fig. 3)

1. Personal protective equipment (PPE) including disposable gowns and/or dedicated scrubs, head cover, shoe covers, surgical mask, and gloves.
2. Digital vet scale (DRE, Louisville, KY, USA).
3. Digital thermometer (ReliOn Inc., Spokane Valley, WA, USA).
4. Stethoscope (3 M Littmann, St. Paul, MN, USA).
5. Handheld pulse oximeter (Masimo Rad-57) (DRE, Louisville, KY, USA).
6. Thermal blanket (100% cotton) (Elite Home Products Inc., Saddle Brook, NJ, USA).
7. Animal work table.
8. Animal hair clipper equipped with a size 50 blade (Oster professional product, Sunbeam products, Inc., Boca Raton, FL, USA).
9. Shielded I.V. Catheter (22G × 1.00 in. needle, or 20G × 1.00 in. needle) (Becton Dickinson company, Franklin Lakes, NJ, USA).

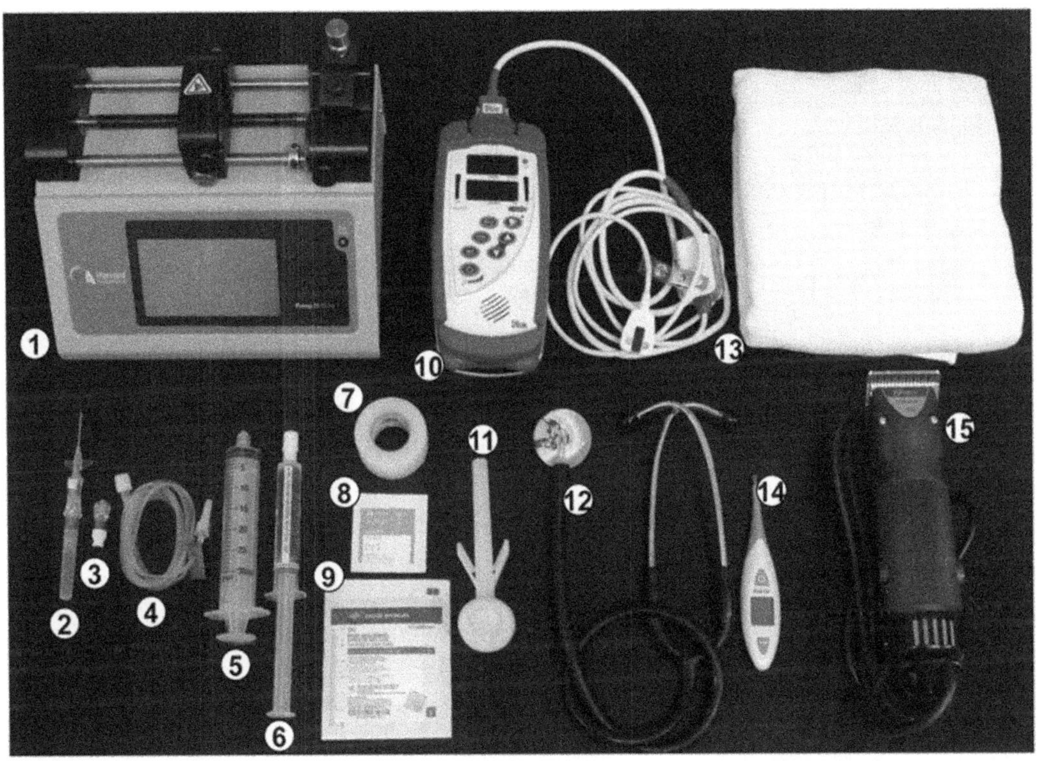

Fig. 3 Materials required for the systemic injection of young/adult dog via the cephalic vein (1) Infusion pump, (2) shielded IV catheter, (3) Luer lock, (4) connector tubing, (5) 30 mL syringe, (6) prefilled saline syringe, (7) medical tape, (8) alcohol wipe, (9) gauze sponge, (10) handheld pulse oximeter, (11) ChloraPrep one step, (12) stethoscope, (13) thermal blanket, (14) digital thermometer, (15) animal hair clipper

10. SmartSite Needle free valve (CareFusion, San Diego, CA, USA).
11. Micro-Bore IV Extension set (A 60 in. tube line with 0.3 mL volume capacity, and contains Luer lock connectors as well as a slide clamp) (NeoChild, Oklahoma City, OK, USA).
12. Medical Tape.
13. Vet wrap (3 M, Maplewood, MN, USA).
14. Webcol Alcohol Prep pad (Covidien, Dublin, Republic of Ireland).
15. Syringe with Luer-Lock Tip (30 mL) (Becton Dickinson Company, Franklin Lakes, NJ, USA).
16. BD PosiFlush Pre-filed saline (0.9% Sodium chloride) syringe for IV injection (Becton Dickinson Company, Franklin Lakes, NJ, USA).
17. Infusion pump (KD Scientific Inc., Holliston, MA, USA).
18. Sterile gauge sponges (4 in. × 4 in.) (Govidien, Mansfield, MA, USA).
19. ChloraPrep one step (2% Chlorhexidine gluconate and 70% Isopropyl alcohol) (CareFusion, San Diego, CA, USA).
20. Recombinant AAV vector (*see* **Note 1**).

3 Methods

3.1 Systemic Delivery of AAV to Neonatal Mice

1. Follow the institutional guidelines for animal care and handling. Obtain the approval from institutional animal care and use committee (ACUC). Wear all PPE before working or handling animals. Use aseptic techniques while preparing the AAV vector for injection and during the injection procedure.
2. Keep the neonates in a pre-warmed cage by keeping a heat pad beneath the cage. Separate the dam from the pups (*see* **Note 2**).
3. Mix the food dye with stock AAV virus at a 1:100 dilution (*see* **Note 3**).
4. Fill the syringe with the AAV vector. Get rid of air bubbles (*see* **Note 4**).
5. Anesthetize the animal using a thermal shock by wrapping the puppy with a piece of gauze and keeping on ice for few seconds (*see* **Note 5**).
6. Secure the neonate body on its side on the Wee Sight transilluminator by placing a gauze pad on the skin and a tape over the gauze (*see* **Note 6**) (Fig. 1b).
7. Use the magnifier to visualize the facial vein (Fig. 1c).
8. Slowly insert the needle into the vein and infuse the AAV vector (*see* **Note 7**) (Fig. 1d). Once all the solution is injected, keep the needle in the vein for additional 15 s (*see* **Note 8**).

9. After removing the needle, gently pressure the injection site to avoid bleeding.
10. Allow 2–3 min for the neonate to recover, and after it is conscious, place the neonate back in the cage. If necessary the animal can be warmed up by keeping on the investigator's hands or a pre-warmed gauze pad.
11. Monitor the neonates for signs of distress.
12. Return the injected pup to the mother after the pup becomes conscious and restores ambulation.

3.2 Systemic Delivery of AAV to Adult Mice

1. Follow the institutional guidelines for animal care and handling. Obtain the approval from institutional animal care and use committee (IACUC). Wear all PPE before working or handling animals. Use aseptic techniques while preparing the AAV vector for injection and during the injection procedure.
2. Weigh the mouse and calculate the volume of the AAV vector needed to reach a certain viral genome (vg) particles/body weight (*see* **Note 9**).
3. Position the mouse in the rodent restrainer (Fig. 4a).

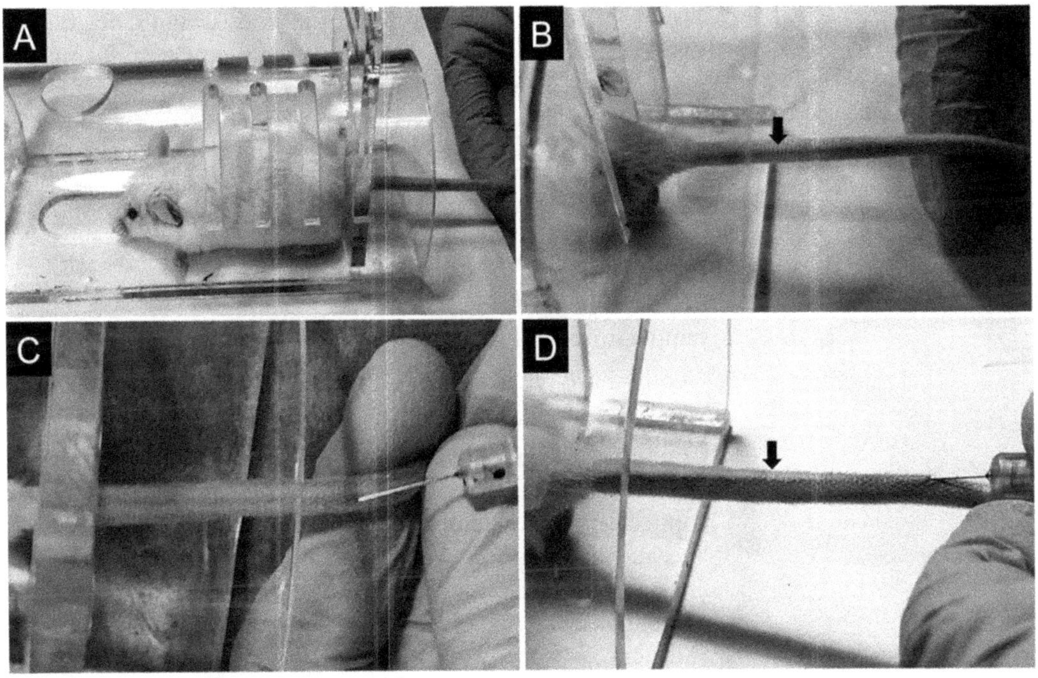

Fig. 4 Systemic injection in adult mice via the tail vein. (**a**) An adult mouse inside the rodent restrainer. (**b**) A photograph showing the prominent lateral tail vein (arrow). (**c**) A photograph showing the placement of the injection needle in the tail vein. (**d**) A photograph showing the color of the tail becomes dark with a successful injection of the AAV. The color change is due to the food dye in the AAV solution

4. Gently warm the tail using a heating pad at its lowest setting to allow vasodilation.
5. Hold the distal end of the tail with the nondominant hand.
6. Slightly rotate the tail to allow clear visualization of the lateral tail vein (Fig. 4b).
7. Clean the injection site with alcohol pad.
8. Insert the needle to the tail vein by holding the tail and needle positioning parallel to the tail (*see* **Note 10**) (Fig. 4c, d).
9. Inject the AAV vector at a steady rate of ~1 mL/min.
10. Once the AAV vector is injected, carefully remove the needle and apply pressure to the injection site to avoid bleeding (*see* **Note 11**).
11. Transfer the mouse to the cage and observe the mouse for at least 4 h.

3.3 Systemic Delivery of AAV to Neonatal Dogs

1. Follow the institutional guidelines for animal care and handling. Obtain the approval from institutional animal care and use committee (ACUC). Wear all PPE before working or handling animals. Use aseptic techniques while preparing the AAV vector for injection and during the injection procedure.
2. Monitor the newborn puppy for the body weight and responsiveness (*see* **Note 12**). Supplement the puppy with formula as needed (*see* **Note 13**). Consult with the veterinary doctor immediately for unexpected changes such as sudden drop of the body weight or temperature.
3. Weight the puppy to determine the injection volume. A normal neonatal puppy can tolerate up to 25 μL liquid/gram body weight (*see* **Note 14**).
4. Measure the vital signs (heart rate, respiration rate, rectal body temperature, and blood oxygen saturation) (*see* **Note 15**).
5. Gently shave the neck hair on the side to allow better observation of the jugular vein for injection.
6. Connect the saline prefilled BD PosiFlush syringe with the IV extension line and attach them to the infusion set. All parts are connected through Luer lock fitting. Prefill the IV extension line and the infusion set with saline to remove air from the lines. Lock the sliding clamp on the extension line. Draw the exact volume of AAV vector (calculated based on body weight and desired dose) into the 3 mL syringe and remove any air bubbles form the syringe. Replace the BD PosiFlush syringe with the AAV syringe. Check the extension line and infusion set for any air bubble.
7. Perform injection procedure with the help of three investigators. The first investigator places a preheated thermal blanket

on the animal working table, positions the puppy supine on the blanket, then gently restrains the forelimbs using both hands (Fig. 2b). The second investigator restrains the head and scrubs the ventral side of the neck with the ChloraPrep, then gently inserts the needle in the jugular vein (Fig. 2c). Once the needle enters the vein, blood will appear in the needle hub. The third investigator unlocks the sliding clamp on the extension line and pulls the plunger back to visualize blood in the infusion line. This will confirm that the needle is correctly placed inside the vein. Inject the AAV vector at the rate of 2 mL/min, then flush the infusion line with saline [12].

8. While pressing the entry site of the needle with a finger, retract the butterfly from the jugular vein. Continue to pressure the injection site to avoid bleeding.
9. Remeasure the vital signs.
10. Return the puppy back to the dam and closely monitor the puppy for the following 24 h. Consult with veterinary doctor immediately with any concerns.

3.4 Systemic Delivery of AAV to Young/Adult Dogs

1. Follow the institutional guidelines for animal care and handling. Obtain the approval from institutional animal care and use committee (ACUC). Wear all PPE before working or handling animals. Use aseptic techniques while preparing the AAV vector for injection and during the injection procedure.
2. Weigh the dog to determine the injection volume. Up to 6.2 mL/kg body weight (6.24×10^{14} viral genome particles/kg body weight) can be tolerated in young adult dystrophic dogs [7].
3. Measure the vital signs (heart rate, respiration rate, rectal body temperature, and blood oxygen saturation).
4. Shave the cranial side of the lower forelimb (Fig. 5a) to allow better observation of the cephalic vein (Fig. 5b) during the injection procedure (*see* **Note 16**).
5. Place a thermal blanket on the animal working table. Place the dog on the table in the sternal recumbency position with the lower forelimb extended. Occlude the cephalic vein at the elbow with your thumb or using a tourniquet by one investigator.
6. Catheterize the cephalic vein by another investigator. Scrub the cranial aspect of the lower forelimb with ChloraPrep. Pull the lower forelimb skin distally to stabilize the vein. Slowly insert the IV catheter needle (bevel up) through the skin and into the vein (Fig. 5c). Blood will appear in the catheter when the needle punctures the vein. Gently slide the catheter into the vein while retracting the needle out. At this time, blood will

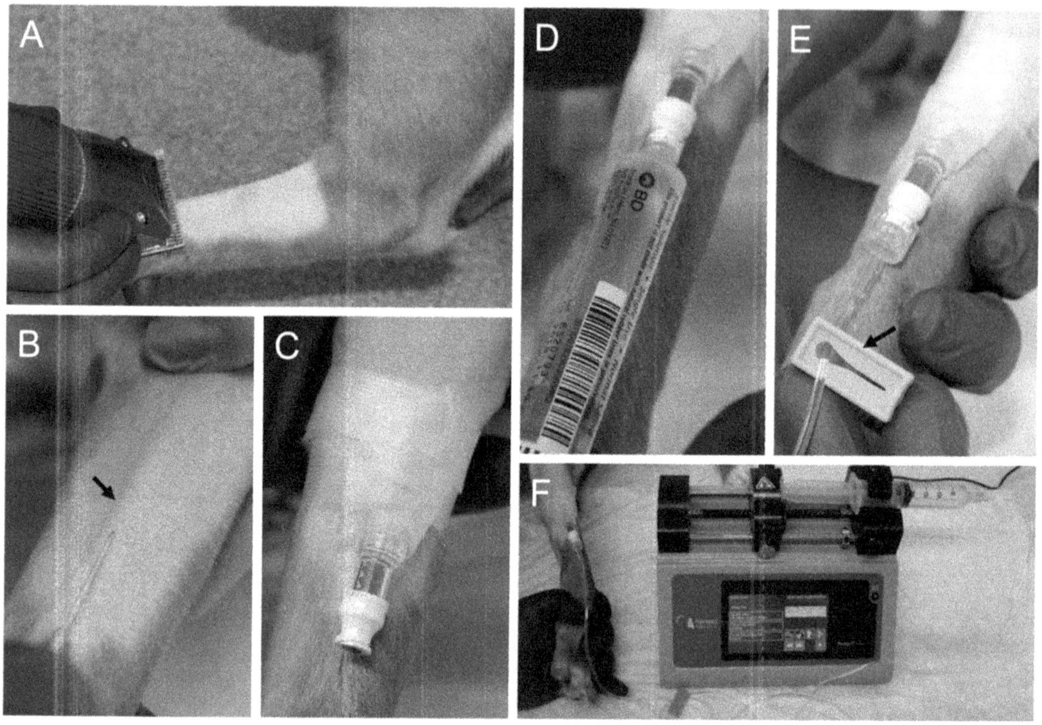

Fig. 5 Systemic injection of young/adult dog via the cephalic vein. (**a**) Shaving of the cranial aspect of the forelimb. (**b**) A photograph showing the placement of the needle in the cephalic vein. Arrow, the cephalic vein. (**c**) A photograph showing the catheterization of the cephalic vein. (**d**) A photograph showing the AAV containing syringe during the injection. (**e**) A photograph showing the catheter inserted with the extension line. Arrow, the sliding clamp. (**f**) A photograph showing injection of AAV using the infusion pump

flow out form the catheter. Quickly occlude the vein at the catheter tip and attach the smart site needle free valve to the catheter. Secure the catheter with tape and warp it with vet wrap. Flush the catheter with saline to ensure there is no leak.

7. Connect the saline prefilled BD PosiFlush syringe with the IV extension line using the Luer lock fitting (Fig. 5d). Prefill the IV extension line with saline to remove air. Lock the sliding clamp on the extension line (Fig. 5e). Draw the exact volume of AAV vector into the 30 mL syringe. Remove any air bubbles form the syringe. Replace the BD PosiFlush syringe with the AAV syringe. Check the extension line for any air bubble. Secure the syringe to the infusion pump and set the infusion rate to 2 mL/min (Fig. 5f).
8. Attach the IV extension line to the valve and unlock the sliding clamp. Start the infusion and monitor the dog's vital signs closely. Once the infusion is done, flush infusion line with saline and disconnect the extension line from the catheter. While pressing the entry site of the needle with your finger, retract the catheter from the cephalic vein. Continue to pressure the puncture site to avoid bleeding.

9. Continue measuring the vital signs for 5, 10, 30, and 60-min post injection and examine the dog for any adverse reaction.
10. Return the dog to its housing and continue monitoring the dog for 72 h. Consult with the veterinary doctor immediately with any concerns.

4 Notes

1. Many different methods have been developed for the production and purification of recombinant AAV vectors (reviewed in refs. 19, 20). For the protocols described in this chapter, AAV vectors were generated using the transient plasmid transfection method [17]. AAV vector titer was determined using quantitative polymerase chain reaction. AAV vector was dialyzed in HEPES buffer (Sodium chloride 8.7 g, HEPES 4.75 g, 10 N Sodium Hydroxide 1.5 mL, Distilled water up to 1 L, pH adjusted to 7.8). AAV-6 and -8 are the first AAV serotypes shown capable of systemic delivery [5, 21]. Since then, a number of AAV serotypes and engineered AAV capsids were found to have the unique systemic delivery property. Some of these AAV serotypes also demonstrated specific tissue tropism following systemic delivery. For example, intravascular injection of AAV-9 results in preferential transduction of the heart in mice [8]. For the protocols described in this chapter, we have used AAV-1, −6, −8, −9 and their tyrosine mutants [22].
2. Neonatal facial vein injections need to be carried out between 24 and 48 h after birth. During this time window, the facial vein is clearly visible.
3. The food dye is added to allow easy visualization and confirmation of injection.
4. A newborn mouse can tolerate up to 100 μL of volume solution.
5. Do not leave the neonate on ice for longer than 1 min to avoid complications related to hypothermia.
6. The newborn mouse should be positioned with its neck turned gently for clear visualization of the facial vein and the head secured by fingers without interfering respiration.
7. Accumulation of solution at the head or neck area (but not in other parts of the body) suggests injection failure. A successful injection should result in the diffusion of the dye throughout the body. The entire puppy will turn the color of the dye.
8. This will allow sufficient time for the AAV solution to flow through the PE-10 tubing and it will also prevent any back flow of the solution.

9. The total AAV vector volume (mL) needed to reach a certain viral dosage (vg/body weight) is calculated using the following formula: Injection volume (mL) = [dosage (vg/body weight) × body weight]/viral titer (vg/mL).
10. Injection should be done approximately 1 in. distal to the midpoint of the tail. Starting at the far end of the tail to allow second or third injection attempt if the first attempt fails.
11. If the injection creates a bleb, the injection at that site should be terminated. In the case of an unsuccessful injection, up to three additional injections can be attempted.
12. The body weight should be monitored every 2–4 h. Bowel movement and/or urination may cause transient body weight fluctuation. Puppies that fail to gain weight should not be used for AAV injection.
13. The Just Born puppy formula produces best results for nutrition supplementation in our hands.
14. We have performed neonatal AAV injection in puppies as early as 24 h after the birth. Successful systemic AAV injection can also be achieved in puppies up to 1 week of the age.
15. Throughout the injection procedure, make sure the puppy is wrapped in a thermal blanket and kept warm. If needed, thermal blanket can be pre-warmed to 37 °C.
16. Besides the cephalic vein, other veins can also be used such as the jugular vein and saphenous vein. For injection through these veins, shave the related area. For example, for saphenous vein injection, we shave the lateral side of the distal tibia above the tarsus joint.

Acknowledgments

The research on systemic AAV delivery in the Duan lab was supported by the National Institutes of Health (NS-90634, AR-70517 and AR-69085), Department of Defense (MD150133), Jesse's Journey-The Foundation for Gene and Cell Therapy, Hope for Javier, Jackson Freel DMD Research Fund and Solid Biosciences Inc. The authors thank Keqing Zhang for the assistance in figure preparation.

Disclosure: D.D. is a member of the scientific advisory board, an equity holder of Solid Biosciences, LLC. D.D. is an inventor on a patent licensed to Solid Biosciences, LLC.

References

1. Russell S, Bennett J, Wellman JA et al (2017) Efficacy and safety of voretigene neparvovec (AAV2-hRPE65v2) in patients with RPE65-mediated inherited retinal dystrophy: a randomised, controlled, open-label, phase 3 trial. Lancet 390(10097):849–860. https://doi.org/10.1016/S0140-6736(17)31868-8
2. FDA News Release (2017) FDA approves novel gene therapy to treat patients with a rare form of inherited vision loss. https://www.fda.gov/NewsEvents/Newsroom/PressAnnouncements/ucm589467.htm. Accessed 19 Dec 2017
3. Duan D (2016) Systemic delivery of adeno-associated viral vectors. Curr Opin Virol 21:16–25. https://doi.org/10.1016/j.coviro.2016.07.006
4. Mendell JR, Al-Zaidy S, Shell R et al (2017) Single-dose gene-replacement therapy for spinal muscular atrophy. N Engl J Med 377 (18):1713–1722. https://doi.org/10.1056/NEJMoa1706198
5. Gregorevic P, Blankinship MJ, Allen JM et al (2004) Systemic delivery of genes to striated muscles using adeno-associated viral vectors. Nat Med 10(8):828–834
6. Yue Y, Ghosh A, Long C et al (2008) A single intravenous injection of adeno-associated virus serotype-9 leads to whole body skeletal muscle transduction in dogs. Mol Ther 16 (12):1944–1952. https://doi.org/10.1038/mt.2008.207
7. Yue Y, Pan X, Hakim CH et al (2015) Safe and bodywide muscle transduction in young adult Duchenne muscular dystrophy dogs with adeno-associated virus. Hum Mol Genet 24 (20):5880–5890
8. Bostick B, Ghosh A, Yue Y et al (2007) Systemic AAV-9 transduction in mice is influenced by animal age but not by the route of administration. Gene Ther 14(22):1605–1609
9. Ghosh A, Yue Y, Long C et al (2007) Efficient whole-body transduction with trans-splicing adeno-associated viral vectors. Mol Ther 15 (4):750–755. https://doi.org/10.1038/sj.mt.6300081
10. Bostick B, Yue Y, Lai Y et al (2008) Adeno-associated virus serotype-9 microdystrophin gene therapy ameliorates electrocardiographic abnormalities in mdx mice. Hum Gene Ther 19(8):851–856. https://doi.org/10.1089/hum.2008.058
11. Ghosh A, Yue Y, Shin J-H et al (2009) Systemic trans-splicing AAV delivery efficiently transduces the heart of adult mdx mouse, a model for Duchenne muscular dystrophy. Hum Gene Ther 20(11):1319–1328
12. Yue Y, Shin JH, Duan D (2011) Whole body skeletal muscle transduction in neonatal dogs with AAV-9. Methods Mol Biol 709:313–329. https://doi.org/10.1007/978-1-61737-982-6_21
13. Bostick B, Shin JH, Yue Y et al (2012) AAV micro-dystrophin gene therapy alleviates stress-induced cardiac death but not myocardial fibrosis in >21-m-old mdx mice, an end-stage model of Duchenne muscular dystrophy cardiomyopathy. J Mol Cell Cardiol 53 (2):217–222. S0022-2828(12)00179-4 [pii]. https://doi.org/10.1016/j.yjmcc.2012.05.002
14. Pan X, Yue Y, Zhang K et al (2013) Long-term robust myocardial transduction of the dog heart from a peripheral vein by adeno-associated virus serotype-8. Hum Gene Ther 24(6):584–594. https://doi.org/10.1089/hum.2013.044
15. Hakim CH, Yue Y, Shin JH et al (2014) Systemic gene transfer reveals distinctive muscle transduction profile of tyrosine mutant AAV-1, -6, and -9 in neonatal dogs. Mol Ther Methods Clin Dev 1:14002. https://doi.org/10.1038/mtm.2014.2
16. Zhang Y, Yue Y, Li L et al (2013) Dual AAV therapy ameliorates exercise-induced muscle injury and functional ischemia in murine models of Duchenne muscular dystrophy. Hum Mol Genet 22(18):3720–3729. https://doi.org/10.1093/hmg/ddt224
17. Pan X, Yue Y, Zhang K et al (2015) AAV-8 is more efficient than AAV-9 in transducing neonatal dog heart. Hum Gene Ther Methods 26 (4):54–61
18. Hakim CH, Pan X, Kodippili K et al (2016) Intravenous delivery of a novel micro-dystrophin vector prevented muscle deterioration in young adult canine Duchenne muscular dystrophy dogs. Mol Ther 24(Suppl 1): S198–S199
19. Kotin RM, Snyder RO (2017) Manufacturing clinical grade recombinant adeno-associated virus using invertebrate cell lines. Hum Gene Ther 28(4):350–360. https://doi.org/10.1089/hum.2017.042
20. Clement N, Grieger JC (2016) Manufacturing of recombinant adeno-associated viral vectors for clinical trials. Mol Ther Methods Clin Dev 3:16002. https://doi.org/10.1038/mtm.2016.2

21. Wang Z, Zhu T, Qiao C et al (2005) Adeno-associated virus serotype 8 efficiently delivers genes to muscle and heart. Nat Biotechnol 23 (3):321–328
22. Zhong L, Li B, Mah CS et al (2008) Next generation of adeno-associated virus 2 vectors: point mutations in tyrosines lead to high-efficiency transduction at lower doses. Proc Natl Acad Sci U S A 105(22):7827–7832. https://doi.org/10.1073/pnas.0802866105

Chapter 19

Intrathecal Delivery of AAV Vectors in Cynomolgus Macaques for CNS Gene Therapy and Gene Expression Analysis in Microdissected Motor Neurons

Florie Borel, Eric Adams, and Christian Mueller

Abstract

This protocol describes a method of delivering adeno-associated viral (AAV) vectors to the intrathecal space of nonhuman primates for CNS-directed gene therapy. It includes the surgical implantation of the catheter, vector infusion, necropsy, laser-capture microdissection of motor neurons, and gene expression analysis. This method allows efficient and reproducible delivery, and would be of interest to test gene therapy vectors for the treatment of disorders of the central nervous system of nonhuman primates. This protocol was tested in cynomolgus macaques and may be adapted for AAV delivery to different species of large animals.

Key words AAV, Adeno-associated virus, Intrathecal, Trendelenburg, CSF, CNS, Laser-capture microdissection, Motor neurons, ddPCR

1 Introduction

The field of AAV-based gene therapy of the central nervous system (CNS) has seen great progress recently, [1] including its first compelling clinical success [2]. However, it is evident that the question of *delivery* is what drives many decisions and pipelines: to treat a target tissue, one must be able to access it, preferably reliably and efficiently. It is crucial to keep in mind that what may work best in small rodents may be completely irrelevant from a translational perspective, whether it is biological (*e.g.*, CSF turnover every hour in rodents vs. every 6 to 8 h in primates), construct-related (*e.g.*, differences in promoters and other regulatory elements), vector-related (*e.g.*, differences in adeno-associated viral (AAV) vector serotypes tropism) [3], or delivery-related. These elements are key, and they need to be taken into consideration early on. Preclinical studies in nonhuman primates are therefore an indispensable element of a gene therapy program life cycle, and such studies need to be carefully designed and implemented to

Fredric P. Manfredsson and Matthew J. Benskey (eds.), *Viral Vectors for Gene Therapy: Methods and Protocols*, Methods in Molecular Biology, vol. 1937, https://doi.org/10.1007/978-1-4939-9065-8_19,

ensure that meaningful data can be derived. Intrathecal delivery of AAV vectors is a delicate procedure that may lead to significant variability [4]. It is essential to rely on efficient, established delivery methods to be able to truly test a vector adequately and generate relevant data. Furthermore, the use of a reproducible delivery method allows for a lower number of animals, ensuring that animals are not used unnecessarily and reducing the animal-related and vector-related costs.

We detail here a reliable and reproducible method for the delivery of AAV vectors to the intrathecal space and downstream tissue processing for transgene expression analysis. In this method, a catheter is surgically pre-implanted intrathecally via hemilaminectomy, and is accessible through a subcutaneous port. After a recovery period of at least ten days, the macaques are then placed in the Trendelenburg position at a 30° angle. The animals are conscious to allow for careful monitoring. A 2.5 mL volume of vector is then infused through the port twice (5 mL total), with a 6-h interval between both infusions to allow for cerebrospinal fluid (CSF) turnover and compliance. This method allows the use of a large volume of vector, and leads to efficient and reproducible vector delivery [5]. We subsequently detail a protocol for laser-capture microdissection of motor neurons and gene expression analysis in the captured cells.

2 Materials

1. Clippers.
2. Atropine sulfate.
3. Ketamine.
4. Anesthesia Machine.
5. Oxygen.
6. Isoflurane.
7. Prednisolone sodium succinate.
8. Flunixin meglumine.
9. Ceftiofur sodium.
10. Catheter, 1.0 mm OD and 0.38 mm ID polyethylene lined polyurethane.
11. Isovue 300.
12. Suture.
13. Tissue adhesive.
14. Subcutaneous titanium access port (such as MIN-LOVOL, Solomon Scientific, Skokie, IL, USA, or equivalent).
15. X-ray detector.

16. Butorphanol tartrate.
17. Restraint table.
18. Transport cart.
19. Calibrated syringe infusion pump with primed extension set.
20. Research grade AAV vector (or other test article of choice).
21. Heparin sodium.
22. 0.001% sodium nitrite in saline.
23. Perfusion pump.
24. Labeled tubes for tissue storage.
25. Cryostat.
26. Silane-treated slides.
27. Arcturus HistoGene LCM Frozen Section Staining Kit (ThermoFisher, Waltham, MA, USA, or equivalent).
28. Xylene.
29. Laser capture microdissection instrument (such as ArcturusXT, ThermoFisher, Waltham, MA, USA, or equivalent).
30. Laser capture microdissection caps (ThermoFisher, Waltham, MA, USA, or equivalent).
31. 42 °C incubator.
32. RNA isolation kit (ThermoFisher, Waltham, MA, USA, or equivalent).
33. 2100 BioAnalyzer (Agilent, Santa Clara, CA, USA) or equivalent.
34. Retrotranscription kit (ThermoFisher, Waltham, MA, USA, or equivalent).
35. Appropriate Taqman probes, one FAM-labeled targeting the gene of interest and one HEX-labeled targeting a normalizer gene.
36. ddPCR machine (Biorad, Hercules, CA, USA) comprising a droplet generator and a reader.
37. ddPCR droplet reader oil (Biorad, Hercules, CA, USA).
38. ddPCR cartridges and gaskets (Biorad, Hercules, CA, USA).
39. ddPCR droplet generating oil (Biorad, Hercules, CA, USA).
40. ddPCR-compatible PCR plates, plate seals, and plate sealer.
41. PCR machine.
42. Pipettes and tips.
43. PCR tubes and/or strips.
44. Vortex.
45. Microcentrifuge.

3 Methods

3.1 Catheter Implantation

1. Shave the animal (*see* **Note 1**) and administer 0.04 mg/kg atropine sulfate subcutaneously; administer 8 mg/kg ketamine intramuscularly approximately 15 min later.
2. Intubate and maintain the animal on approximately 1 L/min of oxygen and approximately 2.0% isoflurane; adjust as needed.
3. Administer 30 mg/kg prednisolone sodium succinate intravenously, and 2 mg/kg flunixin meglumine and 5.0 mg/kg ceftiofur sodium intramuscularly.
4. Make an incision approximately over the L5 vertebra, dissect the muscle, and perform a hemilaminectomy.
5. Incise the dura, insert the catheter, and advance it intrathecally with the tip located near the thoracolumbar junction and secure with suture.
6. Attach the subcutaneous titanium access port and secure to the musculature.
7. Close the skin with sutures and tissue adhesive.
8. Verify placement of the catheter with a myelogram with Isovue 300 (Fig. 1).
9. Upon recovery from anesthesia, administer 0.05 mg/kg butorphanol tartrate and 5.0 mg/kg ceftiofur sodium intramuscularly (*see* **Note 2**).
10. Let the animal recover for at least 10 days.

3.2 Vector Delivery

1. Restrain the animal and transport it to the procedure room (*see* **Note 3**).
2. Immediately prior to the procedure, make sure the animal is appropriately restrained, and adjust it in a prone position with the restraint table tilted approximately 30° head-down (*see* **Note 4**) (Fig. 2).
3. Administer the first dose, which consists of a 2.5 mL infusion of test article followed by approximately 0.3 mL of vector excipient through the IT-L port/catheter system, at a rate of approximately 7.5 mL/h (*see* **Note 5**).
4. Return the animal to its cage for 6 h and monitor.
5. Administer the second dose, as described earlier.
6. Return the animal to its cage and monitor.

3.3 Necropsy

1. Administer 8.0 mg/kg ketamine.
2. Intubate and maintain the animal on approximately 1 L/min of oxygen and approximately 2.0% isoflurane.

5. CSF may be drawn prior to the infusion to verify proper catheter patency or for isovolumetric delivery.
6. Additional tissues may be harvested and additional tissue processing methods may be used, as relevant.
7. It is recommended to avoid sources of RNAse contamination, *i.e.*, change gloves frequently and use RNAse-free tubes and tips; workspace can be cleaned using commercial reagents that remove RNAse activity.
8. We typically discard samples with a RNA Integrity Number (RIN) inferior to 8.
9. We typically use 9 μL of undiluted RNA sample as input for the reaction.
10. We recommend the use of *HPRT* as a normalizer gene for motor neurons.
11. It is recommended to pipet the droplets very slowly from the cartridge to the PCR plate.
12. This step may require optimization.
13. For statistical reasons, each sample must have at least 10,000 droplets of which at least 100 are negative.

Acknowledgments

Florie Borel would like to express her gratitude to Rachel Stock from the Mueller Lab at UMass Medical for generating the images of the laser capture, as well as to Christen Simon and Randy Reed at Northern Biomedical Research for their support.

References

1. Hocquemiller M, Giersch L, Audrain M, Parker S, Cartier N (2016) Adeno-associated virus-based gene therapy for CNS diseases. Hum Gene Ther 27(7):478–496. https://doi.org/10.1089/hum.2016.087
2. Mendell JR, Al-Zaidy S, Shell R, Arnold WD, Rodino-Klapac LR, Prior TW, Lowes L, Alfano L, Berry K, Church K, Kissel JT, Nagendran S, L'Italien J, Sproule DM, Wells C, Cardenas JA, Heitzer MD, Kaspar A, Corcoran S, Braun L, Likhite S, Miranda C, Meyer K, Foust KD, Burghes AHM, Kaspar BK (2017) Single-dose gene-replacement therapy for spinal muscular atrophy. N Engl J Med 377 (18):1713–1722. https://doi.org/10.1056/NEJMoa1706198
3. Gray SJ, Matagne V, Bachaboina L, Yadav S, Ojeda SR, Samulski RJ (2011) Preclinical differences of intravascular AAV9 delivery to neurons and glia: a comparative study of adult mice and nonhuman primates. Mol Ther 19 (6):1058–1069. https://doi.org/10.1038/mt.2011.72
4. Borel F, Gernoux G, Cardozo B, Metterville JP, Toro Cabreja GC, Song L, Su Q, Gao GP, Elmallah MK, Brown RH Jr, Mueller C (2016) Therapeutic rAAVrh10 mediated SOD1 silencing in adult SOD1(G93A) mice and nonhuman primates. Hum Gene Ther 27(1):19–31. https://doi.org/10.1089/hum.2015.122
5. Borel F, Gernoux G, Sun H, Stock R, Blackwood M, Brown RH Jr, Mueller C (2018) Safe and effective superoxide dismutase 1 silencing using artificial microRNA in macaques. Sci Transl Med 10(465)

Chapter 20

Detailed Method for Intrathecal Delivery of Gene Therapeutics by Direct Lumbar Puncture in Mice

Kelsey R. Pflepsen, Cristina D. Peterson, Kelley F. Kitto, Lucy Vulchanova, George L. Wilcox, and Carolyn A. Fairbanks

Abstract

Delivery of viral vectors directly into the central nervous system (CNS) has emerged as an important tool for the refinement of gene therapy. Intrathecal delivery by direct lumbar puncture in conscious rodents offers a minimally invasive approach that avoids tissue damage and/or destruction. Here we describe delivery of small quantities of viral vector product to the intrathecal space of rodents via direct lumbar puncture aided by a catheter.

Key words Intrathecal, Gene therapy, AAV gene therapy, Intrathecal gene delivery, Direct lumbar puncture

1 Introduction

The spinal cord is surrounded by three meningeal layers, which act as protective barriers. These include the dura mater, arachnoid mater, and pia mater. For intrathecal injections, injectate is delivered into the subarachnoid space, the region between the arachnoid and the pia mater, which is filled with cerebrospinal fluid (CSF). Delivery of product directly into the subarachnoid space bypasses tissue and cellular barriers. This approach also minimizes off-target expression observed using other delivery methods, thereby reducing potential limitations to distribution and adverse side effects.

Delivery of morphine by direct lumbar puncture to the intrathecal space was first described in mice by Hylden and Wilcox [1] and later adapted for rats [2]. A detailed description of effective training approaches for acquisition of the intrathecal injection technique was later contributed by Fairbanks [3]. We have since adapted this same procedure to optimize the delivery of gene therapeutics intended to target the CNS and peripheral nervous tissue such as the dorsal root ganglia (DRG) in rodents (Fig. 1) [4–7].

Fredric P. Manfredsson and Matthew J. Benskey (eds.), *Viral Vectors for Gene Therapy: Methods and Protocols*, Methods in Molecular Biology, vol. 1937, https://doi.org/10.1007/978-1-4939-9065-8_20,

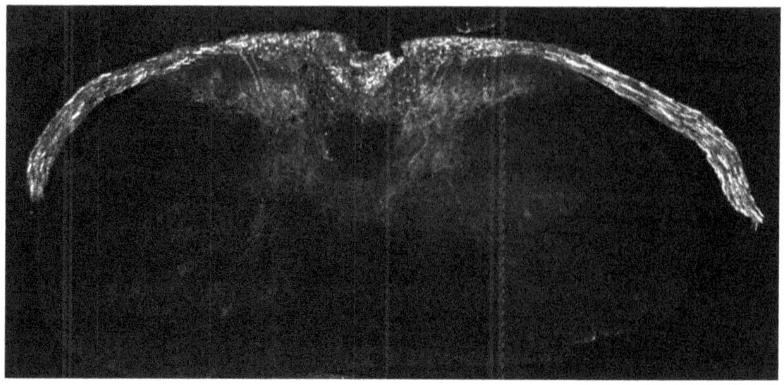

Fig. 1 GFP expression and trafficking in mouse dorsal roots. Adult female ICR-CD1 mice were anesthetized and perfusion-fixed eight weeks following a 10 μL i.t. injection of AAV5-CMV-GFP (~1.0×10^{12} viral particles). Frozen spinal cords were sectioned (14 micron thickness) and prepared for immunohistochemistry. GFP immunoreactivity (−ir) is observed in the dorsal roots. GFP-ir is also present in the dorsal columns and within the dorsal horn

Intrathecal delivery of gene therapeutics offers the potential for increased expression within the spinal cord, both dorsal and ventral horn neurons, and surrounding peripheral tissue. The intrathecal route has been used for modeling gene therapeutic treatment of a variety of neurobiological conditions [8–10] as well as lysosomal storage diseases [11, 12]. The use of the intrathecal route of administration is also of increasing interest to deliver viral vectors to deliver CRE to facilitate tissue-specific and conditional knockdown of floxed genes [13–15].

Viral vectors that have been used in preclinical studies using intrathecal drug delivery include adenovirus (AD) [16, 17], adeno-associated virus (AAV) [4–7, 18], lentivirus (LV) [19], and the herpes simplex virus (HSV) [20, 21]. More recent studies have employed nonvector-based approaches for delivery of gene therapeutics [22, 23].

Intrathecally delivered AAV viral vectors with constitutive promoters and enhancers result in broad distribution and sustained expression profiles throughout the CNS and DRG [4–7, 18]. Further targeting of distinct cell populations can be achieved through use of vectors with cell-specific promoters allowing for controlled gene expression and targeted therapeutic outcomes. In this chapter, we describe a method for directly delivering viral vector product to the intrathecal space of rodents in order to selectively target gene therapeutics to nervous system tissue.

2 Materials

Viral vector injectate should be stored as recommended by manufacturer's instructions prior to and throughout the injection procedure. Freeze-thaw cycles of viral product should be avoided by aliquoting product into volumes likely to be used at the time of injection.

1. Catheter Materials (Fig. 2).
2. 27-gauge, ½″ Needle.

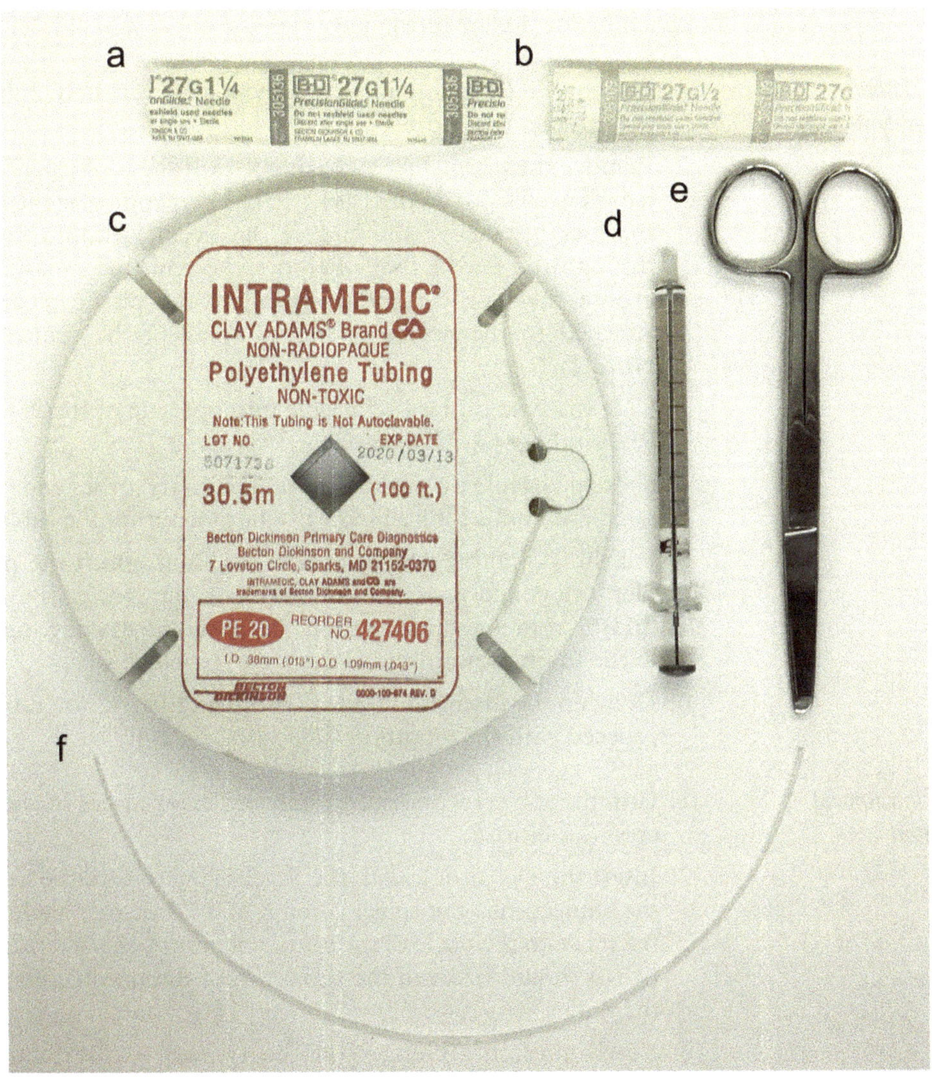

Fig. 2 Catheter materials for intrathecal injection. Catheter is composed of a 27-gauge, 1¼″ needle (**a**), a 27-gauge, ½″ needle (**b**), PE20 tubing (**c**) cut to a length of 250 mm (**f**), 50 μL Hamilton syringe (**d**), and scissors (**e**)

3. 27-gauge, 1¼″ Needle.
4. 50 μL Hamilton Syringe.
5. PE20 Tubing.
6. Scissors.
7. Viral Vector Injectate.

3 Methods

This method of catheter delivery was initially developed by Dr. Alice Larson at the University of Minnesota and has been refined by Mr. Kelley Kitto.

3.1 Catheter Assembly (Fig. 3)

1. Cut the PE20 tubing into lengths of approximately 250 mm per catheter (*see* **Note 1**).
2. Remove the plastic luer lock connector from a 27-gauge, 1¼″ length needle. To do this easily, make a perpendicular cut across the luer just above the end of the needle without cutting through the needle itself. Remove the remaining plastic and expose the base cement point where the plastic piece was attached to the needle as well as some of the cement (*see* **Note 2**).
3. Slide the base of the 1¼″ needle into one side of the 250 mm PE20 tubing (*see* **Note 3**).
4. Slide the needle tip of the ½″ needle into the other end of the same 250 mm PE20 tubing without puncturing the tubing.
5. Fill the Hamilton syringe with diH_2O and attach the plastic luer lock connector of the ½″ needle to the syringe. Push the diH_2O from the syringe into the PE20 tubing, leaving some air in the tubing toward the 1¼″ needle (*see* **Note 4**).
6. Draw up the intended injectate into the catheter. You can now proceed with the intrathecal injection.

3.2 Intrathecal Injection

1. Grip the subject with your thumb and index finger by the iliac crest (*see* **Note 5**).
2. Insert the 1¼″ needle with the bevel facing toward the head of the subject into the spinal column at a 70 degree angle with reference to the plane of the spinal column. A characteristic tail twitch should occur in the subject once the needle punctures the dura.
3. Slowly push the plunger on the syringe to distribute the intended volume of injectate (5–10 μL depending on the dilution of the injectate) (*see* **Note 6**).

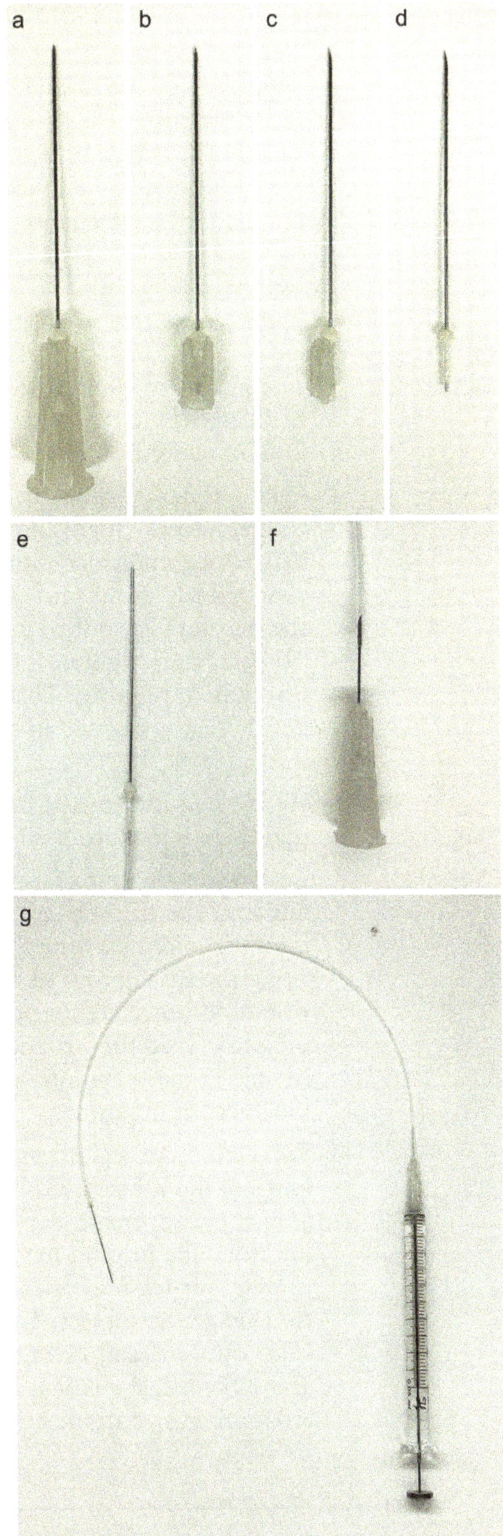

Fig. 3 Catheter assembly. A 27-gauge, 1¼″ needle (**a**) is cut perpendicularly (horizontally) across the luer at the end of the needle (**b**). An additional vertical cut in the plastic is made along the needle (**c**) to release the remaining plastic, thereby allowing it to be removed from the needle (**d**). The newly exposed end of

4. Slightly twist the needle and remove from the spinal column. Place the subject back into its home cage to observe for any variance in behavior. The total injection time should take on the order of a few minutes for a novice injector or less than a minute for an experienced injector.

4 Notes

1. The length of tubing depends on the amount of solution to be injected. We typically inject 5–10 μL of solution, so we use 250 mm length tubing.
2. Half of the plastic luer lock connector is rounded (the end that connects to a syringe) and half is flanged (the needle end). First, make a perpendicular cut across the luer at the point where the rounded part turns into the flanged part (just above the end of the needle). Try not to cut through the needle. Next, make a cut parallel to (right next to) the needle, again trying not to cut through the needle. This cut should expose the end of the needle that is encased in the luer and enable you to pry it out of the plastic.
3. We use a longer needle on the injecting end of the cannula so that there is more to hold onto while performing the injection.
4. Leaving an amount of air inside the tubing between the injectate and the diH_2O creates a bubble that can be monitored prior to and following injection. The bubble should move along the tubing toward the 1¼″ needle proportionally to the volume of injectate delivered intrathecally, thus serving as an indicator that the product was successfully delivered. If the bubble remains stationary, a new catheter must be built and the injection repeated.
5. We often drape a cloth or paper towel over the head and front limbs of the subject prior to handling to minimize stress in the animal and decrease the likelihood of the subject biting the injector. The grip on the iliac crest should be firm enough to prevent any leg movement in the subject but not firm enough to damage the subject. The grip should cause the hind legs to splay outward and downward and cause a plane of taut skin above the spinal column. Handling should be practiced before attempting an intrathecal injection as any movement of the

Fig. 3 (continued) 1¼″ needle is inserted into a 250 mm piece of PE20 tubing (**e**). The needle point of the 27-gauge, ½″ needle is inserted into the opposite end of the same piece of PE20 tubing (**f**). The Hamilton syringe is then attached to the 27-gauge, ½″ needle, completing the assembly of the catheter (**g**)

subject during injection will decrease the likelihood that injectate ends up in the intended space.

6. A rapid injection can cause a temporary hindlimb paralysis that resolves within a minute. While this paralysis resolves, it can be avoided by following proper pacing during injection, leading to a more consistent injection protocol.

Acknowledgments

This work was supported by NIDA grant R01DA035931-01, DoD grant W81XWH-15-1-0494, T32 training grant T32DA07234 to K.P., and T32 training grant T32DA007097 to C.P.

References

1. Hylden JLK, Wilcox GL (1980) Intrathecal morphine in mice: a new technique. Eur J Pharmacol 67:313–316
2. Mestre C, Pelissier T, Fialip J, Wilcox G, Eschalier A (1994) A method to perform direct transcutaneous intrathecal injection in rats. J Pharmacol Toxicol Methods 32(4):197–200
3. Fairbanks CA (2003) Spinal delivery of analgesics in experimental models of pain and analgesia. Adv Drug Deliv Rev 55(8):1007–1041
4. Vulchanova L, Schuster DJ, Belur LR, Riedl MS, Podetz-Pedersen KM, Kitto KF, Wilcox GL, McIvor RS, Fairbanks CA (2010) Differential adeno-associated virus mediated gene transfer to sensory neurons following intrathecal delivery by direct lumbar puncture. Mol Pain 6:31. https://doi.org/10.1186/1744-8069-6-31
5. Schuster DJ, Dykstra JA, Riedl MS, Kitto KF, Honda CN, McIvor RS, Fairbanks CA, Vulchanova L (2013) Visualization of spinal afferent innervation in the mouse colon by AAV8-mediated GFP expression. Neurogastroenterol Motil 25(2):e89–e100. https://doi.org/10.1111/nmo.12057
6. Schuster DJ, Dykstra JA, Riedl MS, Kitto KF, Belur LR, McIvor RS, Elde RP, Fairbanks CA, Vulchanova L (2014) Biodistribution of adeno-associated virus serotype 9 (AAV9) vector after intrathecal and intravenous delivery in mouse. Front Neuroanat 8:42. https://doi.org/10.3389/fnana.2014.00042
7. Schuster DJ, Belur LR, Riedl MS, Schnell SA, Podetz-Pedersen KM, Kitto KF, McIvor RS, Vulchanova L, Fairbanks CA (2014) Supraspinal gene transfer by intrathecal adeno-associated virus serotype 5. Front Neuroanat 8:66. https://doi.org/10.3389/fnana.2014.00066
8. Hardcastle N, Boulis NM, Federici T (2018) AAV gene delivery to the spinal cord: serotypes, methods, candidate diseases, and clinical trials. Expert Opin Biol Ther 18(3):293–307. https://doi.org/10.1080/14712598.2018.1416089
9. Dirren E, Aebischer J, Rochat C, Towne C, Schneider BL, Aebischer P (2015) SOD1 silencing in motoneurons or glia rescues neuromuscular function in ALS mice. Ann Clin Transl Neurol 2(2):167–184. https://doi.org/10.1002/acn3.162
10. Hirai T, Enomoto M, Kaburagi H, Sotome S, Yoshida-Tanaka K, Ukegawa M, Kuwahara H, Yamamoto M, Tajiri M, Miyata H, Hirai Y, Tominaga M, Shinomiya K, Mizusawa H, Okawa A, Yokota T (2014) Intrathecal AAV serotype 9-mediated delivery of shRNA against TRPV1 attenuates thermal hyperalgesia in a mouse model of peripheral nerve injury. Mol Ther 22(2):409–419. https://doi.org/10.1038/mt.2013.247
11. Shyng C, Nelvagal HR, Dearborn JT, Tyynela J, Schmidt RE, Sands MS, Cooper JD (2017) Synergistic effects of treating the spinal cord and brain in CLN1 disease. Proc Natl Acad Sci U S A 114(29):E5920–E5929. https://doi.org/10.1073/pnas.1701832114
12. Watson G, Bastacky J, Belichenko P, Buddhikot M, Jungles S, Vellard M, Mobley WC, Kakkis E (2006) Intrathecal administration of AAV vectors for the treatment of lysosomal storage in the brains of MPS I mice. Gene Ther 13(11):917–925. https://doi.org/10.1038/sj.gt.3302735

13. Liu CC, Huang ZX, Li X, Shen KF, Liu M, Ouyang HD, Zhang SB, Ruan YT, Zhang XL, Wu SL, Xin WJ, Ma C (2018) Upregulation of NLRP3 via STAT3-dependent histone acetylation contributes to painful neuropathy induced by bortezomib. Exp Neurol 302:104–111. https://doi.org/10.1016/j.expneurol.2018.01.011
14. Li YY, Li H, Liu ZL, Li Q, Qiu HW, Zeng LJ, Yang W, Zhang XZ, Li ZY (2017) Activation of STAT3-mediated CXCL12 up-regulation in the dorsal root ganglion contributes to oxaliplatin-induced chronic pain. Mol Pain 13:1744806917747425. https://doi.org/10.1177/1744806917747425
15. Zhang XS, Li X, Luo HJ, Huang ZX, Liu CC, Wan Q, Xu SW, Wu SL, Ke SJ, Ma C (2017) Activation of the RAGE/STAT3 pathway in the dorsal root ganglion contributes to the persistent pain hypersensitivity induced by lumbar disc herniation. Pain Physician 20(5):419–427
16. Morioka N, Zhang FF, Nakamura Y, Kitamura T, Hisaoka-Nakashima K, Nakata Y (2015) Tumor necrosis factor-mediated downregulation of spinal astrocytic connexin43 leads to increased glutamatergic neurotransmission and neuropathic pain in mice. Brain Behav Immun 49:293–310. https://doi.org/10.1016/j.bbi.2015.06.015
17. Morioka N, Fujii S, Kondo S, Zhang FF, Miyauchi K, Nakamura Y, Hisaoka-Nakashima K, Nakata Y (2018) Downregulation of spinal astrocytic connexin43 leads to upregulation of interleukin-6 and cyclooxygenase-2 and mechanical hypersensitivity in mice. Glia 66(2):428–444. https://doi.org/10.1002/glia.23255
18. Towne C, Pertin M, Beggah AT, Aebischer P, Decosterd I (2009) Recombinant adeno-associated virus serotype 6 (rAAV2/6)-mediated gene transfer to nociceptive neurons through different routes of delivery. Mol Pain 5:52. https://doi.org/10.1186/1744-8069-5-52
19. Lu Z, Xu J, Rossi GC, Majumdar S, Pasternak GW, Pan YX (2015) Mediation of opioid analgesia by a truncated 6-transmembrane GPCR. J Clin Invest 125(7):2626–2630. https://doi.org/10.1172/JCI81070
20. Yang X, Liu J, Liu ZJ, Xia QJ, He M, Liu R, Liu W, Wang W, Liu J, Zhou XF, Zhang YH, Wang TH (2014) Reversal of bone cancer pain by HSV-1-mediated silencing of CNTF in an afferent area of the spinal cord associated with AKT-ERK signal inhibition. Curr Gene Ther 14(5):377–388
21. Zou W, Guo Q, Chen C, Yang Y, Wang E (2011) Intrathecal herpes simplex virus type 1 amplicon vector-mediated human proenkephalin reduces chronic constriction injury-induced neuropathic pain in rats. Mol Med Rep 4(3):529–533. https://doi.org/10.3892/mmr.2011.445
22. Hu C, Lu Y, Cheng X, Cui Y, Wu Z, Zhang Q (2017) Gene therapy for neuropathic pain induced by spared nerve injury with naked plasmid encoding hepatocyte growth factor. J Gene Med 19:12. https://doi.org/10.1002/jgm.2994
23. Vanderwall AG, Noor S, Sun MS, Sanchez JE, Yang XO, Jantzie LL, Mellios N, Milligan ED (2018) Effects of spinal non-viral interleukin-10 gene therapy formulated with d-mannose in neuropathic interleukin-10 deficient mice: Behavioral characterization, mRNA and protein analysis in pain relevant tissues. Brain Behav Immun 69:91–112. https://doi.org/10.1016/j.bbi.2017.11.004

Chapter 21

Cerebellomedullary Cistern Injection of Viral Vectors in Nonhuman Primates

Lluis Samaranch, Kousaku Ohno, Waldy San Sebastian, and Krystof Bankiewicz

Abstract

Gene therapy shows great promise for the treatment of neurological disorders, and accessing cerebrospinal fluid (CSF) from the cerebellomedullary cistern through the posterior atlanto-occipital membrane has become a common route of delivery in preclinical studies. Unlike direct brain parenchymal infusions, CSF delivery offers broader coverage to the central and peripheral nervous system. This prospectively increases its translational value, more specially to treat global brain dysfunctions in which the pathology is disseminated throughout the brain and not focalized in one specific brain structure. Also, from the practical point of view, this approach offers a more reliable method for neurological gene replacement in infants, whose immature cranial suture preclude the use of skull-mounted devices. Here we describe a consistent, precise, and safe method for CSF injection.

Key words CSF delivery, Gene therapy, Viral vector, Nonhuman primates, Preclinical research

1 Introduction

The main goal of gene therapy with adeno-associated virus (AAV) is to deliver the transgene of interest to target cells at levels that result in expression that is both safe and effective. Systemic administration of AAV via the peripheral venous route is the most desirable noninvasive approach for patients with global pathological implications. However, systemic administration usually results in restricted or no distribution of most of the serotypes of the AAV to central nervous system (CNS) due to the blood–brain barrier (BBB) [1].

Cerebrospinal fluid (CSF) injection is a quick and straightforward method to disseminate viral particles with minimum invasiveness into the nervous system. Intrathecal (IT) administration is a delivery approach that has shown promise for providing efficient distribution along the spinal cord, but, in our experience, with very limited transduction in the brain. Small volumes, inaccurate

Fredric P. Manfredsson and Matthew J. Benskey (eds.), *Viral Vectors for Gene Therapy: Methods and Protocols*, Methods in Molecular Biology, vol. 1937, https://doi.org/10.1007/978-1-4939-9065-8_21,

delivery, or a strong dilution effect could explain lack of cortical transduction. In contrast, cerebellomedullary cistern route (also known as a cisternal puncture or cisterna magna injection, CM) results in widespread brain distribution with significant levels of transduction in the spinal cord as well [2]. Lumbar puncture is accepted as a standard practice for CSF collection, but lateral cervical puncture (a lateral variant of CM puncture) is also routinely practiced as a safe and useful alternative in cases of spinal dysraphism, spinal fusion, or congenital narrow canal, among others.

The CM is part of the subarachnoid space below cerebellum, and dorsal to the medulla oblongata. CSF is produced by the choroid plexus in the fourth ventricle and drains mainly through the central aperture filling the CM. In mammals, the atlanto-occipital membrane ensures the integrity and preserves CSF flux. There are multiple techniques that the investigator can use to disseminate viral particles into the CSF space throughout the atlanto-occipital membrane, as well as multiple guiding systems like CT-scan, X-Ray, or MRI for preclinical experiments. These imaging techniques differ in sophistication and equipment required, and also differ in accuracy and reproducibility, but all of them follow, basically, the same procedure.

In this chapter we will describe two techniques to inject viral particles into the CSF through the CM in nonhuman primates: stereotaxic-guided delivery and MR-guided delivery.

2 Materials

2.1 Animal Guidelines

1. As for any other experimental investigation involving the use of animals, animal protocols must be submitted and approved by the appropriate Institutional Animal Care and Use Committees prior to acquiring the animals.
2. When working with adult rhesus or cynomolgus monkeys, animals should be housed individually, in rooms set to a 12-h light/dark cycle and with room temperature ranging between 64 and 84 °F.
3. At the time of purchase from an authorized vendor and before animal arrival to the research institution, the investigator must learn if a 31-day quarantine period is required by the competent animal management institution (i.e., Centers for Disease Control in the USA).
4. Upon arrival, and prior to study assignment, animals should undergo a detailed examination by a qualified veterinarian. Animal inclusion/exclusion will be based on the Institutional Animal Care and Use Committees criteria.

2.2 Sedatives, Disinfectants, and Analgesics

1. Ketamine (Ketathesia®, 5 mg/kg, Henry Schein, Melville, NY USA, intramuscular injection), atropine (AtroJect® SA, 0.04 mg/kg, Henry Schein, Melville, NY USA, intramuscular injection), and Isoflurane (IsoThesia®, Henry Schein Melville, NY USA), inhalation anesthetic) are used for animal sedation.
2. Nolvasan® (10% chlorhexidine gluconate solution, Zoetis services LLC, Parsippany, NY, USA) and 70% isopropyl alcohol are used for surgical area cleaning.
3. Meloxicam (EloxiJect®, 0.2 mg/kg, Henry Schein, Melville, NY USA, intramuscular injection) and buprenorphine (Buprenex®, 0.01 mg/kg, Slough, United Kingdom, intramuscular injection) are used for postsurgical care.

2.3 List of Tools and Equipment

2.3.1 Common Tools and Equipment

1. Venous catheter (Sur-Vet® 22Gx1).
2. Min-volume extension set (37″ 0.2 mL priming volume, Smiths Medical, Minneapolis, MN, USA).
3. Medfusion® pump (Smiths Medical, Minneapolis, MN, USA).
4. Three-way stopcock (Intralock®, Abbott Labs, Chicago, IL, USA).
5. Luer lock 3 mL syringe (Becton Dickinson, Franklin Lakes, NJ, USA).
6. Gaymar T/Pump-700 and TP22G warming blanket (Paragon Medical, Pierceton, IN, USA).
7. Sterile gauze, sterile gloves, stopwatch timer, sterile Eppendorf tubes (*see* **Note 1**).

2.3.2 Stereotaxic-Guided Delivery Tools and Equipment

1. Stereotactic monkey frame (Dynatech Machining, Union City, CA, USA).
2. Monkey ear bars (Dynatech Machining, Union City, CA, USA).
3. Stereotactic manipulator arm (David Kopf Instruments, Tujunga, CA, USA).
4. Large probe holder (Stoelting, Kiel, WI, USA).
5. 22G × 1″ spinal needle (0.70 mm × 38 mm, Becton Dickinson, Franklin Lakes, NJ, USA).

2.3.3 MR-Guided Delivery Tools and Equipment

1. MRI contrast agent, gadoteridol (ProHance®), diluted to a final concentration of 2 mM in phosphate-buffered saline (PBS, pH 7.4) and Pluronic F-68 (0.001% v/v; Invitrogen, Carlsbad, CA, USA). Gadoteridol is commercially available (Bracco Diagnostics Inc., Monroe Township, NJ, USA) and each milliliter contains 279.3 mg of gadoteridol with 0.23 mg calteridol calcium and 1.21 mg of tromethamine.

2. MRI scanner, 1.5 or 3.0 Tesla (Signa LX®; GE Medical Systems, Waukesha, WI, USA).
3. Set of two 5-inch circular surface MRI coil (MR Instruments Inc., Hopkins, MN, USA) (*see* **Note 2**).
4. MRI-compatible 22G spinal needle (0.70 mm × 100 mm, MRI Chiba Needle).

3 Methods

3.1 Animal Sedation

1. Animals are sedated in the home cage with an intramuscular injection of ketamine and atropine (*see* Subheading 2.2) and transported to the preparation area.
2. A venous line is established with a 22G × 1″ catheter in the saphenous vein to deliver isotonic fluids during the whole procedure at a rate of 5–10 mL/kg/h.
3. The back of the neck is shaved with a hair clipper and cleaned with Nolvasan and 70% isopropyl alcohol (*see* Subheading 2.2).
4. Animal is intubated with an endotracheal tube after a local anesthetic is applied directly on the larynx according to *IACUC Guidelines and Standard Procedures.* Then, endotracheal tube is secured to animal and connected to the anesthetic circuit.
5. Once endotracheal tube is secured, isoflurane inhalation anesthesia is delivered at 1–3% to maintain a stable plane of anesthesia (*see* Subheading 2.2).
6. Sterile surgical gloves should be worn for the final skin preparation. Supplies, including solutions, must be sterile and non-expired.

3.2 Surgical Setup

3.2.1 Stereotaxic-Guided Delivery

1. Vital signs are monitored continuously during the procedure. When pupil and toe-pinch reflexes are absent, along with low jaw-tone, the animal's head is placed in a stereotactic frame, and the neck is flexed with the animal in a prone position (*see* **Note 3**).
2. Injection pump is mounted in an auxiliary pole or in a supplementary table close to the surgery or exam table (Fig. 1a and *see* **Note 4**).

3.2.2 MR-Guided Delivery

1. Vital signs are monitored continuously during the procedure. When pupil and toe-pinch reflexes are absent, along with low jaw-tone, the animal is placed on its side, and the neck is flexed with the animal in a decubitus position during needle insertion. After the correct location is verified, the animal remains at decubitus but can be returned to a relaxed neck position (*see* **Note 5**).

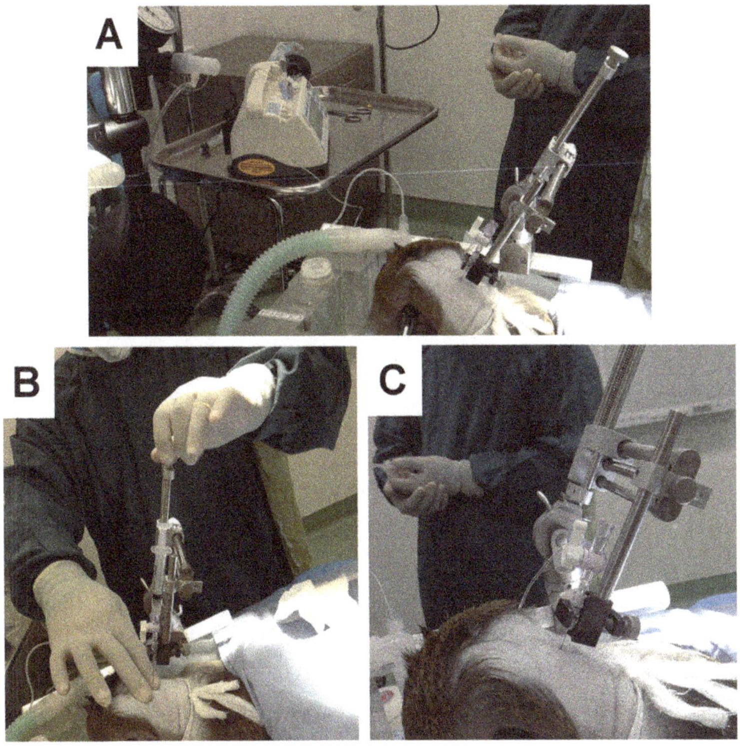

Fig. 1 Stereotactic-guided injection into the cerebrospinal fluid (CSF) through cerebellomedullary cistern (CM). (**a**) Surgical setup for CM injection showing animal in stereotactic frame, stereotactic tower, and infusion pump. Once the injection port is secured in the holder, the atlanto-occipital gap in the cranio-cervical junction is identified by palpation (**b**) and the needle is manually guided to the target. The needle is slowly introduced through the atlanto-occipital membrane until clear CSF comes up to the syringe. Then, the 3-way stopcock is opened to the loading line (**c**) to allow vector infusion by means of a pump

2. Non-MR-compatible injection pumps must remain at secure distances from the magnetic field (5-gauss fringe field line) as determined by Safety protocol, Regulatory Agencies, and MR manufacturer. That distance will condition the length of the tubing required to deliver the vector (*see* **Note 6**).

3.3 Animal Preparation

1. The back of the neck is cleaned with 4 × 4 gauze soaked in 70% isopropyl alcohol and povidone iodine (Betadine®) in an alternate sequence. The skin must be wiped, beginning at the anticipated incision site and working outward. Repeat the sequence three times.
2. Complete the final skin preparation by applying a light coat of antiseptic surgical solution (povidone iodine) with a spray bottle.

3.4 Vector Injection

When flushing the system with sterile saline and/or the infusate, it is important to avoid the introduction and/or formation of air bubbles within the cannula, loading line or secondary line. Although pneumorrhachis and pneumocephalus are asymptomatic, spinal and intracranial collection of air and air that exits the spinal needle tip can impact infusate flux and, therefore, impact the test article distribution. Air bubbles within the infusion system should be avoided.

3.4.1 Stereotaxic-Guided Delivery

Needle Assembly and Tubing Setup

1. An in-line three-way stopcock permits both collection and injection with a single insertion, preserves sample sterility, and minimizes vector-to-CSF cross-contamination during baseline CSF collection.
2. To control the administration into the CM space, a syringe containing test article or vehicle solution is locked into an infusion pump and connected to primed 36-inch high-pressure intravenous tubing.
3. Stopcock open 1 (female luer fitting): Connect the other end of the tubing containing the test article or vehicle solution.
4. Stopcock open 2 (female luer fitting): Connect a 3 mL sterile syringe.
5. Stopcock open 3 (male luer fitting): Connect a 22G × 1″spinal needle, and place it into the large probe holder that is mounted on a stereotactic manipulator arm (*see* **Note 7**).

Delivery

1. After the needle is assembled and infusion system is primed and ready, the atlanto-occipital gap in the craniocervical junction is identified by palpation (Fig. 1b).
2. Once the needle is aligned with the cervical gap with the proper angle, the needle is manually advanced with the manipulator through the skin, atlanto-occipital membrane, and into the cerebellomedullary cistern space (*see* **Note 8**).
3. To verify the correct depth and to avoid damage to the medulla oblongata, negative pressure is applied during needle insertion by slightly retracting the plunger of the attached syringe. It is important to emphasize that the negative pressure has to be maintained during all the needle insertion process until CSF starts to flow inside the syringe. Intermittent peaks of negative pressure while the needle is being inserted should be avoided (*see* **Note 9**).
4. Constant negative pressure while the needle is inserted will ensure to locate needle inside the CM as deep as practicable to avoid medullary injury.
5. Once the needle has passed the atlanto-occipital membrane and entered the CM space, a small volume of CSF (~1 mL) may be

aspirated into the 3 mL syringe and stored at −20 °C as a baseline CSF sample.

6. The position is then secured and the three-way stopcock is opened to the loading line to allow pumped infusion of the vector at a rate of 0.5 mL/min (Fig. 1c and *see* **Note 10**).
7. After vector infusion, the line is flushed with saline buffer by exchanging the syringe containing test article or vehicle solution with a sterile saline syringe (*see* **Note 4**).
8. After 5 min of resting period, the needle is slowly retracted out of the CM space with the manipulator. Hemostasis is achieved by direct pressure for 2–3 min after the needle is completely removed.
9. Finally, the animal's head is removed of the frame, the animal is maintained on a warming blanket decubitus, and vital signs are monitored until its full recovery from anesthesia.
10. Institutional approved postsurgical care protocol should be followed after animal recovery (*see* Subheadings 2.2 and 3.5).

3.4.2 MR-Guided Injection

MR Protocol

The MRI parameters are specific to visualizing the gadoteridol contrast agent during infusion (Fig. 2). Note that this protocol has been written for use in at 1.5T Signa LX scanner and should be modified when using a different type of scanner, hardware and software.

1. The animal is place in the desired position on the MR table and secure to avoid movement. Body mold or bean bags can be used to ensure spinal cord parallel to the acquisition plane (*see* **Note 5**).
2. A 3D high-resolution MP-RAGE scan is first run for multiplanar reconstruction (MPR) to evaluate correct body alignment, neck muscle density, and/or optimal neck position to "open" the CM (*see* **Note 11**).

Delivery

1. After the animal is secure and in an appropriate position, the atlanto-occipital gap in the craniocervical junction is identified by palpation.
2. Once the surgeon identifies the optimal location, MRI-compatible needle (*see* Subheading 2.3.3) is steadily advanced through the skin, atlanto-occipital membrane, and into the cerebellomedullary cistern space (Fig. 2a, b).
3. The needle should be advanced with the bevel down until the dura is punctured.
4. To verify the correct placement, once the dura is perforated, the needle should be turned 90–180° to allow CSF flow. CSF should flow freely from the needle.

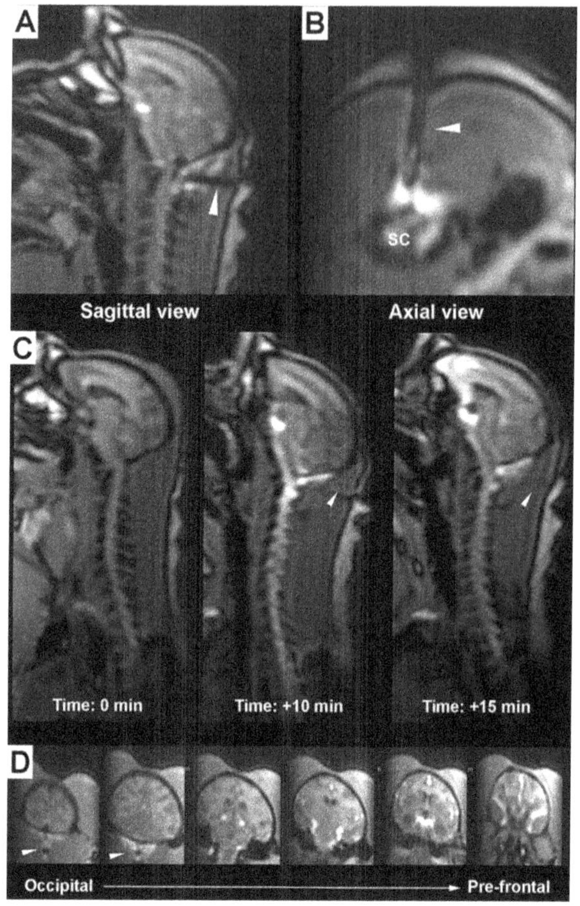

Fig. 2 Gadoteridol-based MR contrast distribution in the CSF after cerebellomedullary cistern (CM) delivery. Representative MR acquisition in the sagittal plane (**a**) and in the axial plane (**b**) showing the MR-compatible spinal needle (white arrowhead) injecting into the CM. (**c**) MR time sequence acquisition showing MR contrast distribution in brain and spinal cord at baseline (0 min), infusion (+10 min), and end of the infusion (+15 min). (**d**) Caudal to rostral MR series showing the distribution of the MR-contrast into the brain from the injection site (CM) to the prefrontal cortex. *SC* spinal cord, *min* minutes

5. Once the needle tip location is confirmed with a good CSF flow, and a baseline CSF sample is collected (if needed), the tubing containing the vector is connected to the needle. Tubing must be secured on the MR table avoiding any tension on the needle.
6. Then, a MR scan will allow the investigator to check the correct position of the needle (white arrowhead Fig. 2).
7. Then, consecutive MR acquisitions during the injection will allow the investigator to monitor the vector distribution into the CSF (Fig. 2c) by the visualization of the gadoteridol signal form the infusate.

8. After vector infusion, ~200 μL of saline can be used to flush any vector left in the injection system.
9. After 5 min of resting period, the needle is slowly retracted out of the CM space with the manipulator. Hemostasis is achieved by direct pressure for 2–3 min after the needle is completely removed. A final scan can be acquired to determine the scope of the gadoteridol signal at the end of the injection (Fig. 2d).
10. Finally, the animal is maintained on a warming blanket decubitus, and vital signs are monitored until its full recovery from anesthesia.
11. Institutional approved postsurgical care protocol should be followed after animal recovery (*see* Subheadings 2.2 and 3.5).

3.5 Postsurgical Care

In order to minimize any discomfort, distress, or pain after the CSF delivery, animals receive an intramuscular injection of meloxicam (NSAID) during cisternal injection and twice the day after the injection. Animal is placed on its side during recovery. Nausea and vomiting can be expected from the animal right after recovering from anesthesia. Once the animal is returned to its cage, it is evaluated twice daily for 5 days by veterinary staff. Standardized forms are completed for each animal that includes evaluations of the surgical-site integrity, edema, infection, balance, locomotion, attitude, food intake, and fecal and urine output. Any abnormalities or signs of discomfort are promptly reported to veterinary staff for evaluation and appropriate treatment.

3.6 Anatomy of the CM Delivery

Pilot experiments with trypan blue dye demonstrated that test article dissemination after CM into the CSF achieve a broad distribution covering not only CM surroundings like cerebellum and the dorsal surface of the medulla oblongata (Fig. 3a, b), but also structures much more distant like prefrontal cortex (upstream) or sacral region of the spinal cord (downstream). This experiment confirmed that CM injection ensures the full coverage of the spinal cord (Fig. 3c) and dorsal root ganglia (Fig. 3d).

The methodology described in this chapter was also applied to different AAV serotypes and the results have been reported in several peer-reviewed publications [2–4].

4 Notes

1. Since the technique described in this chapter will be applied to nonhuman primates, and possibly translational research generating IND-enabling data, sterility in all the steps and material is recommended. The investigator must ensure a sterile environment from the beginning to the end of the procedure.

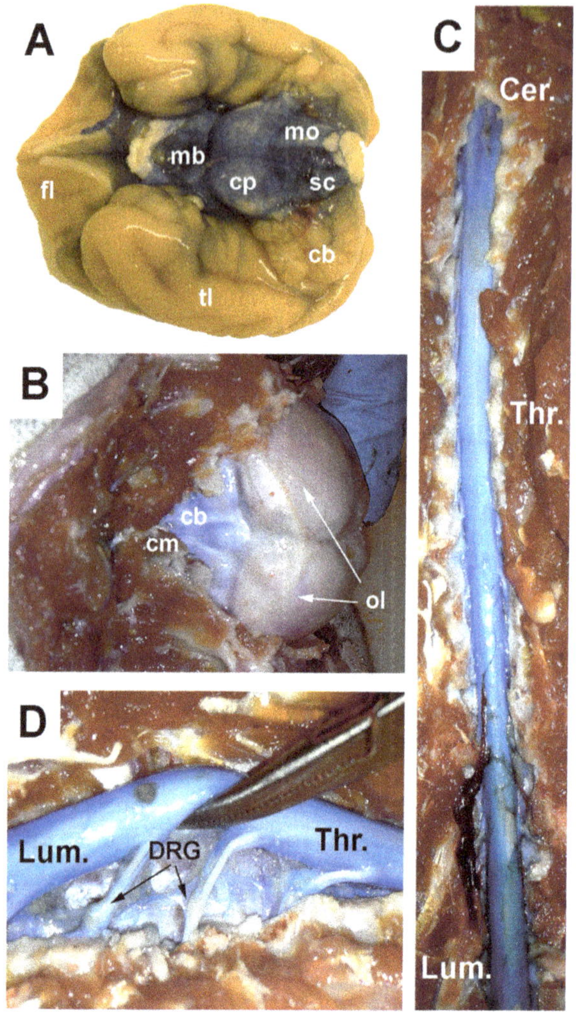

Fig. 3 Macroscopic images from brain and spinal cord after dye injection experiment. (**a**) Image shows ventral areas of the brain after injection of 6 mL of trypan blue dye into the cerebellomedullary cistern (CM). Intense staining was observed in proximal areas of the CM as cervical (C1-C2) regions of the spinal cord (*sc*), medulla oblongata (*mo*), cerebral peduncle (*cp*), and midbrain (*mb*). Other regions like cerebellum (*cb*), ventral temporal lobe (*tl*), or ventral frontal lobe (*fl*) were also stained by the blue dye. Just after CM injection, animal was humanely euthanized and craniotomy (**b**) was performed to expose the brain. Intense blue staining was observed under dura, especially in central and lateral vermis of the cerebellum (*cb*), closest regions of the injection site (*cm*). A total laminectomy (**c**) was also performed to expose the spinal cord. Intense blue staining was observed along the entire cord at cervical (Cer.), thoracic (Thr.), and lumbar (Lum.) regions. Further assessment of the spinal cord (**d**) evinced blue staining along the posterior root of the dorsal root ganglion (DRG), the ganglion, and the spinal nerve

2. A head coil, shoulder coil, or the body coil integrated in the MR table are also feasible options, but the investigator can experience poor resolution and image quality. Loop coils attached to the animal's head, as mentioned in the text, will result in an optimal head image quality. For optimal spinal cord image quality or head-body image quality the investigator should contact the technical support or MR specialist to explore the best options for the scanner.
3. If necessary, the investigator can use the eye bars from the stereotactic monkey frame to maintain the animal's head flexed down by locating them on top of the periorbital bone.
4. The investigator has to take into account the dead volume contained in the tubing after the full delivery when defining the dose. Short tubing will ensure less vector waste, although a small amount of saline (200–650 μL, depending tubing specifications and length) can be used to prime the lines and inject the vector left in the tubing. Larger volumes of saline into the CM are not recommended. In our experience, large volume of saline after vector injection displaces the vector out of the cisterna magna and impairs vector levels for brain transduction.
5. If full body images will be acquired to monitor MR-contrast flow through the spinal cord, a body mold or bean bags are recommended to maintain the spinal cord aligned to the MR acquisition plane. For spinal studies, a prone position can be considered, as well as the lateral cervical puncture.
6. If the volume of the infusate containing the vector can be held in a fraction of the tubing, the rest of the tubing must be preloaded with sterile saline. Briefly, load the vector in the loading line though the distal end and using all sections needed to contain the dose. Lock the tube using the slide clamp. Then, load the rest of the tubing connected in a separate set with sterile saline until a drop shows up in the proximal end of the luer-lock. Then, connect the section with the saline to the section with the vector avoiding any bubble in the female luer-lock. Finally, push the saline into the vector sections until a drop of vector shows up on the distal end of the tubing. This end will connect with the spinal needle. It is recommendable to use a locking male luer-lock to connect with the spinal needle to avoid excessive vibration once the needle is inserted in the CM.
7. The investigator must ensure that there are no air bubbles in the loading line and should prime the side-port (female luer fitting) of the 3-way stopcock with vector. The investigator has to ensure that no vector will be drained to the spinal needle at this initial step by closing the tubing-needle position of the stopcock. Access between 3-mL syringe and spinal needle should be in the open position.

8. Depending on animal's head position in the frame, the investigator may vary the angle of the stereotactic arm to accommodate perpendicular incision into the CM (Fig. 1). Based on our experience, the correct position of the head in the frame and needle alignment will avoid vessel bleeding that would contaminate the CSF sample or cause injury.
9. Because the brain is enclosed in a rigid compartment, vascular pulsations related to cardiac cycle also induce pulsations in flow and pressure in the brain and spinal cord, causing minor tissue movement that the investigator has to take into account when inserting the spinal needle through the atlanto-occipital membrane. Hence, it is strongly recommended that the needle is inserted slowly with the manipulator, locking the position as soon as CSF appears in the syringe in order to locate the needle as far as practicable from the medullary tissue.
10. At this point, if approved IACUC protocol allows the procedure, repeated delivery in delayed intervals or sequential multidrug delivery may be performed.
11. Repetition time (TR): 2110 ms; echo time (TE): 3.6 ms; flip angle: 15°; number of excitations (NEX): 1 (repeated three times); matrix: 240×240; field of view (FOV): $240 \times 240 \times 240$.

Acknowledgments

This work was supported by a grant to K.S.B. from NIH-NINDS (R01NS073940-01).

References

1. Foust KD, Nurre E, Montgomery CL et al (2009) Intravascular AAV9 preferentially targets neonatal neurons and adult astrocytes. Nat Biotechnol 27:59–65. https://doi.org/10.1038/nbt.1515
2. Samaranch L, Salegio EA, San Sebastian W et al (2013) Strong cortical and spinal cord transduction after AAV7 and AAV9 delivery into the cerebrospinal fluid of nonhuman primates. Hum Gene Ther 24:526–532. https://doi.org/10.1089/hum.2013.005
3. Samaranch L, Salegio EA, San Sebastian W et al (2012) Adeno-associated virus serotype 9 transduction in the central nervous system of nonhuman primates. Hum Gene Ther 23:382–389. https://doi.org/10.1089/hum.2011.200
4. Salegio EA, Streeter H, Dube N et al (2014) Distribution of nanoparticles throughout the cerebral cortex of rodents and non-human primates: implications for gene and drug therapy. Front Neuroanat 8:9. https://doi.org/10.3389/fnana.2014.00009

INDEX

Fredric P. Manfredsson and Matthew J. Benskey (eds.), *Viral Vectors for Gene Therapy: Methods and Protocols*, Methods in Molecular Biology, vol. 1937, https://doi.org/10.1007/978-1-4939-9065-8,

P

R

S

T

www.ingramcontent.com/pod-product-compliance
Lightning Source LLC
Chambersburg PA
CBHW081212180926
51613CB00003B/220
* 9 7 8 1 4 9 3 9 9 0 6 4 1 *